LINEAR INVERSE PROBLEMS
The Maximum Entropy Connection
With CD-ROM

Series on Advances in Mathematics for Applied Sciences – Vol. 83

LINEAR INVERSE PROBLEMS
The Maximum Entropy Connection
With CD-ROM

Henryk Gzyl
IESA, Venezuela

Yurayh Velásquez
Universidad Metropolitana, Venezuela

World Scientific

NEW JERSEY · LONDON · SINGAPORE · BEIJING · SHANGHAI · HONG KONG · TAIPEI · CHENNAI

Published by

World Scientific Publishing Co. Pte. Ltd.
5 Toh Tuck Link, Singapore 596224
USA office: 27 Warren Street, Suite 401-402, Hackensack, NJ 07601
UK office: 57 Shelton Street, Covent Garden, London WC2H 9HE

Library of Congress Cataloging-in-Publication Data
Gzyl, Henryk,
    Linear inverse problems : the maximum entropy connection (with CD-ROM) / by Henryk Gzyl & Yurayh Velásquez.
    p. cm. -- (Series on advances in mathematics for applied sciences ; v. 83)
    Includes bibliographical references.
    ISBN-13: 978-981-4338-77-6 (hardcover : alk. paper)
    ISBN-10: 981-4338-77-X (hardcover : alk. paper)
    1. Inverse problems (Differential equations) 2. Maximum entropy method. I. Velásquez, Yurayh. II. Title.
    QA378.5.G98 2011
    515'.357--dc22
                                                                           2010047244

British Library Cataloguing-in-Publication Data
A catalogue record for this book is available from the British Library.

Cover illustration by Stefan Gzyl

Copyright © 2011 by World Scientific Publishing Co. Pte. Ltd.

All rights reserved. This book, or parts thereof, may not be reproduced in any form or by any means, electronic or mechanical, including photocopying, recording or any information storage and retrieval system now known or to be invented, without written permission from the Publisher.

For photocopying of material in this volume, please pay a copying fee through the Copyright Clearance Center, Inc., 222 Rosewood Drive, Danvers, MA 01923, USA. In this case permission to photocopy is not required from the publisher.

Printed in Singapore by World Scientific Printers.

Preface

These lecture notes were originally prepared as backup material for a course on Inverse Problems and Maximum Entropy taught at the Venezuelan School of Mathematics. The event takes place annually in the city of Mérida, a university town in the Venezuelan Andean mountains. The attendance, mainly graduate students are exposed material that goes a bit beyond the standard courses.

The course had three aims. On one hand, to present some basic results about linear inverse problems and how to solve them. On the other, to develop the method of maximum entropy in the mean, and to apply it to study linear inverse problems. This would show the weaknesses and strengths of both approaches. The other aim was to acquaint the participants with the use of the software that is provided along with the book. This consists of interactive screens on which the data of typical problems can be uploaded, and a solution is provided

The present notes eliminate many mistakes and misprints that plague the original version, and hopefully not many new ones crept in the new material that we added for this version. The material was reorganized slightly, new applications were added, but no real effort was undertaken to update the enormous literature on applications of the maximum entropy method.

It is a pleasure to thank the team of TeX experts at World Scientific for their help in bringing the original latex manuscript to its present form.

*"You can't always get what you want
but if you try, sometimes
you may find that you get what you need"*

(M. Jagger & K. Richards)

"Cuando se me termina el azul, uso el rojo"

(P. Picasso)

*"No consigo comprender el significado de la
palabra 'investigación' en la pintura moderna,
a mi manera de ver, buscar no significa nada
en pintura. Lo que cuenta es encontrar"*

(P. Picasso)

*"Desde cuando es el autor de un libro quien
mejor lo comprende"*

(M. de Unamuno)

*"For every problem there is a solution
that is simple, elegant and wrong"*

(H. L. Menken)

The Roman jurists ruled:
*"Concerning evildoers, mathematicians, and the like"
that: "to learn the art of geometry
and to take part in public exercises,
an art as damnable as mathematics, are forbidden"*

(downloaded from the web)

Contents

Preface   v

List of Figures   xv

List of Tables   xxi

1. **Introduction**   1

2. **A collection of linear inverse problems**   5
   - 2.1 A battle horse for numerical computations   5
   - 2.2 Linear equations with errors in the data   6
   - 2.3 Linear equations with convex constraints   8
   - 2.4 Inversion of Laplace transforms from finite number of data points   10
   - 2.5 Fourier reconstruction from partial data   11
   - 2.6 More on the non-continuity of the inverse   12
   - 2.7 Transportation problems and reconstruction from marginals   13
   - 2.8 CAT   15
   - 2.9 Abstract spline interpolation   20
   - 2.10 Bibliographical comments and references   21

3. **The basics about linear inverse problems**   25
   - 3.1 Problem statements   25
   - 3.2 Quasi solutions and variational methods   30
   - 3.3 Regularization and approximate solutions   31

3.4	Appendix	35
3.5	Bibliographical comments and references	36

4. Regularization in Hilbert spaces: Deterministic and stochastic approaches — 37

4.1	Basics	37
4.2	Tikhonov's regularization scheme	40
4.3	Spectral cutoffs	44
4.4	Gaussian regularization of inverse problems	46
4.5	Bayesian methods	48
4.6	The method of maximum likelihood	49
4.7	Bibliographical comments and references	51

5. Maxentropic approach to linear inverse problems — 53

5.1	Heuristic preliminaries	53
5.2	Some properties of the entropy functionals	58
5.3	The direct approach to the entropic maximization problem	59
5.4	A more detailed analysis	62
5.5	Convergence of maxentropic estimates	64
5.6	Maxentropic reconstruction in the presence of noise	67
5.7	Maxentropic reconstruction of signal and noise	70
5.8	Maximum entropy according to Dacunha-Castelle and Gamboa. Comparison with Jaynes' classical approach	72
	5.8.1 Basic results	72
	5.8.2 Jaynes' and Dacunha and Gamboa's approaches	77
5.9	MEM under translation	79
5.10	Maxent reconstructions under increase of data	80
5.11	Bibliographical comments and references	82

6. Finite dimensional problems — 87

6.1	Two classical methods of solution	87
6.2	Continuous time iteration schemes	90
6.3	Incorporation of convex constraints	91
	6.3.1 Basics and comments	91
	6.3.2 Optimization with differentiable non-degenerate equality constraints	95

	6.3.3	Optimization with differentiable, non-degenerate inequality constraints	97
6.4		The method of projections in continuous time	98
6.5		Maxentropic approaches	99
	6.5.1	Linear systems with band constraints	100
	6.5.2	Linear system with Euclidean norm constraints	102
	6.5.3	Linear systems with non-Euclidean norm constraints	104
	6.5.4	Linear systems with solutions in unbounded convex sets	105
	6.5.5	Linear equations without constraints	109
6.6		Linear systems with measurement noise	112
6.7		Bibliographical comments and references	113

7. Some simple numerical examples and moment problems — 115

7.1		The density of the Earth	115
	7.1.1	Solution by the standard $L_2[0,1]$ techniques	116
	7.1.2	Piecewise approximations in $L_2([0,1])$	117
	7.1.3	Linear programming approach	118
	7.1.4	Maxentropic reconstructions: Influence of *a priori* data	120
	7.1.5	Maxentropic reconstructions: Effect of the noise	122
7.2		A test case	125
	7.2.1	Standard $L_2[0,1]$ technique	126
	7.2.2	Discretized $L_2[0,1]$ approach	127
	7.2.3	Maxentropic reconstructions: Influence of *a priori* data	128
	7.2.4	Reconstruction by means of cubic splines	131
	7.2.5	Fourier versus cubic splines	135
7.3		Standard maxentropic reconstruction	141
	7.3.1	Existence and stability	144
	7.3.2	Some convergence issues	146
7.4		Some remarks on moment problems	146
	7.4.1	Some remarks about the Hamburger and Stieltjes moment problems	149
7.5		Moment problems in Hilbert spaces	152
7.6		Reconstruction of transition probabilities	154

	7.7	Probabilistic approach to Hausdorff's moment problem	156
	7.8	The very basics about cubic splines	158
	7.9	Determination of risk measures from market price of risk	159
		7.9.1 Basic aspects of the problem	159
		7.9.2 Problem statement	161
		7.9.3 The maxentropic solution	162
		7.9.4 Description of numerical results	163
	7.10	Bibliographical comments and references	164

8. Some infinite dimensional problems 169

	8.1	A simple integral equation	169
		8.1.1 The random function approach	170
		8.1.2 The random measure approach: Gaussian measures	173
		8.1.3 The random measure approach: Compound Poisson measures	174
		8.1.4 The random measure approach: Gaussian fields	176
		8.1.5 Closing remarks	177
	8.2	A simple example: Inversion of a Fourier transform given a few coefficients	178
	8.3	Maxentropic regularization for problems in Hilbert spaces	179
		8.3.1 Gaussian measures	179
		8.3.2 Exponential measures	182
		8.3.3 Degenerate measures in Hilbert spaces and spectral cut off regularization	183
		8.3.4 Conclusions	184
	8.4	Bibliographical comments and references	184

9. Tomography, reconstruction from marginals and transportation problems 185

	9.1	Generalities	185
	9.2	Reconstruction from marginals	187
	9.3	A curious impossibility result and its counterpart	188
		9.3.1 The bad news	188
		9.3.2 The good news	190
	9.4	The Hilbert space set up for the tomographic problem	192
		9.4.1 More on nonuniquenes of reconstructions	194

9.5	The Russian Twist	194
9.6	Why does it work	195
9.7	Reconstructions using (classical) entropic, penalized methods in Hilbert space	198
9.8	Some maxentropic computations	201
9.9	Maxentropic approach to reconstruction from marginals in the discrete case	203
	9.9.1 Reconstruction from marginals by maximum entropy on the mean	204
	9.9.2 Reconstruction from marginals using the standard maximum entropy method	207
9.10	Transportation and linear programming problems	209
9.11	Bibliographical comments and references	211

10. Numerical inversion of Laplace transforms — 215

10.1	Motivation	215
10.2	Basics about Laplace transforms	216
10.3	The inverse Laplace transform is not continuous	218
10.4	A method of inversion	218
	10.4.1 Expansion in sine functions	219
	10.4.2 Expansion in Legendre polynomials	220
	10.4.3 Expansion in Laguerre polynomials	221
10.5	From Laplace transforms to moment problems	222
10.6	Standard maxentropic approach to the Laplace inversion problem	223
10.7	Maxentropic approach in function space: The Gaussian case	225
10.8	Maxentropic linear splines	227
10.9	Connection with the complex interpolation problem	229
10.10	Numerical examples	230
10.11	Bibliographical comments and references	236

11. Maxentropic characterization of probability distributions — 241

11.1	Preliminaries	241
11.2	Example 1	243
11.3	Example 2	244
11.4	Example 3	245

11.5	Example 4	245
11.6	Example 5	246
11.7	Example 6	246

12. Is an image worth a thousand words?   249

12.1	Problem setup	249
	12.1.1 List of questions for you to answer	251
12.2	Answers to the questions	251
	12.2.1 Introductory comments	251
	12.2.2 Answers	251
12.3	Bibliographical comments and references	258

Appendix A   Basic topology   261

Appendix B   Basic measure theory and probability   265

B.1	Some results from measure theory and integration	265
B.2	Some probabilistic jargon	272
B.3	Brief description of the Kolmogorov extension theorem	275
B.4	Basic facts about Gaussian process in Hilbert spaces	276

Appendix C   Banach spaces   279

C.1	Basic stuff	279
C.2	Continuous linear operator on Banach spaces	281
C.3	Duality in Banach spaces	283
C.4	Operators on Hilbert spaces. Singular values decompositions	289
C.5	Some convexity theory	290

Appendix D   Further properties of entropy functionals   293

D.1	Properties of entropy functionals	293
D.2	A probabilistic connection	297
D.3	Extended definition of entropy	301
D.4	Exponetial families and geometry in the space of probabilities	302
	D.4.1 The geometry on the set of positive vectors	304
	D.4.2 Lifting curves from G^+ to G and parallel transport	306
	D.4.3 From geodesics to Kullback's divergence	307

		D.4.4	Coordinates on $\mathbb{P}$	308
	D.5	Bibliographical comments and references		310

Appendix E   Software user guide   313

	E.1	Installation procedure		313
	E.2	Quick start guide		316
		E.2.1	Moment problems with MEM	317
		E.2.2	Moment problems with SME	318
		E.2.3	Moment problems with Quadratic Programming	318
		E.2.4	Transition probabilities problem with MEM . . .	319
		E.2.5	Transition probabilities problem with SME . . .	320
		E.2.6	Transition probabilities problem with Quadratic Programming	320
		E.2.7	Reconstruction from Marginals with MEM	320
		E.2.8	Reconstruction from Marginals with SME	321
		E.2.9	Reconstruction from Marginals with Quadratic Programming	321
		E.2.10	A generic problem in the form $Ax = y$, with MEM	322
		E.2.11	A generic problem in the form $Ax = y$, with SME	323
		E.2.12	A generic problem in the form $Ax = y$, with Quadratic Programming	323
		E.2.13	The results windows	323
		E.2.14	Messages that will appear	324
		E.2.15	Comments	326

List of Figures

2.1	Basic scheme.	16
2.2	Basic geometry of the CAT process.	17
2.3	Tomographic image formation.	18
2.4	Actual CAT setup.	19
6.1	$C \cap A^{-1}y = \emptyset$.	94
6.2	$C \cap A^{-1}y = \emptyset$. $x^* \neq x_0$.	94
6.3	$C \cap A^{-1}y \ni x^* = x_0$.	95
7.1	Standard L_2 solution to (7.1).	116
7.2	Standard L_2 with penalization.	117
7.3	Reconstruction with $\lambda = 10^{-7}$, $\varepsilon = 6.0063 \times 10^{-5}$.	118
7.4	Penalized least square reconstruction. Constrained $a = 0$, $b = 1$; parameters $\lambda = 10^{-5}$, 10^{-7}. Reconstruction errors $\varepsilon = 5.23 \times 10^{-3}$, 6.39×10^{-5}.	119
7.5	Penalized least square reconstruction. Constrained $a = 0.1$, $b = 0.9$; parameters $\lambda = 10^{-5}$, 10^{-7}, $\varepsilon = 7.13 \times 10^{-5}$.	119
7.6	Reconstructing using linear programming.	120
7.7	Maxentropic reconstruction with uniform *a priori* on different $[a, b]$.	121
7.8	Maxentropic reconstruction with uniform *a priori* on different $[a, b]$.	121
7.9	*A priori* distribution of Bernoulli type masses $p = q = \frac{1}{2}$ and different intervals.	122
7.10	*A priori* distribution of Bernoulli type on a fixed interval but different masses at the ends.	123
7.11	Reconstructions with uniform *a priori* distribution on $[0, 1]$, fixed $\sigma_1 = \sigma_2 = 0.01$ and varying T.	124

7.12 Reconstructions with uniform *a priori* distribution on $[0, 4]$, varying σs and Ts. 124

7.13 Reconstructions with fixed Bernoulli distribution on $[0, 10]$ with $p = q = \frac{1}{2}$, $T = 1$ and $\sigma_1 = \sigma_2$ varying. 125

7.14 Reconstructions under Bernoulli *a priori* distribution on $[0, 10]$ with $p = 1 = \frac{1}{2}$, fixed $\sigma_1 = \sigma_2 = 0.1$, but T varying as indicated. 125

7.15 Reconstructions from first list of moments. Different λs and its respectively error. 127

7.16 Reconstructions from second list of moments. Different λs and its respectively error. 128

7.17 Reconstructions for first list moments and several λ. 129

7.18 Reconstructions for second list of moments and several λ. .. 129

7.19 Reconstructions for first list of moments and several λ, projected onto constraint space. 130

7.20 Reconstructions for second list of moments and several λ, projected onto constrained space. 130

7.21 Maxentropic reconstructions for different constraints and uniform *a priori* measure. 131

7.22 Maxentropic reconstruction for different constrained and Bernoulli *a priori* measure with $p = q = \frac{1}{2}$. 132

7.23 Maxentropic reconstruction for $[a, b] = [-2, 2]$ and $p = 0.5, 0.2$, 0.8 with $q = 1 - p$. 132

7.24 Maxentropic reconstruction with cubic splines from $m = 2$, $w = 1$ with error $\varepsilon = 1.9830 \times 10^{-10}$. 133

7.25 Maxentropic reconstruction with cubic splines from $m = 2$, $w = \pi$ with error $\varepsilon = 1.8340 \times 10^{-8}$. 134

7.26 Maxentropic reconstruction with cubic splines from $m = 4$, $w = \pi$ with error $\varepsilon = 4.5339 \times 10^{-8}$. 134

7.27 Maxentropic reconstruction with cubic splines from $m = 4$, $w = \pi$ with error $\varepsilon = 7.2104 \times 10^{-9}$. 135

7.28 Maxentropic reconstruction with cubic splines from $m = 4$, $w = 2\pi$ with error $\varepsilon = 5.2992 \times 10^{-8}$. 135

7.29 Maxentropic reconstruction with cubic splines from $m = 7$, $w = 4\pi$ with error $\varepsilon = 6.2902 \times 10^{-6}$. 136

7.30 Maxentropic reconstruction with cubic splines from $m = 10$, $w = 4\pi$ with error $\varepsilon = 9.9747 \times 10^{-6}$. 136

7.31 Maxentropic reconstruction with cubic splines from $m = 4$, in $[-5, 5]$, $w = \pi$, $\sigma = 0.1$, $T_1 = 0.5$, $T_2 = 1$, $T_3 = 1.5$ and $T_4 = 2$ with error $\varepsilon_1 = 0.05$, $\varepsilon_2 = 0.10$, $\varepsilon_3 = 0.20$ and $\varepsilon_4 = 0.20$. 137

7.32 Maxentropic reconstruction with cubic splines from $m = 4$, in $[-5, 5]$, $w = \pi$, $\sigma = 0.01$, $T_1 = 1.5$, $T_2 = 2$, $T_3 = 3$, $T_4 = 4$ and $T_5 = 5$ with the respective error $\varepsilon_1 = 0.020$, $\varepsilon_2 = 0.020$, $\varepsilon_3 = 0.030$, $\varepsilon_4 = 0.040$ and $\varepsilon_5 = 0.050$. 137

7.33 Maxentropic reconstruction with cubic splines from $m = 8$, in $[-5, 5]$, $w = \pi$, $\sigma = 0.1$, $T_1 = 0.5$, $T_2 = 2$, $T_3 = 3$ and $T_4 = 4$ with the respective error $\varepsilon_1 = 0.20$, $\varepsilon_2 = 0.20$, $\varepsilon_3 = 0.30$ and $\varepsilon_4 = 0.40$. 138

7.34 Maxentropic reconstruction with cubic splines from $m = 8$, in $[-5, 5]$, $w = \pi$, $\sigma = 0.01$, $T_1 = 0.5$, $T_2 = 1$, $T_3 = 1.5$ and $T_4 = 2$ with the respective error $\varepsilon_1 = 0.0131$, $\varepsilon_2 = 0.0135$, $\varepsilon_3 = 0.020$ and $\varepsilon_4 = 0.020$. 138

7.35 Maxentropic reconstruction with cubic splines from $m = 8$, in $[-1, 1]$, $w = 2\pi$, $\sigma = 0.01$, $T_1 = 0.5$, $T_2 = 1$, $T_3 = 2$ and $T_4 = 3$ with the respective error $\varepsilon_1 = 0.0118$, $\varepsilon_2 = 0.01$, $\varepsilon_3 = 0.02$ and $\varepsilon_4 = 0.03$. 139

7.36 Maxentropic reconstruction with cubic splines from $m = 8$, in $[-2, 2]$, $w = 2\pi$, $\sigma = 0.01$, $T_1 = 0.5$, $T_2 = 1$, $T_3 = 1.5$ and $T_4 = 4$ with the respective error $\varepsilon_1 = 0.005$, $\varepsilon_2 = 0.010$ and $\varepsilon_3 = 0.020$. 139

7.37 Maxentropic reconstruction with cubic splines and method No. 2, in $[-25, 25]$, from $m = 12$, $w = 2\pi$, $\sigma = 0.01$, $\mu_0 = 0.01$, $\mu_1 = 0$, $\mu_2 = 0.005$ and $\mu_3 = 0.02$ with the respective error $\varepsilon_1 = 0.0274$, $\varepsilon_2 = 0.0326$, and $\varepsilon_3 = 0.745$. 140

7.38 Maxentropic reconstruction with cubic splines and method No. 2, in $[-80, 80]$, from $m = 12$, $w = 4\pi$, $\sigma = 0.01$, $\mu_0 = 0.01$, $\mu_1 = 0$, $\mu_2 = 0.005$ and $\mu_3 = 0.02$ with the respective error $\varepsilon_1 = 0.0246$, $\varepsilon_2 = 0.030$, and $\varepsilon_3 = 0.0734$. 140

7.39 Reconstructions from $m = 8$, the respective L_1 errors are $\sigma_1 = 0.5272$ and $\sigma_2 = 0.6786$ and the reconstruction error is $\varepsilon_1 = 6.1374 \times 10^{-8}$. 141

7.40 Reconstructions from $m = 14$, the respective L_1 errors are $\sigma_1 = 0.3811$ and $\sigma_2 = 0.5654$ and the reconstruction error is $\varepsilon_1 = 8.2644 \times 10^{-8}$. 141

7.41 Reconstructions from $m = 6$, $\sigma = 0.1$, $T_1 = 0.5$, $T_2 = 1$ and $T_3 = 1.5$ the respective L_1 errors are $\delta_1 = 0.6088$, $\delta_2 = 0.6269$, $\delta_3 = 0.6502$, and $\delta_4 = 0.6340$, and the reconstruction errors for maximum entropy method are respectively $\varepsilon_1 = 0.05$, $\varepsilon_2 = 0.1$, $\varepsilon_3 = 0.15$. 142

7.42 Reconstructions from $m = 6$, $\sigma = 0.01$, $T_1 = 0.25$, $T_2 = 1$ and $T_3 = 3$ the respective L_1 errors are $\delta_1 = 0.5852$, $\delta_2 = 0.5895$, $\delta_3 = 0.5999$, and $\delta_4 = 0.6340$, and the reconstruction errors for maximum entropy method are respectively $\varepsilon_1 = 0.00255$, $\varepsilon_2 = 0.01$, $\varepsilon_3 = 0.03$. 142

7.43 Original function and three reconstructions using standard ME applied to 7, 8 or 9 givens moments. 143

7.44 Graph of ϕ^* reconstructed from four different prices. 164

9.1 Difference between data and reconstruction, using uniform a priori measure. 206

9.2 Difference between data and reconstruction, using Bernoulli a priori measure, with $p = 0.85$, $q = 0.15$. 206

9.3 Difference between data and reconstruction. 208

9.4 Maxentropic approach to a linear programming problem, with $p = 0.75$, $q = 0.25$, $\gamma = 50.2199$. The reconstruction error $\varepsilon = 1.7831 \times 10^{-11}$. 210

10.1 Reconstruction of $\exp(-t)\sin(2\pi t)$ using sine, Legendre and Laguerre bases. 231

10.2 Reconstruction of $\exp(-t)\sin(8\pi t)$ using sine, Legendre and Laguerre bases. 232

10.3 Reconstruction of $\sin(2\pi t)$ using sine, Legendre and Laguerre bases. 232

10.4 Reconstruction of $\sin(8\pi t)$ using sine, Legendre and Laguerre bases. 233

10.5 Reconstruction of $N[1 + \sin(2\pi t)]\exp(-t)$ and $N[1 + \sin(8\pi t)]\exp(-t)$ using standard maximum entropy. ... 233

10.6 Reconstruction of $N[1 + \sin(2\pi t)]$ and $N[1 + \sin(8\pi t)]$ using standard maximum entropy. 234

10.7 Maxentropic reconstruction with first order splines from $m = 5$ (0,2,4,6,8) $\beta = 1$, $t_0 = 1$, $\sigma_1 = \sigma_2 = 0.6931$, for different intervals of reconstruction $I_1 = [0, 1]$, $I_2 = [-1, 1]$ with errors $\varepsilon_1 = 1.4438 \times 10^{-6}$, $\varepsilon_2 = 1.7652 \times 10^{-6}$. 234

10.8 Maxentropic reconstruction with first order splines from $m = 5$ (0,2,4,6,8) $\beta = 1$, $t_0 = 1, \sigma_1 = \sigma_2 = 0.6931$, for different intervals of reconstruction $I_1 = [0, 1.5]$, $I_2 = [-2, 2]$ with errors $\varepsilon_1 = 1.2994 \times 10^{-4}$, $\varepsilon_2 = 1.4573 \times 10^{-6}$. 235

10.9 Maxentropic reconstruction with first order splines from $m = 10$ (0,1,2,3,4,5,6,7,8,9), $\omega = 1$, $\sigma = 0.6931$ with error $\varepsilon = 1.7567 \times 10^{-5}$. 235

10.10 Multiplying by $\exp(\beta t)$ the functions showed in figure 10.2. . . 236

10.11 Maxentropic reconstructions with first order splines from $m = 15$ $(0, 2, 4, ..., 26, 28)$, $\omega = 2\pi$, with different σ's values. Respectively values of σ's and errors are listed: $\sigma_1 = 0.009$, $\varepsilon_1 = 3.28 \times 10^{-6}$; $\sigma_2 = 0.09$, $\varepsilon_2 = 0.0014$; $\sigma_3 = 0.9$, $\varepsilon_3 = 0.0025$. 237

10.12 Maxentropic reconstructions with first order splines from $m = 15$ (0,2,4,...,26,28), $\omega = 8\pi$, with different σ's values. Respectively values of σ's and errors are listed: $\sigma_1 = 0.009$, $\varepsilon_1 = 1.22 \times 10^{-7}$; $\sigma_2 = 0.09$, $\varepsilon_2 = 4.68 \times 10^{-5}$; $\sigma_3 = 0.9$, $\varepsilon_3 = 6.25 \times 10^{-5}$. 237

12.1 The original picture. 250

12.2 The latitude (LAT) is equal to the height of the pole P above the horizon. 252

12.3 Here we define variables for the analysis. We identified some of the stars in the picture. The South Pole is identified with a letter P. The star pointed with the arrow is the most weak start that the author could identify on the original photo. . . 254

12.4 Southern Hemisphere's Map, used to determine the photo's date. The date is displayed in the external bound (Norton, 1978). 256

List of Tables

7.1	List of moments.	126
7.2	Transition matrix obtained with maxentropic reconstruction, $\varepsilon = 2.3842 \times 10^{-10}$.	155
7.3	Transition matrix obtained with maxentropic reconstruction, $\varepsilon = 1.1230 \times 10^{-12}$.	156
7.4	Error of reconstruction risk price.	163
9.1	Data matrix.	204
9.2	Reconstruction using uniform *a priori* measure. $\varepsilon = 2.1544 \times 10^{-9}$.	205
9.3	Reconstruction using Bernoulli *a priori* measure, with $p = 0.85$, $q = 0.15$. $\varepsilon = 1.6444 \times 10^{-9}$.	205
9.4	Reconstruction using standard maximum entropy.	208

Chapter 1

Introduction

In this volume we examine some basic aspects of linear inverse problems, examine some standard methods of solutions, and present the method of maximum entropy in the mean (MEM for short) as a powerful alternative method to solve constrained linear inverse problems. We shall see that the method takes care of the constraints in a natural way, and competes reasonably with the standard methods. As we shall eventually see below, MEM includes some standard methods as particular cases. We shall illustrate the range of applications with examples from many different fields.

The generic, and most general form of a linear inverse problem with convex constraints consists of solving for x in

$$Ax \in B_M(y, T), \quad x \in C \subset V \qquad (1.1)$$

where V and W are given Banach spaces, $A : V \to W$ is a linear bounded operator, C is some specified convex set in V. These are the convex constraints imposed on the solution.

Usually the data vector y is known up to some experimental error, time and again it does not even belong to $A(V)$ and we have to be flexible as to what we admit as solution. The right-hand side of (1.1) suggests that we will admit any $x \in C$, such that Ax is inside some ball, not necessarily in the same norm as in W, but certainly related to it and to the experimental errors, the radius T of which measures the "tolerance" with which we accept solutions.

When the data is certain, we can be strict and intolerant, set $T = 0$ to obtain, instead of (1.1):

$$Ax = y, \quad x \in C \subset V. \qquad (1.2)$$

So if linear inverse problems basically consist of solving a linear equation, why are they not simply called linear equations? Probably because of the interpretation of (1.2) or (1.1). Quite often x is interpreted as an input or as an initial condition, the datum y as an output or as actual observed state of the system and A as the transfer operator describing the evolution of the system. Thus, finding the input or stimulus given the output or response is an inverse problem.

Or you could follow Keller in saying: computing y from A and x is called a direct problem, the solving for x in $Ax = y$ is called the inverse problem.

The difficulties in solving (1.1) or (1.2) are of two kinds:

First, it usually happens that A is neither injective nor surjective. The lack of surjectivity is easy to deal with by restricting the range of A. When A is not injective, then non-uniqueness is present and infinitely many solutions exist. The problem is to devise algorithms that produce "meaningful" solutions. Here meaningful is tied to the interpretation of the solution, and is not an obviously mathematizable concept.

The second issue, important from the practical point of view, is that the "inversion" or "reconstruction" or more simply, the solution algorithms, should be robust, that is, small changes in the datum vector y should produce small changes in the reconstructed x. This amounts to saying that if $y \to x = B(y)$ is to satisfy (1.2) say, then B must be continuous in y.

Different aspects of these issues dealing with either functional analysis and/or linear algebra on one hand, and with the computational side of these problems have been extensively studied. Many of the references to Chapter 2, 3, 4 or 5 are but entrance gates to these different aspects.

One of the aims of these notes is to briefly review some basic notions related to solving (1.1) or (1.2). We carry out this in Chapter 3 and 4. Chapter 2 is devoted to listing some standard linear inverse problems.

A standard way of approaching (1.2) is to convert it in a variational problem that exploits the metric nature of the spaces V and W: one searches for points x in C, which minimize a penalized error functional

$$F(x) = \|Ax - y\|_W + \lambda \|x\|_V \qquad (1.3)$$

defined on the constraint set C. In this fashion the machinery of convex analysis can be brought in to bear on theoretical (and practical) aspects of the problem. Even though the interpretation of (1.3) is rather direct, we must comment on the role of the penalization parameter λ: Among those x's $\in C$ yielding similar reconstruction error $\|Ax - y\|_W$, we have to choose one with a norm as small as possible.

Why? Well in some cases $\|x\|_V$ may have an interpretation like "energy" or cost, and it makes sense to choose among the many solutions the one with smallest energy. In other instances one may be forced to shrug and say: well it works and I don't have anything else.

The basic heuristic behind the MEM, developed in Chapter 5, consists of considering the set C of possible reconstructions as values of a random variable X. Instead of devising methods to find an explicit solution to the given equation, we search for a probability distribution P on C such that $AE_P[X]$ satisfies either (1.1) or (1.2). Here $E_P[X] = \int X dP$ denotes the mean value of X with respect to P.

The problem is how to choose one among all possible P's that do the job. It is here that the concept of entropy enters: in analogy with (1.3), the entropy $S(P)$ will be a concave functional, defined on the set of all probability measures which is a convex set. Thus, the problem becomes one of maximization of a concave functional on some convex set. Hence the name of the method.

To be fair, we have to ask why the specific functional $S(P)$ chosen in Chapter 5 works. Despite the effort in making things appear "natural" (not yet in a functional sense), all we can say is that the procedure works.

Originally, the variational method was proposed by Jaynes to lay the foundations of statistical physics, where it provided an astounding bridge between micro- and macro-physics. The list of problems in which the method works is still growing. See Section 6.5 of Chapter 6 and take a look at references [12]-[27] for an interesting list of a variety of applications. Or look at the journal "Entropy" for more.

Now, since the: Why does it work? is tied to assembling a whole framework in which it does, we could as well as ourselves: Why that framework? To avoid longer regressions, the best answer so far is that it works.

It will be up to you, dear reader to help explaining why. Or at least, to have fun applying it.

Chapter 6 is devoted to the finite dimensional problems arising from the discretization of continuous problems or the truncation of infinitely many dimensional ones, and related issues. Some basic formulae are obtained there. We devote Chapter 7 to present some numerical examples.

In Chapter 8 we carry out the program outlined in Chapter 5 in a truly infinite dimensional setup. In particular, we shall see that some regularization methods in Hilbert space can be obtained from the maxentropic approach.

We shall devote Chapter 9 to review some results about reconstructing a function in the plane when its integral along a few lines is known. We end the chapter examining the reconstruction of the entries in a table from their row and column sums, using two maxentropic approaches: the classical maximum entropy method and the method of maximum entropy in the mean. After that, instead of solving a transportation problem, we approach maxentropically a linear programming problem, arising in Chapter 7 when reconstructing a function from its moments.

In Chapter 10 we review some basic stuff about Laplace transforms as well as some of the work done to deal with a vexing inverse problem in applied mathematics: that of reconstructing a function from a few values of its Laplace transform. We try a few maxentropic shots at it as well.

The material in Chapter 11 is just for fun. It provides a maxentropic characterization of some non-exponential families, but it involves a departure from the traditional or standard maximum entropy method.

We provided a few appendices, where some basics on topology, measure and probability theory, (very) elementary stuff on Banach spaces and convexity is gathered. It is really a commented list of definitions for those unfamiliar with that terminology.

To finish we would like to thank Aldo Tagliani for contributing material to Chapters 7 and 10 as well as Michael Hazewinkel for making the resources of the nice library of the CWI in Amsterdam available to us.

Special thanks go to Ignacio Ferrín for writing the last chapter. There he presents a "detectivesque inverse" problem consisting of interpreting a picture.

Chapter 2

A collection of linear inverse problems

In this chapter we collect some examples of linear inverse problems. The comments about each item are to reflect typical issues described in the next chapters.

Some of these we shall solve by the method of maximum entropy in the mean. Others we solve by the method of maximum entropy in the mean and at least one other method. This is just to compare solutions, and to emphasize the need of para-mathematical or meta-mathematical criteria of model building and problem solving.

2.1 A battle horse for numerical computations

The problem consists of solving

$$Ax = y \tag{2.1}$$

where A is an $n \times m$ matrix, y is a given vector in $\mathbb{R}^n$ and x is an unknown vector in $\mathbb{R}^m$ with $n \neq m$ is so ubiquitous that we should just state it, describe some variants on the problem as well as some methods of solution. Even when $m = n$, A^{-1} may no exists and the system may not have a unique solution or no solution at all.

Suppose to begin with that $m \geq n$ so that $Ker(A)$ is not $\{0\}$. Two standard ways of choosing one among the infinitely many solutions depend on whether $(A^*A)^{-1}$ or $(AA^*)^{-1}$ exists. In the first case multiply both sides of (2.1) with A^* and then solve for x in $(A^*A)x = A^*y$ obtaining $x = (A^*A)^{-1}A^*y$.

In the second case, and when $Ker(A) \neq \{0\}$ but AA^* is invertible, one way of choosing among the infinitely many possible solutions, is to consider the x_0 that solves the variational problem:

$$\inf\left\{\frac{1}{2}\|Ax-y\|^2\right\}. \qquad (2.2)$$

We can apply either the standard Lagrange multiplier method or duality techniques to obtain:

$$x_0 = A^*(AA^*)^{-1}y.$$

When, for example either A^*A is not invertible, or if it is invertible, its eigenvalues are so small that possible measurement errors in y are amplified too much, one recurs to searching for x_0 realizing

$$x_0 = \arg\inf\left\{\frac{\lambda}{2}\|x\|^2 + \|Ax-y\|^2 : x \in \mathbb{R}^m\right\}. \qquad (2.3)$$

Here λ is a penalization parameter by means of which we assign "different weights" to the two terms in (2.3). An easy computation yields

$$x_0 = (\lambda I + A^*A)^{-1} A^*y \qquad (2.4)$$

as the solution to (2.3). Here the role of λ becomes clear: if the eigenvalues of the symmetric matrix are too small, we push them up with λ. If they are all large, we take λ to be small.

2.2 Linear equations with errors in the data

This time, instead of (2.1), we are supposed to have an equation like:

$$Ax = y_0 + \varepsilon = y$$

where y_0 is supposed to be a true value and ε is supposed to model a random error in the measurement of y_0.

Again suppose $Ker(A) \neq \{0\}$. Sometimes we may have enough measurements of y, in which the errors may add up to zero. Since the methods deployed in Section 2.1 yield a solution linear in y, when we average over solutions, we end up washing away the effect of the measurement errors.

But usually one only has one value of y plus a mathematical or physical model for the error ε.

In this case, a procedure that exploits both Euclidean Geometry and Gaussian distribution of errors suggests considering to solve for x in

$$Ax \in B_M(y_0, T) \qquad (2.5)$$

where
$$B_M(y_0, T) = \left\{ \eta \in \mathbb{R}^n : \sum m_{ij}(y_j - y_j^0)(y_i - y_i^0) \leq T \right\}$$
is the ball of radius T with center y_0 in a (positive definite) metric m_{ij}. When the measurements of the components of y_0 are independent of each other, it is natural to assume that M is diagonal matrix.

Again, when $Ker(A) \neq \{0\}$, there are infinitely many solutions to (2.5), or solutions are unique up to addition of elements in $Ker(A)$. To determine a solution, we do as in section 2.1, namely we look for
$$x_0 = \arg\inf \left\{ \frac{1}{2} \|x\|^2 : \|Ax - y\|_M \leq T \right\}.$$

Actually, even if the drawing below is not exact, but it suggests the proof of the fact that we may replace the former characterization by:
$$x_0 = \arg\inf \left\{ \tfrac{1}{2} \|x\|^2 : \|Ax - y\|_M = T \right\}. \qquad (2.6)$$

The drawing is:

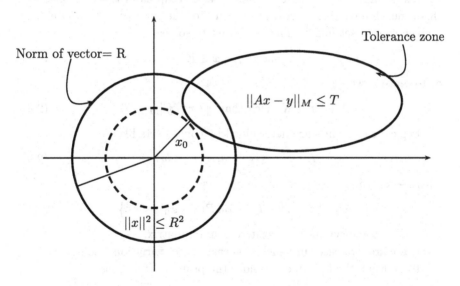

Lemma 2.1. *The infimum of* $\left\{ \tfrac{1}{2} \|x\|^2 : \|Ax - y\|_M \leq T \right\}$ *is achieved at some x_0 for which*
$$\|Ax_0 - y\|_M = T.$$

Proof. Assume that x_0 realizes the minimum but $\|Ax_0 - y\|_M = T_1 < T$. The value of the tolerance at βx_0, $0 < \beta < 1$ is given by

$$\|A(\beta x_0) - y\|^2 = \|\beta(Ax_0 - y) - (1-\beta)y\|^2$$
$$= \beta T_1^2 + 2\beta(1-\beta)(y, Ax_0 - y) + (1-\beta)^2 \|y\|^2 .$$

A simple continuity argument shows that as β gets closer to 1, we can keep the right-hand side of the last expression less than T^2, i.e. βx_0 is within the tolerance zone, but its norm is less than that of x_0, contrary to the assumption. □

The importance of this lemma is that to find x_0 it suffices to apply Lagrange multipliers to (2.6). But this leads to a non-linear system. Let us drop this issue for the time being.

2.3 Linear equations with convex constraints

In this example we consider the same basic setup as in the first one, but this time we have convex constraints on x of the type: $x \in K$, where K is a given convex set in $\mathbb{R}^m$. Thus, we want to solve:

$$Ax = y, \quad x \in K \tag{2.7}$$

or to make it worse

$$\text{Find } x \in K \text{ such that } \|Ax - y\|_M \leq T. \tag{2.8}$$

Typical constraints are determined by convex sets like:

$$K = \{x \in \mathbb{R}^m : a_i < x_i < b_i\}, \tag{2.9}$$

where $-\infty \leq a_i < b_i \leq +\infty$, or we may have

$$K = \{x \in \mathbb{R}^m : \langle x - x_0, B(x - x_0)\rangle < b^2\}, \tag{2.10}$$

where B is some symmetric, positive definite matrix.

There are two cases in which it is easy to get variational solutions. To describe the first one, consider the problem of finding:

$$x_0 = \arg\inf \left\{ \tfrac{\lambda}{2} \|x\|_X^2 + \tfrac{1}{2} \|Ax - y\|_Y^2 : x \in K \right\} \tag{2.11}$$

where K is described by (2.10).

Note that if we did not have the restriction $x \in K$, the following iterative procedure would lead to x_0 : Define:

$$x_{n+1} = x_n - \varepsilon \left\{\lambda x_n - A^*(y - Ax_n)\right\}, \tag{2.12}$$

i.e., take steps of size ε again, the gradient of the convex function

$$F(x) = \tfrac{\lambda}{2} \|x\|_X^2 + \|Ax - y\|_Y^2.$$

If we define the projection operator $P : \mathbb{R}^m \to K$ by:

$$(Px)_i = \begin{cases} x_i & \text{if } a_i < x_i < b_i \\ a_i & \text{if } x_i \leq a_i \\ b_i & \text{if } b_i \leq x_i \end{cases}$$

for $i = 1, 2, ..., m$; then instead of (2.12), to find the x_0 solving (2.11) we have

$$x_{n+1} = (1 - \lambda\varepsilon)Px_n + \varepsilon PA^*(y - Ax_n). \qquad (2.13)$$

This procedure already appears in some books about numerical recipes.

The other variant appears when it makes sense to impose an L_1 norm on $\mathbb{R}^m$. We should consider

$$\|x\|_1 = \sum_{i=1}^m \omega_i |x_i|.$$

Note that $|x_i| = x_i$ or $-x_i$ depending on whether $x_i > 0$ or $x_i < 0$. We shall then write $x = u - v$ with u_i and v_i both ≥ 0 and recast problem (2.1) as

$$\tilde{A}\tilde{x} = \begin{bmatrix} A & -A \end{bmatrix} \begin{pmatrix} u \\ v \end{pmatrix} = y, \quad \tilde{x} \in \mathbb{R}^{2m}_+ \qquad (2.14)$$

where the notational correspondences are obvious.

The variational method for producing a solution to (2.14) becomes a standard linear programming problem: Find

$$\tilde{x}_0 = \arg\inf\left\{ \sum_{i=1}^m \omega_i u_i + \sum_{i=1}^m \omega_i v_i : \tilde{A}\tilde{x} = y, \tilde{x} \in \mathbb{R}^{2m}_+ \right\}. \qquad (2.15)$$

And it may be a physical requirement that instead of $\tilde{x} \in \mathbb{R}^{2m}_+$, we actually have a more general convex constraint as above, for example, instead of (2.15) we may have to find:

$$\tilde{x}_0 = \arg\inf\left\{ \sum_{i=1}^{2m} \omega_i \tilde{x}_i : \tilde{A}\tilde{x} = y, a_i \leq \tilde{x}_i \leq b_i \right\}. \qquad (2.16)$$

Of course $0 \leq a_i < b_i < +\infty$, as consistency requires.

2.4 Inversion of Laplace transforms from finite number of data points

This example and the next are typical of situations in which one has an integral transform like:

$$\tilde{f}(s) = \int K(s,t) f(t) m(dt), \qquad (2.17)$$

where $f(t)$ is in some class of functions X defined on some measure space $(E, \mathcal{E}, m)$, and the transform maps X onto Y in a bijective function. Y is another class of functions on some other measure space (S, Σ, σ).

But it may now happen that $\tilde{f}(s)$ can be observed, measured or determined only for a finite number of values of s. The problem is to solve (2.17) for $f(t)$ when one is given $\tilde{f}(s_1), ..., \tilde{f}(s_n)$.

A tough, particular case of (2.17), consists of numerically inverting the Laplace transform of $f(t)$ given finitely many values $\tilde{f}(s_i)$ of the transformed function. .

A candidate for X is for example the class of function defined on $[0, \infty)$ that grow not faster than $p(t) \exp(\alpha_0 t)$ for some fixed α_0.

Then

$$\tilde{f}(s) = \int_0^\infty \exp(-st) f(t) dt \qquad (2.18)$$

is well defined for complex s with $\mathcal{R}]s > \alpha_0$. Actually it is continuous for $\mathcal{R}]s \geq \alpha_0$, and analytic in $\mathcal{R}]s > \alpha_0$ if $f(t)$ is integrable.

Assume (2.18) is known only for finitely many numbers in $\mathcal{R}]s > \alpha_0$, and it is known that $f(t)$ can be regarded to be in a smaller class; say linear combinations of products of exponentials times polynomials times elementary trigonometric functions (say sines and cosines). This happens when for example, $f(t)$ is the response of a linear system.

As started above, the problem belongs to the class of generalized moment problems.

There are two approaches to follow to solve for $f(t)$ in (2.18) given $\tilde{f}(s_1), ..., \tilde{f}(s_n)$.

First approach: Just solve for $f(t)$ given the finite data.

Second approach: Try to find an $\tilde{f}_l(s)$ in a certain class such that

$$\tilde{f}_l(s_i) = \tilde{f}(s_i), \ i = 1, ..., n;$$

then invert the Laplace transform assuming that $\tilde{f}_l(s)$ is the right Laplace transform of the unknown f.

We shall have more to say about both approaches. To finish note that a conformal mapping transforming the right-hand-complex half-plane onto the unit circle allows us to relate the second approach to the famous:

Pick-Nevanlina interpolation problem: Given two finite sets of complex numbers $z_1, ..., z_n$ and $\tilde{f}_1, ..., \tilde{f}_n$; all of absolute value less than 1, find a holomorphic function $h : U \to U$ on the unit disk such that:

$$h(z_i) = \tilde{f}_i \quad i = 1, ..., n.$$

2.5 Fourier reconstruction from partial data

Let $f(t)$ be a real-valued, function defined on an interval $[0, T]$. It is known that

$$\int_0^T \exp(i\omega t) f(t) dt = \hat{f}(\omega) \qquad (2.19)$$

for ω in the finite set $\{\pm\omega_1, \pm\omega_2, ..., \pm\omega_n\}$. The problem is to find $f(t)$. Sometimes one may want to think of $\hat{f}(\omega)$ as being the Fourier transform of a measure $dF(t)$ not necessarily absolutely continuous with respect to dt. For example:

$$dF(t) = \sum p_n \delta(t - t_n) dt$$

where p_n are "weights" and t_n are the points at which the weights are concentrated.

This type of problem appears when one studies second order stationary processes. Here one determines correlations $R(k)$ from the data and one searches for a (spectral) measure $dF(t)$ on $[0, 2\pi]$ such that:

$$R(k) = \int_{[0,2\pi]} \exp(ik\lambda) dF(\lambda), \quad |k| \leq N. \qquad (2.20)$$

Another problem in which one knows a Fourier transform occurs when reconstructing velocity profiles in stratified models of the Earth. We refer the reader to the references quoted at the end of the chapter. The problem we are interested in is to find $a(x)$ of compact support in $[0, \infty)$ such that:

$$\hat{a}(\omega) = \int_0^\infty \exp(i\omega x) a(x) dx \qquad (2.21)$$

is known. Here the interest lies in finding $a(x)$'s satisfying $a_1 \leq a(x) \leq a_2$. These bounds can be justified since the speed of propagation of sound is bounded below (by the speed of propagation of sound in the air) and above (by the speed of propagation in some very rigid material).

Different types of constraints appear when one reconstructs probability densities $\rho(r)$ from their Fourier transforms. Here one knows $\hat{F}(p)$ such that:

$$\hat{F}(p) = \int_{\mathbb{R}^5} \exp(ipr)\rho(r)dr \qquad (2.22)$$

and one wants $\rho(r) \geq 0$ such that $\int \rho(r)dr = 1$. Of course, if $\hat{F}(p)$ is known for all p in $\mathbb{R}^3$, one would just invert the Fourier transform. But usually $\hat{F}(p)$ is known only for finitely many values of p, and measurement errors must be dealt with on top of it.

A classical example relating the issues of the choice of norms and the summation of Fourier series is the following. Consider the mapping:

$$A : C[0,1] \to l_2 \quad \text{defined by} \quad f(t) \to c_n = \frac{1}{2\pi} \int_0^1 \exp(2i\pi nt)f(t)dt.$$

Let $f(t) \in C[0,1]$ and let $\sum c_n \exp(-2\pi i n t)$ be its Fourier representation.

Let $\{d_n\} \in l_2$ be given by $d_n = c_n + \frac{\varepsilon}{|n|}$, thus

$$\|d - c\| = \left(\sum (d_n - c_n)^2\right)^{\frac{1}{2}} = \varepsilon \sqrt{\frac{\pi^2}{6}}$$

which can be made as small as desired, but if $g(t) = \sum d_n \exp(-2i\pi nt)$ then

$$\|f - g\|_\infty = \varepsilon \left\|\sum \tfrac{1}{n} \exp(-2\pi i n t)\right\|_\infty.$$

Since at $t = 1$ we obtain the harmonic series, $\|f - g\|_\infty = \infty$. If you (rightly) feel we seem to be cheating, just consider a sequence $d_n^N = c_n + \frac{\varepsilon}{n}$ for $|n| \leq N$, and zero otherwise. Then $\{d_n^N\}$ is as close as you want to $\{c_n\}$ and $g^N(t)$ is well defined, but $\|g^N - f\|_\infty$ is as large as you want.

The moral is, even though A is continuous, A^{-1} is not.

2.6 More on the non-continuity of the inverse

A classical example of non-continuity of inverses is the following. On $C[0,1]$ provided with its usual l_∞ norm define $T : C[0,1] \to C[0,1]$ by:

$$(Tx)(t) = \int_0^t x(s)ds.$$

Certainly the range of T is the class $C^1[0,1]$ of continuously differentiable functions on $(0,1)$.

If $y_1(t)$ is continuously differentiable, let $y_2(t) = y_1(t) + A\sin(\omega t)$. Note that $\|y_1 - y_2\|_\infty = |A|$ but $\|x_2 - x_1\|_\infty = \|y_2' - y_1'\|_\infty = |A\omega|$ which can be made as large as desired keeping $|A|$ as small as desired as well.

But note that if we consider $T : C^1[0,1] \to C[0,1]$ and on $C^1[0,1]$ we put the norm:

$$\|y_1 - y_2\|^* = \|y_1 - y_2\|_\infty + \|y_1' - y_2'\|_\infty$$

then T has a continuous inverse T^{-1} (when restricted to $C^1[0,1]$!).

Comment: If on $C[0,1]$ we define the σ-algebra $\mathcal{F}$ that makes the coordinate mapping $X_t : C[0,1] \to \mathbb{R}$, $X_t(\omega) = \omega(t)$ measurable, and on $(C[0,1], \mathcal{F})$ we define the usual Wiener measure, then $C^1[0,1]$ has measure zero in $C[0,1]$!

2.7 Transportation problems and reconstruction from marginals

The simplest transportation problem consists of shipping gods from m sources, in amounts s_i from the i-th source. Goods are to be received at n destinations, in required amounts d_j at the j-th one.

If there is a cost c_{ij} involved in the shipping from origin i to destination j, the simple transportation problem consists of finding:

$$x_{ij}^* = \inf \left\{ \sum_{i,j=1,1}^{m,n} c_{ij} x_{ij} : x_{ij} \geq 0, \sum_{j=1}^{n} x_{ij} = s_i, \sum_{i=1}^{m} x_{ij} = d_j \right\}.$$

This problem is usually solved in a few seconds by many software packages. Actually it usually takes longer to translate a given problem into that set up, to key it in at the keyboard, than what it takes the PC to provide the solution.

What is important for us here is the issue of the existence of feasible solutions, i.e., given: $\{s_i : i = 1, 2, ..., m\}$ and $\{d_j : j = 1, 2, ..., n\}$ find $x_{ij} \geq 0$, or $a_{ij} \leq x_{ij} \leq b_{ij}$ for preassigned a_{ij}, b_{ij}, such that:

$$\sum_{i=1}^{m} x_{ij} = d_j, \quad \sum_{j=1}^{n} x_{ij} = s_i$$

and consistency requires that $\sum d_j = \sum s_i$.

To consider explicitly an example in which the non-uniqueness of the solution appears look at the empty 3 × 4 array:

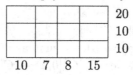

which we have to fill up with positive numbers whose sums by rows and columns adds up to the indicated amounts.

A started method of filling the array is called the northwest corner method, which consists of:

i) Fill the most northwest (empty) box with a number not exceeding the smallest constraint.
ii) If it equals the column (row) constraint, delete the column (or row) and update the corresponding column (row) contained the filled box.
iii) Repeat with the remaining array.
iv) In case of a tie, break it *a piacere*.

If instead of considering the most northwest, we had taken the most northeast, southwest or southeast corners we would have ended with different reconstructions. Here are two of them:

10	5	3	2	20
0	0	5	5	10
0	2	0	8	10
10	7	8	15	

0	0	5	15	20
0	7	3	0	10
10	0	0	0	10
10	7	8	15	

By taking convex combinations one may generate infinitely many solutions.

This problem is a particular case of a much more general problem. To state it consider two measure spaces $(X, \mathcal{F}, \mu)$ and (Y, Ψ, u) and form

$$Z = X \times Y \quad \Gamma = \mathcal{F} \otimes \Psi.$$

The problem consists of finding a measure ρ on (Z, Γ) such that

$$\rho(A \times Y) = m(A), \quad \rho(X \times B) = n(B) \qquad (2.23)$$

for all $A \in \mathcal{F}$, and $B \in \Psi$.

And if possible, categorize all measures ρ on (Z, Γ) such that (2.23) holds! As in the linear programming problem, one hopes that looking for:

$$\rho^* = \arg\inf \left\{ \int c(x, y) \rho(dx, dy) : (2.23) \text{ holds} \right\}$$

leads to an easier characterization of solutions to (2.23).

Actually, since the class:

$$\wp(m,n) = \{\rho \text{ measure on } (Z,\Gamma) : (2.23) \text{ holds}\}$$

is a convex set, trivially non-empty for $m \otimes n$ belongs to it, any lower semicontinuous, convex function $F : \wp(m,n) \to \mathbb{R}$ is likely to provide us with candidates solving (2.23) when some compactness is available.

A variation on the theme of (2.23) and usually called the problem of tomographic reconstruction consists of finding a positive function ρ on some compact K of a metric space from the knowledge of:

$$\int \rho(\xi)\mu_i(d\xi) = d_i \qquad (2.24)$$

where the measures $\mu_i(d\xi)$ are usually concentrated on sets of measure zero with respect to a given measure $m(d\xi)$ on (K,β), β denoting the Borel sets in K.

2.8 CAT

CAT stands for either computer axial tomography or computer aided tomography. Either name describes a complex process involves obtaining X-ray images of an object.

The technique consists of a combination of hardware-software by means of which an image of a series of consecutive slices (tomos) of an object, usually part of a human body, are obtained. The image consists of an X-ray plaque of each given section (slice). Different aspects of the historic development of the tomographic process are contained in [10]-[13]. In [11] you will find a brief account of the story from the point of view of the research-development-commercialization process, whereas the other two emphasize the connection with the evolution of the development of the mathematics of the inverse problem that is needed to present the images of the slices.

Let us begin by describing the phenomenological model of light absorption by tissue. Everything is contained in the following simple, one-dimensional model: consider a slab of material, of linear density $\rho(x)$ extending from 0 to L. Let the radiation incident at $x = 0$ be of intensity I_0 and denote by $I(x)$ the intensity of the radiation reaching point x. Thus the radiation reaching $x + dx$ is

$$I(x + dx) = I(x) - \lambda \rho(x) I(x) dx$$

from which we obtain that the radiation at $x = L$ is

$$I(L) = I_0 \exp\left(-\int_0^L \lambda \rho(x) dx\right). \tag{2.25}$$

We shall follow [10] for the description of conventional tomography and to present the reader with one more set of references to the subject. According to [13] it seems that the basics were patented by the Frenchman Bocade around 1921. Consider two parallel planes, separated by a distance L.

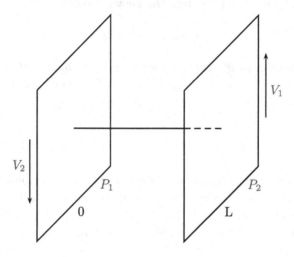

Fig. 2.1 Basic scheme.

Consider two points 1 and 2 on each plane. After some time Δ they will be at $1'$ and $2'$ respectively, with $\overline{11'} = V_1 \Delta$ and $\overline{22'} = V_2 \Delta$.

Using this as starting point, we leave it to the reader to prove that there is a plane Pf which stays fixed. Write $V_2 = \alpha V_1$ and determine the position of the plane in terms of α.

Imagine now a source of X-rays on plane 1 and a photographic plate on plane 2. Since the motion of the planes is much, much slower than the speed of light, we might imagine that situation depicted in figure 2.2 holds instantaneously, and in a "more realistic" schematic depiction of an X-raying process we will have something like figure 1 in [10].

Clearly, the details in the fixed plane get more (negatively) illuminated than objects in non-fixed planes, so they will appear less blurred in the exposed plate. If this process is repeated for different relative source-plate po-

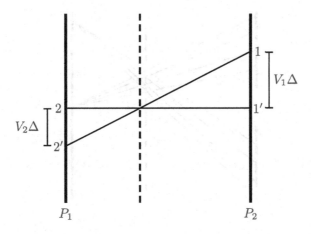

Fig. 2.2 Basic geometry of the CAT process.

sitions and for different source-plate speeds we obtain a sequence of images which are the basis for the inverse problem of tomographic reconstruction.

For this we need to put in some detail into figure 2.3 with respect to figure 2.4, in which both the edges of the photographic plate and the center of the photographic plate are joined to the X-ray source by solid lines, denote by θ the angle of this last line and the horizontal axis.

Let η be the vertical distance measured from the center of the plate, and let $\theta(\eta)$ be the angle between the ray from the source to η and the horizontal axis.

If we assume that the object is located in the central area between the plate (which slides along the $x = -L$ vertical axis) and the X-ray source (which slides along $x = L$ vertical axis), and if we denote by γ the dotted line in figure 2.4, assuming that this configuration is held during a time δ during which the intensity of the radiation collected at η is

$$\delta I(\eta,\theta) = \delta I_0(\eta,\theta) \exp\left(-\int_{-L}^{L} \lambda \rho(x, x\tan\theta + \frac{L-x}{2L}\eta) \frac{dx}{\cos\theta}\right). \quad (2.26)$$

Here $y = x\tan\theta + \frac{L-x}{2L}\eta$ is the vertical coordinate of a point in the object whose horizontal coordinate is x. We are approximating $\theta(\eta)$ by θ and the length element along the line γ is approximated by $\frac{dx}{\cos\theta}$. It is usually assumed that the error incurred in this approximation is negligible.

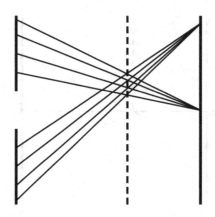

Fig. 2.3 Tomographic image formation.

In (2.26) $dI_0(\eta,\theta)$ would be the intensity recorded in the plate if there were no object in the path of the X-rays. If you do not find the approximation in (2.26) satisfactory, the exponent should be replaced by the integral

$$\int_\gamma \rho\, dl(\theta) \tag{2.27}$$

of the density $\rho(x,y)$ of the object along the ray $\gamma(\theta)$.

For a given pair (V_1, V_2) of source-plate velocities, the central ray rotates in angle from θ_0 to θ_1, the total radiation collected at η in the plate is

$$I(\eta) = \int_{\theta_0}^{\theta_1} d\theta \frac{dI_0}{d\theta}(\eta,\theta) \exp\left(-\lambda \int_\gamma \rho\, dl(\theta)\right) \tag{2.28}$$

$$I(\eta) \approx \int_{\theta_0}^{\theta_1} d\theta\, C \exp\left(-\lambda \int_{-L}^{L} \rho(x, x\tan\theta + \frac{L-x}{2L}\eta) \frac{dx}{\cos\theta}\right). \tag{2.29}$$

It would be reasonable to assume that for not too long an exposition time $\frac{dI_0(\eta,\theta)}{d\theta} = C$ is a constant. Notice now that when $\theta_1 - \theta_0 << \theta_0$, which has been brought about by the improved technology, we can rewrite (2.28) as

$$\ln\left(\frac{dI(\eta)}{dI_0(\eta)}\right) \propto \int_\gamma \rho\, dl(\theta). \tag{2.30}$$

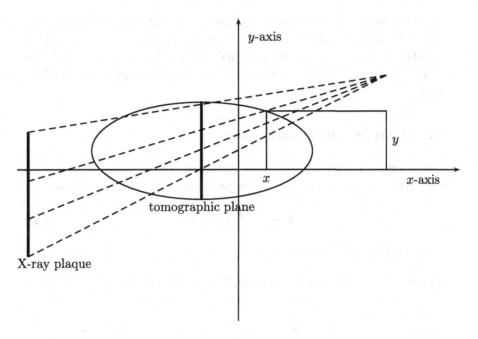

Fig. 2.4   Actual CAT setup.

If we consider the image to be formed by a parallel stream of X-rays impinging on a plate (actually a linear array of photosensitive detectors) lying along the line characterized by the unit vector. The X-ray image of the two-dimensional density $\rho(x,y)$ at a point t along the plate is

$$(R_\theta \rho)(t) = \int_{-\infty}^{\infty} \rho(t\cos\theta - s\sin\theta, t\sin\theta + s\cos\theta)ds.$$

After this lengthy digression, we are ready to state what is commonly known as the **tomographic reconstruction** problem.

Let V be a given Banach space of functions defined in a unit disk in $\mathbb{R}^2$ (or in some bounded set in $\mathbb{R}^n$). For a given family of lines with unit normals $\eta_i = (-\sin\theta_i, \cos\theta_i)$ (or hyperplanes with normals η_i) for $i = 1, 2, ..., N$ define the projection operators

$$(R_{\theta(i)}\rho)(x) = \int_{-\infty}^{\infty} \rho(x + s\eta_i)ds \qquad (2.31)$$

where x denotes a point in the line (or in the hyperplane).

The problem consists in finding $\rho \in V$ from the knowledge of (2.31).

Comment: It is left to the reader to convince himself/herself that (2.31) describes marginalization of ρ along successively rotated axes. Thus standard reconstructions from marginals fall within the class of problems described above. We will examine some aspects of these problems in Chapter 9.

2.9 Abstract spline interpolation

The classical interpolation problem consists of finding a $\sigma \in C^2[0,1]$ such that $\sigma(t_i) = z_i$ for $i = 1, 2, ..., n$. The procedure consists in minimizing:

$$F(\sigma) = \int_0^1 (\sigma''(t))^2 dt \qquad (2.32)$$

and the result is contained in:

Theorem 2.1. *There is a unique $\sigma \in C^2[0,1]$ minimizing (2.32) such that:*

a) $\sigma(t_i) = z_i$.
b) σ *is a cubic polynomial in* $[t_i, t_{i+1}]$, $i = 1, 2, ..., n-1$.
c) $\sigma''(t_1) = \sigma''(t_n) = 0$.
d) σ *is linear on* $[0, t_1]$ *and* $[t_n, 1]$.

Instead of the interpolation conditions $\sigma(t_i) = z_i$ for $i = 1, 2, ..., n$; one could search for a function σ satisfying:

$$\int_0^1 \sigma(t) P_l(t) \omega(t) dt = z_i \qquad (2.33)$$

where $\omega(t)$ is some weight function and the $\{P_i(t), i \geq 1\}$ are the orthogonal polynomials with respect to $\omega(t)$, i.e., they are orthogonal in $\ell_2([0,1], \omega)$.

The problem can be abstracted into the following setup: Let X, Y, Z be Hilbert spaces. Let $A : X \to Y$ and $B : X \to Z$ be two bounded, linear surjections. Let $y_0 \in Y$ and the *problem consists of finding* $x_0^* \in X$ such that

$$x_0^* = \arg\inf \{\|Bx\|_Z : Ax = y_0\}. \qquad (2.34)$$

The existence and uniqueness of x_0^* is covered by:

Theorem 2.2. *If*

a) $B(Ker(A))$ *is closed in* Z.

b) $Ker(B) \cap Ker(A) = \{0_X\}$. ($0_X$ denotes the zero element of X.)
c) $y_0 \in Y$.

Then there is a unique x_0 such that $Ax_0 = y_0$ and
$$\|B(x_0)\|_Z = \inf \{\|Bx\|_Z : Ax = y\}.$$

Proof. A being surjective, implies the existence of $x_1 \in X$ such that
$$A^{-1}\{y_0\} = x_1 + Ker(A).$$
Thus $B(A^{-1}\{x_0\}) = B(x_1) + B(Ker(A))$.

From the continuity of A and B we obtain the closedness of $B(Ker(A))$ and hence, $B(A^{-1}\{x_0\})$, which being a translate of a subspace is convex.

Let z_0 denote the projection of 0_Z onto $B(A^{-1}\{x_0\})$, thus
$$z_0 \in B(x_1) + B(A^{-1}\{x_0\})$$
which means that there exists $x \in Ker(A)$ such that $z_0 = B(x_1) + B(x) = B(x_1 + x)$ or there exists $x \in Ker(A)$ such that $x_1 + x \in B^{-1}\{z_0\}$. But,
$$A(x + x_1) = A(x_1) + A(x) = y_0 + 0_X = y_0$$
or $x + x_1 \in A^{-1}\{y_0\} \cap B^{-1}\{z\}$, i.e., this set is not empty. Now let x_1, x_2 be two different elements in
$$A^{-1}\{y_0\} \cap B^{-1}\{z\}.$$
Then,
$$B(x_1 - x_2) = 0_Z, \ A(x_1 - x_2) = 0_Y$$
plus assumption b) implies that $x_1 = x_2$ or $A^{-1}\{y_0\} \cap B^{-1}\{z\}$ is a singleton, which we denote by x_0. □

2.10 Bibliographical comments and references

<u>Generic comment</u>: References will be cited as [n] or [m-n]. The first is a reference n at the end of the present chapter, whereas [m-n] denotes the nth reference in the list at the end of Chapter m.

A couple of new (as of 1998) books in which finite dimensional inverse problems are treated in full are: [1] and [2]. A nice collection of inverse problems and the basic theory to deal with them is exposed in [2] and in [3]. Even more theory and applications, both linear and nonlinear appear in [4], [5] and [6]. Two related references on the trigonometric moment problem are [7], [7-7] and [7-8]. On the transportation problem take a look at any book on Operations Research. For information about tomography, its history, and further references, consult with [8]-[13], and with many of the papers described in Chapter 9.

References [14]-[16] contain historical remarks and description of the evolution of attempts to solve the inverse problem of CAT and related imaging problems in various fields as well as a list of further references.

For related work on transportation problems consider references [9-18] and [9-27] and for a nice and short expose on abstract splines take a look at [17] from which we took material for Section 9.10.

References

[1] Mansen, P. C. "Rank-Deficient and Discrete Ill-Posed Problems". SIAM, Philadelphia, 1998.

[2] Bertero, M. and Bocacci, P. "Introduction to Inverse Problems in Imaging". IOP Publishing, Bristol, 1998.

[3] Parker, R. "Geophysical Inverse Theory". Princeton Univ. Press, Princeton 1994.

[4] Ramm, A. G. *"Scattering by Obstacles"*. Kluwer Acad. Pub., Dordrecht, 1986.

[5] Ramm, A. G. *"Multidimensional Inverse Problems"*. Longman Scientific, 1992.

[6] Laurentiev, M. M. "Some Improperly Posed Problems of Mathematical Physics". Springer-Verlag. 1967.

[7] Romanov, V. G. *"Inverse Problems of Mathematical Physics"*. VNU Science Press, Utrecht, 1987.

[8] Orphanoudakis, S. and Strohben, J. *"Mathematical model of conventional tomography"*. Med. Phys., Vol. 3 (1976), pp. 224-232.

[9] Adams, J. L. *"Flying Buttresses, Entropy and O. Rings"*. Harvard Univ. Press, Cambridge, 1991.

[10] Shepp, L. A. and Kruskal, J. B. *"Computerized Tomography: the new medical X-ray technology"*. Ann. Math. Monthly, Vol. 85 (1978) pp. 420-439.

[11] Panton, D. *"Mathematical reconstruction techniques in computer axial tomography"*. Math. Scientist, Vol. 6 (1981), pp. 87-102.

[12] Fuchs, J., Mast, K., Hermann, A. and Lackner, K. *"Two-dimensional reconstruction of the radiation power density in ASDEX upgrade"*. 21st. E.P.S. Conf. on Controlled Fusion and Plasma Physics. Joffrin, E., Platz, P. and Scott, P. (eds), Publ. E.P.S., Paris, 1994.

[13] Friedman, A. *"Image reconstruction in oil refinery"* in *Mathematics in Industrial Problems*, IMA Volumes in Mathematics and its Applications, Vol. 16, Springer-Verlag, 1988.

[14] Hounsfield, G. *"Historical notes on computerized axial tomography"*. The J. of the Canadian Assos. of Radiologists, Vol. 27 (1976), pp.135-191.

[15] Brooks, R and Di Chiro, G. *"Principles of computer assisted tomography (CAT) in radiographic and radioisotopic imaging"*. Phys. Med. Biol., Vol. 21 (1976), pp. 689-732.

[16] Gordon, R. and Herman, G. *"Three dimensional reconstruction from projections: a review of algorithms"*. Inter. J. Cytology, Vol. 39 (1974), pp. 111-151.

[17] Champion, R., Lenard, T. C. and Mills, T. M. *"An introduction to abstract splines"*, Math. Scientist. Vol.21, (1996), pp. 8-26.

Chapter 3

The basics about linear inverse problems

3.1 Problem statements

Much of what we shall say below would apply to linear as well as to nonlinear problems but, since the maxentropic methods we shall develop in the next chapters are designed to be applied to linear problems, we shall concentrate on those. This chapter is to present some basic issues in inverse problem theory.

Let V and W be two Banach spaces with norms $\|...\|_V$ and $\|...\|_W$ respectively. The subscripts shall be dropped when no ambiguity arises.

Let $A : V_1 \subset V \to W$ be a linear operator defined on a linear subset V_1 of V. We do not assume that V_1 is the largest domain on which A may be defined.

In actual modeling, the points of V_1 are the possible "inputs" or "sources" or "impulses". The operator A models some physical effect or propagation and the points of W represent "data" or "observations" or "outputs" of the system.

In the equation:

$$Ax = y \tag{3.1}$$

the computation of y when A and x are known is called a *direct problem*, and solving for x when A and y are known is called an *inverse problem*.

Exercise 3.1. Formulate the problem of determining A in (3.1), when y is given for a countable (usually finite) class of x's.

From the purely analytical point of view, the following problems may arise when solving (3.1) for x :

i) The inverse operator A^{-1} is not defined.

ii) $A^{-1}\{y\}$ has more than one element for each $y \in R(A)$.
iii) A^{-1} may not be continuous.

Item (i) means that A is not surjective. But even if it were, the map A, may not be $1:1$ so there are many inverses. But even if it were, item (ii) means that A may not be $1:1$ so there are (infinitely) many inverses.

But even if the operator A is a bijection, its inverse may not be continuous. This means that small errors in the measured data may produce large indetermination in the possible inputs.

To describe the solution to the inverse problem we may introduce the following

Definition 3.1. An operator $B : W_1 \subset W \to V_1 \subset V$ solves (3.1) whenever:

$$A(By) = y. \qquad (3.2)$$

Comments: The class W_1 should be thought of as the "available" data and the fact that $By \in V_1$ is just a consistency requirement.

With this notation, cases (i), (ii) and (iii) can be rephrased as:

a) There may exist no B such that (3.2) holds.
b) There exist many B's such that (3.2) holds.
c) Given a B such that (3.2) holds, B is not continuous.

Definition 3.2. We say that problem (3.1) is well posed when, there exists a unique $B : R(A) \to V$ which is continuous and (3.2) holds.

Comment: It is clear that the continuity of B is relative to the topology induced on $R(A)$ by the topology on W associated with $\|\ldots\|_W$. The notion of well posedness may be relativized to pairs V_1, W_1 without problem.

It is easy to define an inverse when A is bijective. When it is not, and $K = Ker(A)$ is known, then $\hat{A} : \hat{V} = V/K \to W_1 = Im(A)$ is a bijection. When A is bounded, W_1 is a closed subspace of W and a Banach space on its own.

As a consequence of the open mapping theorem, see Appendix C, $\hat{A}^{-1}$ exists and is continuous.

Theorem 3.1. Let $A : V \to W$ be compact and suppose that the factor space $\hat{V} = V/K$ has infinite dimension. Then there exists a sequence $\{x_n : n \geq 1\} \subset V$ such that $\|Ax_n\| \to 0$, but $\{x_n\}$ does not converge. Even worse, the $\{x_n\}$ can be chosen so that $\|x_n\| \to \infty$. In particular if A is 1:1, the inverse $A^{-1} : W_1 = A(V) \subset W \to V$ is unbounded.

Proof. The operator $\hat{A} : \hat{V} \to Y$ induced by A on the factor space $\hat{V} = V/K$, defined by $\hat{A}[x] = Ax$ is compact and 1:1. If $\hat{A}^{-1} : W_1 \to \hat{V}$, then $\hat{A}^{-1}$ cannot be bounded.

If it were, $I = \hat{A}^{-1}\hat{A}$ and $\hat{V} \to \hat{V}$ would be compact which is impossible if $\hat{V}$ is infinite dimensional.

Since $\hat{A}^{-1}$ is unbounded, there exists a sequence $\{x_n : n \geq 1\}$ in V and a corresponding sequence $\{[x_n] : n \geq 1\}$ in $\hat{V}$ such that $\|Ax_n\| \to 0$ but $\|[x_n]\| = 1$. Let $\xi_n \in K$ be such that $\|x_n + \xi_n\| \geq a > 1$ and put $x'_n = \dfrac{x_n + \xi_n}{\sqrt{\|Ax_n\|}}$, $\|Ax'_n\| \to 0$ but $\|x'_n\| \geq \dfrac{a}{\sqrt{\|Ax_n\|}} \to \infty$ as $n \to \infty$. $\square$

Comments: To avoid contradiction with the comments above the statement of the Theorem, either

i) $\hat{V}$ is finite dimensional (and $\hat{A}^{-1}$ is bounded) or
ii) $\hat{V}$ is infinite dimensional and W_1 is of first category in W. It can be represented as union of a sequence of nowhere dense sets.

Besides the issues described above, there are two more classes of issues that are involved in solving (3.1).

The first class is related to the fact that both, the model A and the data y, depend on actual measurements and our problem does not merely consist in solving (3.1) but in giving meaning to the problem.

(P)   Find $x \in V_1$ such that $Ax = y$ where $A \in N(A_0, \delta)$ and $y \in B(y_0, \varepsilon)$

Here $y \in B(y_0, \varepsilon)$ is clear: it denotes the ball of radius ε around y_0 in the $\|...\|_W$ distance. The indeterminacy in A may be due to indeterminacy in some physical parameters, and thus $N(A_o, \delta)$ denotes some appropriate neighborhood of A_0. Tomographic reconstruction, see [9-24] is a typical situation of this kind.

The other issue compounding the difficulties is related to the fact that in applications, both x and y lie in infinite dimensional spaces, and that to describe B and By one has to resort to finite dimensional setups. To be more precise, one has to be able to produce families of projections $P_n : V \to V_n$ and $Q_n : W \to W_n$, where V_n and W_n are of appropriate finite dimension, n, and such that $\|x - P_n x\|_V \to 0$ for every x, $\|y - Q_n y\|_W \to 0$

for every y, and then for each n, m solve

$$A_{mn}(P_n x) = Q_m y \qquad (3.3)$$

where A_{mn} denotes the restriction of $Q_m A$ to $P_n V$.

Reasonable definitions of convergence of this projection method are not hard to conceive. Here is one.

Definition 3.3. Assume that we are given a projection scheme $S(P_n, Q_n)$. We shall say that

i) The scheme solves $Ax = y$ if for every $m > 0$, for any $n \geq m$,

$$A_{mn}\xi = y_m$$

has solution $\xi = x_n^{(m)}$ in $P_n V$. Here $y_m = Q_m y$.

ii) The scheme is convergent for equation (3.1) if it is solvable for all y in $R(A)$ and there exist:

    a) $x^{(m)} \in V$ such that $x_n^{(m)} \to x^{(m)}$ and $Ax^{(m)} = y_m$.
    b) $x \in V$ such that $x^{(m)} \to x$ and $Ax = y$.

Comment: Actually $(ii) - a)$ is important in applications, for most of the time data is collected in such a way which is describable by projection on finite dimensional space. Condition $(ii) - b)$ is important for we need to know we are near a solution to the full problem, even when forced to describe it in finite dimensional terms.

As the following example shows, (i) and $(ii) - a)$ can hold without $(ii) - b)$ taking place.

Example 3.1. Take $V = L_1[0, 1]$, $W = C[0, 1]$ and let us define:

$$(Ax)(t) = \int_0^t x(s)ds.$$

Let

$$P_n x(t) = \sum_{j=1}^{2^n} \left(2^n \int_{\frac{j-1}{2^n}}^{\frac{j}{2^n}} x(s)ds \right) I_{[\frac{j-1}{2^n}, \frac{j}{2^n})}(t).$$

Note that if $n \geq m$, $P_n P_m = P_m$.
For $y \in W$, $Q_n y(t) = \sum_{k=1}^{2^n} y\left(\frac{k}{2^n}\right) I_{[\frac{k-1}{2^n}, \frac{k}{2^n})}(t)$.

The convergence of $Q_n y(t)$ is easy, but that of $P_n x(t)$ to $x(t)$ in L_1 is not. (Probabilists relate it to the Martingale Convergence Theorem.) But let us not worry about that here.

Note that $P_n V$ and $Q_n W$ can be identified with finite dimensional spaces.

Note that given y, and therefore y_n, the solution to $A_{mn}\xi = y_m$ is $x_m^{(m)}(t) = \sum_{j=1}^{2^m} 2^m \left(y(\frac{j}{2^m}) - y(\frac{j-1}{2^m})\right) I_{[\frac{k-1}{2^m}, \frac{k}{2^m})}(t)$, and clearly $x_n^{(m)}(t) = x_{(m)}^{(m)}$ for $n \geq m$. Thus (i) and (ii) $-$ a) are satisfied. But (ii) $-$ b) is not if $y(.)$ is not differentiable.

Comment: A result asserting consistency of solution schemes is contained in:

Theorem 3.2. *Let $A : V \to W$ be a bounded operator and let (P_n, Q_n) be a projection scheme.*

Suppose that the V_n are increasing and that $y_m = Q_m A x \in Q_m A X_n$. Assume as well that there exist linear maps

$$R_{nm} : Q_m A(V_n) \to V_n, \quad n \geq m$$

such that $R_{nm}(y_n) = x_n^{(m)}$ solves $A_{mn}\xi = y_m$ for $\xi \in V_n$. Let $x^{(m)}$ solve $Q_m A x = y_m$. Then $x_n^{(m)} \to x^{(m)}$ if and only if $\sup_n \|R_{nm} Q_m A\| < c(m) < \infty$ for some positive $c(m)$.

Proof. $x_n^{(m)} \to x^{(m)}$ means that $\|R_{nm}(y_n)\| = \|R_{nm} Q_m A x^{(m)}\| = \|R_{nm} Q_m A x\|$ is bounded in n for every x in V. By the uniform boundedness principle

$$\sup_n \|R_{nm} Q_m A\| = c(m) < \infty.$$

Conversely, since $R_{nm} Q_m A$ is a projection operator on X_n

$$\left(x^{(m)} - x_n^{(m)}\right) = (I - R_{nm} Q_m A) x^{(m)} = (I - R_{nm} Q_m A)\left(x^{(m)} - \xi\right),$$

for any $\xi \in X_n$. Thus,

$$\left\|x^{(m)} - x_n^{(m)}\right\| \leq (1 + c(m)) \left\|x^{(m)} - \xi\right\|$$

and since $P_n\left(x^{(m)}\right) \to x^{(m)}$, by setting $\xi = P_n x^{(m)}$ for large enough n, we obtain the desired convergence of $x_n^{(m)}$ to $x^{(m)}$. $\square$

Comments: Along the proof we used the fact that both x and $x^{(m)}$ solve $Q_m A \xi = y_m$. Even if A were 1:1, since Q_m is a projection, $Q_m A$ ceases being 1:1.

To see what happens when the data collection improves, i.e., when $y_m \to y$, consider

Theorem 3.3. *With notation and assumptions as above, assume furthermore that $Q_m A(V) \subset A(V)$ and that $A(V)$ is a second Baire category set, then $y_m \to y$ implies that there exist $\{x_n : n \geq 1\}$ and x such that $Q_m A x_m = y_m$, $Ax = y$ and $x_m \to x$.*

Proof. If we pass to the factor space $\hat{V}$ and consider the bijection $\hat{A} : \hat{V} \to A(V)$, we know that a continuous $\hat{A}^{-1}$ exists. This means that there exists $\hat{x}_m, \hat{x} \in \hat{V}$ such that $\hat{A}^{-1} y_m = \hat{x}_m \to \hat{x} = \hat{A}^{-1} y$. Now, if we choose representatives x_m, x in $\hat{x}_m$, and $\hat{x}$ such that $\|\hat{x}_m - \hat{x}\| \to 0$ implies that $\|x_m - x\| \to 0$ and we are done. □

3.2 Quasi solutions and variational methods

The following definition proves useful for numerical procedures as well as for theoretical analysis, specially when y is not in $R(A)$.

Definition 3.4. The point (or vector) x_0 in V is a **quasi solution** of 1 for $y \in W$, if it realizes $\inf \{\|Ax - y\|_W : x \in V\}$. This is usually restated as $x_0 = \arg\inf \{\|Ax - y\|_W : x \in V\}$.

Comment: This definition can clearly be relativized to $x_0 \in V_1$, $y \in W_1$, for appropriate classes V_1 and W_1.

Useful as it is, Definition 3.4 must still be supplemented to pin down candidates to solve (3.1). When A restricted to V_1 ($\subseteq V$) is 1:1, one should expect (if all goes well) a unique x_0. But this does not occur when A is not 1:1, even when $y \in R(A)$. The problem is that if $Ker(A) = \{x : Ax = 0\} \neq \{0\}$, an infinity of solutions arise. In this case two variational methods are usually proposed to provide a choice.

Definition 3.5. We say that x_0 is a minimum norm solution of (3.1) if

$$x_0 = \arg\inf \left\{ \frac{1}{2} \|x\|_V^2 : Ax = b \right\}.$$

Comment: If we pass to the quotient map $\hat{A} : \hat{V} \to W$, and denote by $\hat{x}_0$ the solution to $\hat{A}x = y$, then the x_0 in Definition 3.5, is the element in V such that $\|\hat{x}_0\|_{\hat{V}} = \|x_0\|_V$.

To overcome the troublesome search process, and to be able to accommodate for the possibility of having data vectors y not in $R(A)$, a penalization method is proposed. Instead of the previous definition we have:

Definition 3.6. Let $A : V \to W$ be given. For $b \in W$ we say that x_0 is a minimum norm best approximate solution to $Ax = b$ if

$$x_0 = \arg\inf \left\{ \frac{1}{2} \|x\|_V^2 + \frac{1}{2} \|Ax - b\|_W^2 : x \in V \right\}.$$

A first result ascertaining existence of quasi solutions to (3.1) is contained in:

Theorem 3.4. *Let $A : V \to W$ be a bounded injective operator. Let K be a convex compact subset of V and let W be strictly convex space. Then a unique quasi solution x^0 exists for any $y^0 \in W$. Also $x^0 = B(y^0)$ is a continuous function of y^0 $\{x_n : n \geq 1\}$.*

Proof. Notice that $f(x) = \|Ax - y^0\|_Y^2$ is strictly convex. Since K is compact, $f(x)$ reaches a minimum at a unique x^0 in K. Set $B(y^0) = x^0$. If $\{y_n : n \geq 1\}$ is such that $y_n \to y^0$ in Y, and x_n is the unique point in K minimizing $f(x)$ for each y_n, then there exists a subsequence, denote it by x_n again, converging to a point $\hat{x}$ in K. Since A is bounded $Ax_n \to A\hat{x}$, and therefore $\hat{x}$ minimizes $f(x) = \|Ax - y^0\|_Y$ as well. The uniqueness of minima takes care of the remaining detail. □

Corollary 3.1. *Under the assumptions of Theorem 3.4, if y varies in $R(A)$, then $B(y)$ is a continuous inverse of A.*

Comment: The compactness assumption on $K \subset V$ can be relaxed. We can search for solutions in domains D whose bounded subsets are compact. Such D can be written as $K \oplus V_1$, where $K \cap V_1 = \{0\}$, K being compact and V_1 a finite dimensional subspace of V.

3.3 Regularization and approximate solutions

When solving an equation like (3.1), even if the model, A were known exactly, one may be confronted with the fact that the right-hand side y is not known exactly. What one has instead is some data η which is known to lie within an experimental error δ of y. Thus whatever procedure to assign an element in V to each η, we must be able to apply it to every element in $B(y, \delta)$, and even better to any $\eta \in W$.

Definition 3.7. Let $B_\delta : W \to V$ be a family of continuous mappings defined for every $\delta \in [0, \delta_0]$. We shall say that it provides us with **consistent approximate solution** scheme to (3.1) whenever

$$\sup \{\|A(B_\delta(\eta)) - \eta\|_W : \eta \in B(y, \delta)\} \downarrow 0 \qquad (3.4)$$

as $\delta \downarrow 0$.

We now consider the notion of regularization of an operator. Our presentation is slightly more restrictive, but useful to prove it equivalent to Definition 3.7.

Definition 3.8. Let $A : V \to W$ and $A^{-1} : W \to V$ be everywhere defined, inverse to each other and continuous. We say that a family $R_\delta : W \to V$ of continuous operators, defined on $[0, \delta_0]$, **regularizes** $B = A^{-1}$ whenever

$$\sup \{\|R_\delta(\eta) - x\|_V : \eta \in B(y, \delta)\} \downarrow 0 \qquad (3.5)$$

as $\delta \downarrow 0$. In (3.5) $y = Ax$.

We can now prove:

Theorem 3.5. *With the notations and under the assumptions in Definitions 3.7 and 3.8, we have: a consistent approximation scheme is a regularization family as well and viceversa.*

Proof. Let R_δ be a regularization family. Then

$$\|A(R_\delta(\eta)) - \eta\|_Y \leq \|A(R_\delta(\eta)) - Ax\|_Y + \|y - \eta\|_Y$$
$$\leq \|A\| \, \|R_\delta(\eta) - x\|_X + \|y - \eta\|_Y$$

which certainly tends to zero as $\delta \downarrow 0$. Conversely, let B_δ be a consistent approximation scheme. Then

$$\|B_\delta(\eta) - x\|_X = \|A^{-1}(AB_\delta(\eta)) - A^{-1}y\|_X$$
$$\leq \|A^{-1}\| (\|A(B_\delta(\eta)) - \eta\|_Y + \|\eta - y\|_Y). \qquad \square$$

The potential importance or usefulness of (3.4) relative to (3.5) is that in the former no explicit reference to the unknown point x is made.

Hidden in Definition 3.8 is the fact that $R_\alpha y \to A^{-1}y$ for $y \in A(V)$, but this convergence cannot be uniform as ascertained in

Theorem 3.6. *Let $A : V \to W$ be a linear, compact and injective operator. Assume that $\dim V = +\infty$. If R_α is a regularization scheme, the operator R_α cannot be uniformly bounded with respect to α, and the operators $R_\alpha A$ cannot be norm convergent as $\alpha \downarrow 0$.*

Proof. Assume to begin with that there exists a positive K such that $\|R_\alpha\| \leq K < \infty$. Therefore from $\|R_\alpha f - A^{-1}f\| \to 0$ we deduce that $\|A^{-1}f\| \leq c\|f\|$ which implies that A^{-1} is bounded. This contradicts Theorem 3.1. To obtain the second statement, notice that norm convergence of $R_\alpha A$ implies that $\|R_\alpha A - I\| < \frac{1}{2}$ for some $\alpha > 0$. Let now $y \in A(V)$ we have
$$\|A^{-1}y\| \leq \|A^{-1}y - R_\alpha A A^{-1}y\| + \|R_\alpha y\|$$
$$\leq \tfrac{1}{2}\|A^{-1}y\| + \|R_\alpha\|\|f\|$$
or $\|A^{-1}y\| \leq 2\|R_\alpha\|\|f\|$ which again is impossible according to Theorem 3.1. □

Now a definition which we shall implement in Sections 4.2 and 4.3 of the next chapter.

Definition 3.9. Given a regularization scheme R_α for an operator A, a choice of regularization parameter $\alpha = \alpha(\varepsilon)$ depending on the error level ε is called regular if for any $y \in A(V)$ and any $y(\varepsilon)$ with $\|y(\varepsilon) - y\| \leq \varepsilon$ we have $R_{\alpha(\varepsilon)}y(\varepsilon) \to A^{-1}y$ as $\varepsilon \downarrow 0$.

Comment: In other words, for a regular scheme, the approximate solution is near the true solution when the approximate data is near the true data.

Actually Definition 3.8 can be relaxed considerably if we do not care to relate it to Definition 3.7, and we do not demand that A^{-1} be continuous (if it exists).

Definition 3.10. The problem $Ax = y$ is conditionally correct in a correctness class $V_1 \subset V$ whenever

a) The solution x is unique in V_1, i.e., if $x_1, x_2 \in V_1$ and $Ax_1 = Ax_2 = y$, then $x_1 = x_2$.
b) The solution is stable in V_1 : if solutions to $Ax = y_1$ and $Ax = y_2$ exist in X_1 and $\|y_1 - y_2\| \to 0$ then $\|x_1 - x_2\| \to 0$ (Interpret correctly!).

Comment: Item (a) does not necessarily mean that A restricted to V_1 is injective. It will be so if a solution exist in V_1 for every $y \in \mathcal{A}(V)$.

For the next definition let $\mathcal{A}(V)$ denote the class of closed subsets of V. If S_1 and $S_2 \in \mathcal{A}(V)$ put
$$d(S_1, S_2) = \sup_{x_2 \in S_2} \inf_{x_1 \in S_1} d(x_1, x_2) + \sup_{x_1 \in S_1} \inf_{x_2 \in S_2} d(x_1, x_2)$$
$$= \sup_{x_2 \in S_2} d(S_1, x_2) + \sup_{x_1 \in S_1} d(x_1, S_2).$$

This defines a distance on $\mathcal{A}(V)$.

Definition 3.11. A set valued collection

$$R_\alpha : \mathcal{N}(A(V_1)) \to \mathcal{A}(V)$$

defined at least on a neighborhood of $A(V_1)$ for all $0 < \alpha < \alpha_0$ is called a **regularizer** whenever

$$\lim_{\alpha \downarrow 0} d(R_\alpha(Ax), \{x\}) = 0.$$

Exercise 3.2. As in Theorem 3.6, verify that if A does not have a continuous inverse, then R_α cannot be uniformly continuous.

Consider now

$$\Phi_\alpha(x, y) = \|Ax - y\|_W^2 + \alpha \|x\|_V^2 \qquad (3.6)$$

where clearly $\Phi_\alpha(x, y)$ is convex, but not necessarily strictly convex, continuous (and lower semi continuous) in x.

Theorem 3.7. *Assume that the sets* $M(\tau) := \left\{x \in V : \|x\|_V^2 \leq \tau\right\}$ *are compact in V for any $\tau \geq 0$. Then*

$$R_\alpha(y) := \arg\inf\{\Phi_\alpha(x, y) : x \in V\} \qquad (3.7)$$

with $\Phi_\alpha(x, y)$ as in (3.6), is a regularizer.

Proof. Take any $x_0 \in V_1$ and put $\tau = \|Ax_o - y\|^2 + \alpha \|x_0\|^2$. Then

$$V_0 = \left\{\|x\|_V^2 \leq \frac{\tau}{\alpha}\right\} \cap V_1$$

is compact, thus $\Phi_\alpha(x, y)$ reaches its minimum at some x^* on V_0. Let $R_a(y)$ be the minimizing set, as in (3.7). (See Theorem A.2 in Appendix A) Since $\Phi_\alpha(x, y)$ is lower semi continuous, $R_\alpha(y)$ is a closed set. Let us first verify that for fixed α, $y \to R_\alpha(y)$ is continuous. If it were not, there would exist a sequence $y_k \to y$ in $A(V_1)$ and $\varepsilon > 0$, such that $d(R_\alpha(y_k), R_\alpha(y)) > \varepsilon$ for all k. Thus, there exist $x_k \in R_\alpha(y_k)$ and $x \in R_\alpha(y)$ such that $d(x_k, x) > \varepsilon$. Invoking the compactness of V_0 we would obtain a contradiction. (This takes up a few lines, do it!)

To verify that $R_\alpha(Ax) \to x$ as $\alpha \downarrow 0$, assume the contrary: That for $\alpha_k \leq \frac{1}{k}$ there exist $x_k \in R_{\alpha_n}(y)$ such that $d(x_k, x) > \varepsilon$ for all k. Since $y = Ax$ and $\Phi_\alpha(x_k, y) \leq \Phi_\alpha(x, y)$, then

$$x_k \in V_* = \left\{\xi \in V : \|\xi\|_V^2 \leq \|x\|_V^2\right\} \cap V_1$$

which is a compact set. Thus, there is $x^* \in V_*$ such that $x_k \to x^*$ (or at least a subsequence of x_k does). But then, $d(x, x^*) = \|x - x^*\| > \varepsilon$.

By definition, the minimizers x_k satisfy $\|Ax_k - Ax\|^2 \leq \frac{1}{k} \|x\|^2$. Since A is continuous, by passing to the limit we obtain $Ax^* = Ax$, and since V_1 is a conditional correctness class, we obtain $x^* = x$, which contradicts $\|x - x^*\| > \varepsilon$. $\square$

Comment: When $\|x\|_V$ and $\|y\|_W$ are strictly convex functions on V and W respectively, $R_\alpha(y)$ becomes a one-point set, and things are slightly nicer.

We shall stop here and refer the reader to the references at the end of the chapter, where the topics mentioned here are dealt with in full depth. We shall come back to them in the context inverse problems in Hilbert spaces only to relate them to the maxentropic approaches.

3.4 Appendix

The following material explains why one should use $\|\cdot\|^2$ instead of $\|\cdot\|$ in the variational methods introduced above to define quasi solutions and regularized solutions.

Definition 3.12. A Banach space V is said to be rigorously convex whenever $\|x + y\| = \|x\| + \|y\|$ for $y \neq x$, implies that $y = \lambda x$ for some $\lambda \in \mathbb{R}$.

The standard examples of rigorously convex spaces are the L_p and ℓ_p spaces for $p > 1$, whereas L_1, ℓ_1 and $C[K]$ where K is a compact topological space (see Appendices A and C) are non rigorously convex spaces.

Comment: Verify that in a rigorously convex space V, the norm $x \to \|x\|$ is not a strictly convex function. But nevertheless, we have

Proposition 3.1. *Let V be a rigorously convex Banach space. Then $x \to \|x\|^2$ is rigorously convex.*

Proof. Assume the opposite, and let $t \in (0,1)$ such that for $x \neq y$ we have

$$\|tx + (1-t)y\|^2 = t\|x\|^2 + (1-t)\|y\|^2.$$

From the triangle inequality we have

$$t\|x\|^2 + (1-t)\|y\|^2 \leq t^2 \|x\|^2 + 2t(1-t)^2 \|x\| \|y\| + (1-t)^2 \|y\|^2$$

which in turn amounts to

$$t(1-t)\left\{\|x\|^2 + 2\|x\|\|y\| + \|y\|^2\right\} < 0$$

which is a contradiction, for the left-hand side is always positive. □

3.5 Bibliographical comments and references

Besides the books [2-2], [2-3], [2-5], [2-6], [2-9] in which the generic statements presented above appear, scattered in the following references (which are listed chronologically) you will find material on which this chapter is based and which goes well beyond our introductory remarks.

A way of regularizing inverse problems using entropy functionals appears in [6] and [7]. A variation on this theme is presented in Section 9.7.

References

[1] Tikhonov, A and Arsenin, V. *"Solutions of Ill-Posed Problems"*. John Wiley, N.Y., 1977.
[2] Glasko, V. *"Inverse Problems of Mathematical Physic"*. Am. Inst. Phys. Publication, 1988.
[3] Anger, G. *"Inverse Problems in differential equations"*. Akademie-Verlag, Berlin, 1990.
[4] Bakushinsky, A and Goncharsky, A. *"Ill-Posed Problems Theory and Applications"*, Kluwer Acad. Pubs., Dordrecht, 1994.
[5] Tikhonov, A., Goncharsky, A, Stephanov, V and Yagola, A. *" Numerical Methods for the Solution of Ill-Posed Problems"*. Kluwer Acad. Pub., Dordrecht, 1995.
[6] Eggermont, P. *"Maximum entropy regularization of Fredholm integral equations of the first kind"*. SIAM J. Math. Anal., Vol. 24 (1993) pp. 1557-1576.
[7] Amato, U. and Hughes, W. *"Maximum entropy regularization of Fredholm integral equations of the first kind"*. Inv. Problems, Vol. 7 (1991) pp. 793-808.

Chapter 4

Regularization in Hilbert spaces: Deterministic and stochastic approaches

4.1 Basics

This chapter is devoted to regularization of ill-conditioned linear inverse problems in Hilbert spaces. The methods that we present exploit the geometric structure of the setting of the problem, namely that in Hilbert spaces it is easy to talk about eigenvalues and eigenvectors. Let us begin by recalling some basics.

Let $A : V \to V$ be a self adjoint, compact operator on a Hilbert space V, i.e. $\langle A\varphi, \psi \rangle = \langle \varphi, A\psi \rangle$ for any pair of vectors φ, ψ in V. All eigenvalues of A are real, all nonzero eigenvalues have finite multiplicity and the eigenvectors corresponding to different eigenvalues are orthogonal. Also at most a countable set has zero as accumulation point.

If we denote by ξ_n a sequence of eigenvectors with associated eigenvalues listed as $|\lambda_1| \geq |\lambda_2| \geq \cdots$ then each $x \in V$ can be expanded as

$$x = \sum \langle \xi_n, x \rangle \xi_n + P_0 x, \qquad (4.1)$$

where $P_0 : V \to Ker(A)$ denotes the orthogonal projection onto the kernel of A. Also

$$Ax = \sum \lambda_n \langle \xi_n, x \rangle \xi_n \qquad (4.2)$$

describes the action of A on any $x \in V$.

If V and W are Hilbert spaces, if we identify (as it must be done) their duals with themselves, then

$$\ker(A^*) = A(V)^\perp \text{ and } Ker(A^*)^\perp = \text{cl}(A(V)) = A(V) \qquad (4.3)$$

where for any subset $V_1 \subset V$, $V_1^\perp = \{\eta : \langle \eta, x \rangle = 0 \ \forall \ x \in V_1\}$.

Let $A : V \to W$ be a compact operator and let $A^* : W \to V$ be its dual (adjoint) which is also compact.

Definition 4.1. The non-negative square roots of the eigenvalues of A^*A (which are positive) are called the singular values of A.

The next result is standard.

Theorem 4.1. *Let λ_n denote the nonzero singular values of A ordered as decreasing sequence, and repeated according to multiplicity, $\lambda_1 \geq \lambda_2 \geq \cdots$. (The multiplicities are the dimensions of $Ker A^*A - \lambda_n^2 I$).) Then there exist $\{\xi_n : n \geq 1\} \subset V$ and $\{\eta_n : n \geq 1\} \subset W$ such that*

$$A\xi_n = \lambda_n \eta_n, \quad A^*\eta_n = \lambda_n \xi_n \qquad (4.4)$$

for every $n \geq 1$. For each $x \in V$ we have (the singular value decomposition)

$$x = \sum \langle \xi_n, x \rangle \xi_n + P_0 x \qquad (4.5)$$

as well as

$$Ax = \sum \lambda_n \langle \xi_n, x \rangle \eta_n. \qquad (4.6)$$

Comment: The system $(\lambda_n, \xi_n, \eta_n)$ is called a singular system of A.

The next result is a famous theorem by Picard asserting when and how can the equation $Ax = y$ be solved. The representation it provides for the solution shows clearly where the instabilities of the inverse problem come from.

Theorem 4.2. *Let $A : V \to W$ be as above and let $(\lambda_n, \xi_n, \eta_n)$ denote a singular system for it. The equation $Ax = y$ has a solution if $y \in Ker(A^*)^\perp$ and satisfies*

$$\sum \frac{1}{\lambda_n^2} |\langle y, \eta_n \rangle|^2 < \infty. \qquad (4.7)$$

The solution is then given by

$$x = \sum \frac{1}{\lambda_n} \langle y, \eta_n \rangle \xi_n. \qquad (4.8)$$

Proof. The necessity of $y \in Ker(A^*)^\perp$ follows from (4.3).
If x is a solution, then

$$\lambda_n \langle x, \xi_n \rangle = \langle x, A\eta_n \rangle = \langle Ax, \eta_n \rangle = \langle y, \eta_n \rangle \quad \text{or} \quad \langle x, \xi_n \rangle = \frac{\langle y, \eta_n \rangle}{\lambda_n}.$$

Thus
$$\sum \frac{1}{\lambda_n^2} |\langle y, \eta_n \rangle|^2 = \sum |\langle x, \xi_n \rangle|^2 \leq \|x\|^2$$
which is (4.7).

Assume now that (4.7) holds and that $y \in Ker(A^*)^\perp$. Then the sum (4.8) converges in V. From the continuity of A and for (4.4) we see that
$$A\left(\sum \frac{1}{\lambda_n} \langle y, \eta_n \rangle \xi_n\right) = \sum \langle y, \eta_n \rangle \eta_n.$$

Apply (4.5) to y, this time with $Q_0 =$ orthogonal projection on $Ker(A^*)$ and use $y \in Ker(A^*)^\perp$ to conclude that the last identity above means that the right of (4.8) is a solution to the equation. $\square$

Comment: The usual argument to show the instability of (4.8) goes as follows: replace y by $y + a\eta_n$ for some real a. Then the solution changes by $\frac{a}{\lambda_n}\xi_n$ and $\frac{\|\delta x\|}{\|\delta y\|} = \frac{1}{\lambda_n}$ can be made as large as desired by choosing a small enough eigenvalue.

Before passing on to the regularization schemes, let us consider a generic result.

Theorem 4.3. *Let $A : V \to W$ be compact and injective. Let $\gamma(\alpha, \lambda) : (0, \infty) \times (0, \|A\|] \to \mathbb{R}$ be a bounded function, such that for each fixed α there is a positive constant $k(\alpha)$ such that*
$$|\gamma(\alpha, \lambda)| \leq k(\alpha)\lambda \tag{4.9}$$
for $\lambda \in (0, \|A\|]$. Assume also that for each $\lambda \in (0, \|A\|]$
$$\lim_{\alpha \to 0} \gamma(\alpha, \lambda) = 1. \tag{4.10}$$
Then the bounded linear operators $R_\alpha : W \to V$, defined by
$$R_\alpha y = \sum_{n=1}^\infty \frac{1}{\lambda_n} \gamma(\alpha, \lambda_n)(y, \eta_n)\xi_n \tag{4.11}$$
for each $y \in W$ describe a regularization scheme with
$$\|R_\alpha\| \leq k(\alpha). \tag{4.12}$$

Proof. Notice to begin with that
$$\|R_\alpha y\|^2 = \sum_{n=1}^\infty \frac{1}{\lambda_n^2} |\gamma(\alpha, \lambda_n)|^2 |\langle y, \eta_n \rangle|^2 \leq k(\alpha)^2 \|y\|^2$$

on account of (4.9) and the orthonormality of η_n. This implies (4.12). Notice also that for any $x \in V$

$$\|R_\alpha Ax - x\|^2 = \sum_{n=1}^\infty |\langle R_\alpha Ax - x, \xi_n\rangle|^2 = \sum_{n=1}^\infty |\gamma(\alpha, \lambda_n) - 1|^2 |\langle x, \xi_n\rangle|^2.$$

The injectivity of A was implicitly brought in the chain above: we forgot about $\|P_0 x\|$. If we denote by K a bound for $\gamma(\alpha, \lambda_n)$. For $\varepsilon > 0$ there exists N_0 such that

$$\sum_{n=N_0+1}^\infty |\langle x, \xi_n\rangle|^2 < \frac{\varepsilon}{2(K+1)^2}.$$

From (4.10) we infer the existence of $\alpha_0 > 0$ such that $|\gamma(\alpha, \lambda_n) - 1|^2 \le \frac{\varepsilon}{2\|x\|^2}$ for $0 < \alpha < \alpha_0$ and $n = 1, 2, ..., N_0$. Therefore

$$\|R_\alpha Ax - x\|^2 \le \sum_{n=1}^{N_0} |\gamma(\alpha, \lambda_n) - 1|^2 |\langle \xi_n, x\rangle|^2 + \sum_{n=N_0}^\infty |\langle \xi_n, x\rangle|^2 (K+1)^2$$

$$< \frac{\varepsilon}{2\|x\|^2} \sum |\langle \xi_n, x\rangle|^2 + \tfrac{\varepsilon}{2} < \varepsilon$$

for all $0 < \alpha < \alpha_0$. In other words $R_\alpha Ax \to x$ as $\alpha \to 0$ for all $x \in V$. □

4.2 Tikhonov's regularization scheme

The simple, but powerful, idea consists of noticing that the eigenvalues of A^*A can be pushed away from zero by adding αI to it.

Theorem 4.4. *For each $\alpha > 0$, $A^*A + \alpha I$ has a bounded inverse. If A happened to be injective, then $R_\alpha = (A^*A + \alpha I)^{-1} A^*$ provides a regularization scheme with $\|R_\alpha\| \le \frac{\sqrt{\alpha}}{2}$.*

Proof. The inequality $\langle (A^*A + \alpha I) x, x\rangle \ge \alpha \|x\|^2$ implies that $A^*A + \alpha I$ is injective. If $(\lambda_n, \xi_n, \eta_n)$ denotes a singular system for A, clearly $T: V \to V$ defined by

$$Tx = \sum \frac{1}{\alpha + \lambda_n^2} \langle x, \xi_n\rangle \xi_n + \frac{1}{\alpha} P_0 x$$

is bounded and satisfies $(A^*A + \alpha I)T = T(A^*A + \alpha I) = I$.

If A is injective and $x(\alpha)$ is the unique solution of

$$(A^*A + \alpha I)x(\alpha) = A^*y$$

then $P_0 x$ does not appear in

$$x(\alpha) = \sum \frac{1}{\alpha + \lambda_n^2} \langle A^* y, \xi_n \rangle \xi_n = \sum \frac{\lambda_n}{\alpha + \lambda_n^2} \langle y, \eta_n \rangle \xi_n$$

since $\langle A^* y, \xi_n \rangle = \lambda_n \langle y, \eta_n \rangle$. Therefore

$$\begin{aligned} x(\alpha) := R(\alpha) &= \sum \frac{1}{\lambda_n} \frac{\lambda_n^2}{\alpha + \lambda_n^2} \langle y, \eta_n \rangle \xi_n \\ &= \sum \frac{1}{\lambda_n} \gamma(\alpha, \lambda_n) \langle y, \eta_n \rangle \xi_n \end{aligned} \quad (4.13)$$

with $\gamma(\alpha, \lambda) = \dfrac{\lambda^2}{\alpha + \lambda^2}$. Certainly $\gamma(\alpha, \lambda) < 1$, and for $k(\alpha) = \dfrac{1}{2\sqrt{\alpha}}$ we have $\gamma(\alpha, \lambda) \leq k(\alpha) \lambda$ from $2\sqrt{\alpha} \lambda \leq (\alpha + \lambda^2)$.

To complete, invoke Theorem 4.3. $\square$

Since now norms come from scalar products, it is clear that the following holds.

Theorem 4.5. *For each $\alpha > 0$ and each $y \in W$, $x(\alpha)$ satisfying (4.13) solves the variational problem*

$$x(\alpha) = \arg\inf \left\{ \|Ax - y\|^2 + \alpha \|x\|^2 : x \in V \right\}.$$

The solution depends continuously on y.

Proof. For the reader. $\square$

Consider now

Theorem 4.6. *Assume now that $A(V)$ is dense in W. Let $y \in W$ be such that $0 < \varepsilon < \|y\|$. Then there is a value of α such that*

$$\|AR_\alpha y - y\| = \varepsilon. \quad (4.14)$$

Proof. Suffices to show that $F(\alpha) := \|AR_\alpha y - y\|^2 - \varepsilon^2$, defined on $(0, \infty)$, has a unique zero. From

$$F(\alpha) = \sum \frac{\alpha^2}{(\alpha + \lambda_n^2)^2} |\langle y, \eta_n \rangle|^2 - \varepsilon^2$$

we note that $F(\alpha)$ is continuous, that $\lim_{\alpha \to 0} F(\alpha) = -\varepsilon^2 < 0$, and $\lim_{\alpha \to +\infty} F(\alpha) = \|y\|^2 - \varepsilon^2 > 0$, thus the result. $\square$

Comment: The assertion of the previous theorem is known as the discrepancy principle. It can be further extended to

Theorem 4.7. *Assume that A is injective, (and the rest is as in Theorem 4.6). Let $y \in A(V)$ and $y(\varepsilon) \in W$ with $\|y(\varepsilon) - y\| \leq \varepsilon < \|y(\varepsilon)\|$ for $\varepsilon > 0$. Then there exists $\alpha(\varepsilon)$ such that*

$$\|AR_{\alpha(\varepsilon)}y(\varepsilon) - y(\varepsilon)\| = \varepsilon \tag{4.15}$$

and furthermore

$$R_{\alpha(\varepsilon)}y(\varepsilon) \to A^{-1}y \tag{4.16}$$

as $\varepsilon \downarrow 0$.

Proof. Clearly (4.15) is covered by the preceding theorem. To obtain (4.16) note that $x(\varepsilon) = R_{\alpha(\varepsilon)}y(\varepsilon)$ minimizes the functional described in Theorem 4.5, thus

$$\begin{aligned}\varepsilon^2 + \alpha \|x(\varepsilon)\|^2 &= \|Ax(\varepsilon) - y(\varepsilon)\|^2 + \alpha \|x(\varepsilon)\|^2 \\ &\leq \|AA^{-1}y - y(\varepsilon)\|^2 + \alpha \|A^{-1}y\|^2 \\ &\leq \varepsilon^2 + \alpha \|A^{-1}y\|^2\end{aligned}$$

therefore

$$\|x(\varepsilon)\| \leq \|A^{-1}y\|.$$

Now let $\eta \in W$. Then

$$|\langle Ax(\varepsilon) - y, \eta\rangle| \leq \{\|Ax(\varepsilon) - y(\varepsilon)\| + \|y(\varepsilon) - y\|\} \|\eta\| \leq 2\varepsilon \|\eta\|$$

which tends to zero as $\varepsilon \downarrow 0$, which amounts to the weak convergence of $x(\varepsilon)$ to $A^{-1}y$ as $\varepsilon \downarrow 0$. This is so because of the injectivity of A, the range $A^*(W)$ is dense in V (see (4.3)), and we already saw that $\|x(\varepsilon)\|$ is bounded (by $\|A^{-1}y\|$).

Thus

$$\begin{aligned}\|x(\varepsilon) - A^{-1}y\|^2 &= \|x(\varepsilon)\|^2 - 2\Re\langle x(\varepsilon), A^{-1}y\rangle + \|A^{-1}y\|^2 \\ &\leq 2\{\|A^{-1}y\| - \Re\langle x(\varepsilon), A^{-1}y\rangle\} \to 0\end{aligned}$$

as $\varepsilon \to 0$. $\square$

Instead of worrying about the data vector y, we narrow the class of potential approximate solutions according to

Theorem 4.8. *Let $A: V \to W$ be compact and let $\rho > 0$. For each $y \in W$ there exists a unique $x_0 \in C = \{x \in V : \|x\| \leq \rho\}$ satisfying*

$$\|Ax_0 - y\| \leq \|Ax - y\| \quad \forall x \in C.$$

Comment: We call x_0 the *quasi solution* of $Ax = y$ with constraint ρ (or in the constraint set C).

Proof. Let x_n be a minimizing sequence, i.e., $x_n \in C$ and $\lim_{n \to \infty} \|Ax_n - y\| = \inf \{\|Ax - y\| : x \in C\}$. Then x_n converges weakly to some $x_0 \in X$. Since A is compact this implies convergence $\|Ax_n - Ax_0\| \to 0$ as $n \to \infty$, i.e.,

$$\|Ax_0 - y\| = \inf \{\|Ax - y\| : x \in C\}.$$

Since $\|x_0\|^2 = \lim_{n \to \infty} \langle x_n, x_0 \rangle \leq \rho \|x_0\| \Rightarrow x_0 \in C$. Consider $g(x) = \|Ax - y\|^2$. Then $g(x)$ is strictly convex and V is strictly normed, then $g(x)$ attains its minimum at a unique point in the convex set C. (See Chapter 3.) Thus x_0 is unique. $\square$

To finish, let us relate quasi solutions to Tikhonov's regularization method.

Theorem 4.9. *Let $A : V \to W$ be a compact operator with dense range in W. Let $y \notin A(C) = \{Ax : x \in C\}$. Then the quasi solution x_0 assumes the constraint, i.e., $\|x_0\| = \rho$. Also, these exists a unique $\alpha > 0$ such that*

$$(A^*A + \alpha I)x_0 = A^*y.$$

Proof. If $A^*Ax_0 + \alpha x_0 = A^*y$, then

$$\|Ax - y\|^2 \geq \|Ax_0 - y\|^2 + 2\Re \langle x_0 - x, x_0 \rangle + \|A(x - x_0)\|^2$$

and if $\|x_0\| = \rho$ the middle term can be bounded from below by $2\alpha (\rho^2 - \rho \|x\|) > 0$ for $x \in C$. Thus x_0 is a quasi solution of $Ax = y$ with constraint set C.

To establish the theorem we must show the existence of x_0 such that $\|x_0\| = \rho$ and $A^*Ax_0 + \alpha x_0 = A^*y$.

For each $\alpha > 0$, let $x(\alpha)$ be a solution to $(A^*A + \alpha I)x(\alpha) = A^*y$ and define $F(\alpha) : (0, \infty) \to \mathbb{R}$ by

$$F(\alpha) = \|x(\alpha)\|^2 - \rho^2$$

and let us verify that it has a unique zero. We know that

$$F(\alpha) = \sum_{n=1}^{\infty} \frac{\lambda_n^2}{(\alpha + \lambda_n^2)^2} |\langle y, \eta_n \rangle|^2 - \rho^2$$

from which the continuity and strict monotonicity of $F(\alpha)$ drops out: Notice that

$$F(\alpha) \to -\rho^2 \quad \text{as} \quad \alpha \to +\infty$$

as well as

$$F(\alpha) \to \sum_{n=1}^{\infty} \frac{1}{\lambda_n^2} |\langle y, \eta_n \rangle|^2 - \rho^2 > 0 \text{ or } +\infty.$$

If not, since $Ker(A^*) = \{0\}$ and Theorem 4.2, the condition

$$\sum_{n=1}^{\infty} \frac{1}{\lambda_n^2} |\langle y, \eta_n \rangle|^2 \leq \rho^2 < \infty$$

would imply that $x = \sum_{n=1}^{\infty} \frac{1}{\lambda_n} |\langle y, \eta_n \rangle| \xi_n$ satisfies $Ax = y$ which contradicts the assumptions. Thus an unique zero of $F(\alpha)$, and we are done. □

4.3 Spectral cutoffs

As the name suggests, the method consists of a reduction of the size of the involved spaces by neglecting the very small singular values and forgetting about the null spaces. The regularization scheme is now

$$R_\alpha y = \sum_{\lambda_n \geq \alpha} \frac{1}{\lambda_n} |\langle y, \eta_n \rangle| \xi_n$$

which maps $Ker(A^*)^\perp$ into $Ker(A)^\perp$. Notice that we can bring this into the form (4.11) by putting $r(\alpha, \lambda) = 1$ or 0 if $\alpha \leq \lambda$ or $\lambda < \alpha$.

Now we have the following analogue of Theorem 4.7.

Lemma 4.1. *Assume $A : V \to M$ is compact, injective, with singular system $(\xi_n, \eta_n, \lambda_n)$. Then the spectral cutoff*

$$R_\alpha y = \sum_{\lambda_n \geq \alpha} \frac{1}{\lambda_n} |\langle y, \eta_n \rangle| \xi_n$$

provides us with a regularization scheme and $\|R_\alpha\| \leq \dfrac{1}{\lambda_{n(\alpha)+1}}$ where

$$\lambda_{n(\alpha)} = \inf \{\lambda_n : \lambda_n \geq \alpha\}.$$

Also $\|R_{\lambda_m}\| = \frac{1}{\lambda_m}$ for any m.

Proof. Certainly $\|R_\alpha y\|^2 \leq \dfrac{1}{\lambda_{n(\alpha)+1}^2} \sum_{\lambda_n \geq \alpha} |\langle y, \eta_n \rangle|^2 \leq \dfrac{1}{\lambda_{n(\alpha)+1}^2} \|y\|^2$.
Note that $R_{\lambda_m} \eta_m = \dfrac{1}{\lambda_m} \xi_m$. □

How well does $R_\alpha y$ solve $Ax = y$ is contained in

Theorem 4.10. Let $A : V \to W$ be compact and injective, with $R(A) = A(V)$ dense in W. Let $\varepsilon > 0$. Then there exists a largest α such that
$$\|AR_\alpha y - y\| \leq \varepsilon.$$

Proof. If $\text{cl}(A(V)) = W$, then $Ker(A^*) = \{0\}$, then the singular eigen-expansion for $y \in W$ is $y = \sum \langle y_n, \eta_n \rangle \xi_n$ which means that $F(\alpha)$ defined by
$$F(\alpha) := \|AR_\alpha y - y\| = \|y\|^2 - \sum_{\lambda_n \geq \alpha} |\langle y, \eta_n \rangle|^2 = \sum_{\lambda_n < \alpha} |\langle y, \eta \rangle|^2$$
satisfies: $F(\alpha)$ is nondecreasing in α,

i.) $F(\alpha) \to 0$ as $\alpha \downarrow 0$.
ii.) $F(\alpha) \to \|y\|^2$ as $\alpha \uparrow \|A\|$.

Thus, for $0 < \varepsilon \, (< \|A\|)$ the result drops out. □

Comment: When $\overline{A(V)}$ is not W, but $y \in A(V)$ the result still holds. When A is not injective $R_\alpha y \in Ker(A)^\perp$ and something must be done to eliminate the non-uniqueness.

Theorem 4.11. Let $A : V \to W$ be compact, injective and such that $\sum \lambda_n^{-2} = M < \infty$. Assume that $A(V)$ is dense in W. For $y \in A(V)$ and $\varepsilon > 0$, let $y(\varepsilon) \in W$ be such that $\|y(\varepsilon) - y\| \leq \varepsilon$. Then these exists $\alpha(\varepsilon) \downarrow 0$ as $\varepsilon \downarrow 0$ such that
$$\|AR_{\alpha(\varepsilon)} y(\varepsilon) - y(\varepsilon)\| \leq \varepsilon$$
and
$$R_{\alpha(\varepsilon)} y(\varepsilon) \to A^{-1} y \text{ as } \varepsilon \downarrow 0.$$

Proof. By Theorem 4.10 let $\alpha_1(\varepsilon)$ be the largest α such that the first claim holds. To complete the proof note that for any $\alpha > 0$
$$\|R_{\alpha(\varepsilon)} y(\varepsilon) - A^{-1} y(\varepsilon)\|^2 = \sum_{\lambda_n \geq \alpha} \frac{1}{\lambda_n^2} |\langle y(\varepsilon) - y, \eta_n \rangle|^2 + \sum_{\lambda_n \geq \alpha} \frac{1}{\lambda_n^2} |\langle y, \eta_n \rangle|^2.$$

Let $x \in V$ be such that $Ax = y$, then as in the proof of Theorem 4.3, we can rewrite the last term as $\sum_{\lambda_n \geq \alpha} |\langle x, \xi_n \rangle|^2$. Therefore there exists $\alpha_2(\varepsilon)$ such that
$$\|R_{\alpha(\varepsilon)} y(\varepsilon) - A^{-1} y(\varepsilon)\|^2 \leq \|y(\varepsilon) - y\| M + \varepsilon \leq \varepsilon(M + 1).$$

Take now $\alpha(\varepsilon) = \alpha_1(\varepsilon) \wedge \alpha_2(\varepsilon)$ to complete the proof. □

4.4 Gaussian regularization of inverse problems

Since later we are going to be working in a probabilistic setup, in which the treatment of the measurement errors can be dealt with in a natural way, at this point we might as well consider the problem of solving for x in

$$Ax + n = y \qquad (4.17)$$

where $A : V \to W$ is a bounded operator, V and W are Hilbert spaces and n, the noise process, is a W-valued Gaussian random variable. Instead of considering deterministic x and y in (4.17), we shall think of it as realization of a stochastic identity

$$AX + n = Y \qquad (4.18)$$

where now $X : \Omega \to V$ is a V-valued Gaussian random variable and $Y : \Omega \to W$ is a W-valued Gaussian random variable. By modifying the mean of Y we may assume that both X and n are centered, i.e., $E[X] = 0$, and $E[n] = 0$.

We shall assume that the input process X and the noise process n are independent, and that our modeling abilities allow us to prescribe

$$R_{XX} = E[X \otimes X^*], \quad R_{nn} = E[n \otimes n^*]$$

that is, the autocorrelation operators. (See the appendix for these matters.) From the independence of X and n we obtain

$$R_{Xn} = E[X \otimes n^*] = 0, \quad R_{nX} = E[n \otimes X^*] = 0.$$

We shall assume that R_{YY} is positive definite, which amounts to saying that there exists $c > 0$ such that for any $\eta \in W$ $E\left[|(AX+n)^*\eta|^2\right] \geq c \langle \eta, \eta \rangle$.

From the definition of Y, its autocorrelation operator R_{YY} is given by

$$R_{YY} = E\left[(AX + n) \otimes (AX + n)^*\right] = AR_{XX}A^* + R_{nn} \qquad (4.19)$$

from (4.19) and the independence of X and n. Notice that if the autocorrelation of the noise is positive definite, so is that of Y. Usually noise is white and $R_{nn} = \sigma^2 I$ for some positive σ.

Let us now consider the best linear estimation problem. If the data Y, the noise n and the input X are related by (4.18) and if the correlation operators R are known, we want to estimate the input X by a linear operation L on the data Y.

We measure the quality of the estimation by $E\left[|\langle (X - LY), \xi \rangle|^2\right]$ for any $\xi \in V$.

Since $\langle LY, \xi \rangle = \langle Y, L^*\xi \rangle$ we look for an element η, which is a linear function F^* of ξ (with F^* bounded and independent of ξ) such that

$$E\left[|\langle X, \xi\rangle - \langle Y, \eta\rangle|^2\right] = \text{minimum}. \qquad (4.20)$$

If there exists a unique $\eta = L^*\xi$ solving this problem, we define FY to be our best reconstruction of X.

In this case we have

$$E\left[|\langle X, \xi\rangle - \langle LY, \xi\rangle|^2\right] \geq E\left[|\langle X, \xi\rangle - \langle FY, \xi\rangle|^2\right] \qquad (4.21)$$

for all bounded linear operators L.

We can phrase the minimization problem in (4.21) as: Which random variable $a_2 = \langle Y, \eta\rangle$ obtainable from the data minimizes the stochastic distance to $a_1 = \langle X, \xi\rangle$ obtained from the input?

To begin with note the identity

$$E\left[|\langle X, \xi\rangle - \langle Y, \eta\rangle|^2\right] = \langle \xi, R_{XX}\xi\rangle - \langle \xi, R_{XY}\eta\rangle - \langle \eta, R_{YX}\xi\rangle + \langle \eta, R_{YY}\eta\rangle$$

$$= \langle (\eta - R_{YY}^{-1}R_{YX}\xi), R_{YY}(\eta - R_{YY}^{-1}R_{YX}\xi)\rangle + \langle \xi, (R_{XX} - R_{YX}^* R_{YY}^{-1} R_{YX})\xi\rangle$$

and since R_{YY} is positive definite we conclude that (4.21) is minimized by η given by

$$\eta = R_{YY}^{-1} R_{YX} \xi = F^*\xi$$

and clearly F^* is linear and independent of the particular ξ.

Summing up we have

Theorem 4.12. *The linear bounded operator satisfying (4.22) for all bounded linear L on V and all $\xi \in V$ is given by $F = (R_{YY}^{-1} R_{YX})^* = R_{XY} R_{YY}^{-1}$.*

Proof. Check through the computations sketched above. □

Denote by x_{op} the best estimate of X, then

$$x_{op} = Fy. \qquad (4.22)$$

If we use (4.18) to compute R_{XY} and bring in identity (4.19), we can recast (4.22) as

$$x_{op} = FY = R_{XX} A^* (A R_{XX} A^* + R_{nn})^{-1} y. \qquad (4.23)$$

Observe that when both X and n are white noise, with $R_{XX} = a^2 I$ as well as $R_{nn} = \sigma^2 I$, we can rewrite (4.24) as

$$x_{op} = A^*(AA^* + \lambda I)^{-1} y$$

which solves
$$Ax = [I - \lambda(AA^* + \lambda I)^{-1}]y$$
or filtering away the noise amounts to shifting the data vector. When $\sigma = 0$ and $R_{XX} = a^2 I$ and $(AA^*)^{-1}$ is invertible, then we reobtain
$$x_{op} = A^*(AA^*)^{-1}y$$
which was related to a variational problem in example 1 of Chapter 2.

To compute the lowest error and lowest relative error consider for $\xi \in V$
$$E\left[|\langle \xi, R_{XX} R_{YY}^{-1} Y - X \rangle|^2\right]$$
which after some elementary arithmetics becomes
$$\langle \xi, \hat{R}_{op} \xi \rangle = \langle \xi, (R_{XX} - R_{XY} R_{YY}^{-1} R_{YX}) \xi \rangle$$
the variance at ξ of the true input being $E\left[|\langle \xi, X \rangle|^2\right]$, then the relative error in variance (at ξ) is
$$\varepsilon^2 = \frac{\text{error in approximation}}{\text{variance of data}} = \frac{\langle \xi, \hat{R}_{op} \xi \rangle}{\langle \xi, R_{xx} \xi \rangle}$$
which we leave for the reader to play with.

4.5 Bayesian methods

The following approach is halfway between the deterministic approaches presented in the first sections of this or the former chapters, and the approach developed in the following chapter.

It essentially consists in considering
$$y = Ax + \varepsilon \tag{4.24}$$
as a particular realization of
$$Y = AX + n \tag{4.25}$$
which is (4.18), except that this time we are not requiring that X and n be both distributed according to a Gaussian distribution. But we still do assume that X, and n are random variables with a known distribution.

Let $Q(d\xi, d\eta)$ be some given distribution on $(V \times W, \mathcal{B}(V \times W))$, where $\mathcal{B}(V \times W)$ stands for the Borel sets in $V \times W$. In the finite dimensional case $V = \mathbb{R}^n$, $W = \mathbb{R}^m$.

Assume that (X,Y) has a *known* probability density with respect to $Q(d\xi, d\eta)$, let us denote it by $\rho_{XY}(\xi, \eta)$. Actually the available datum is the *(known)* density of $\rho_X(\xi)$, obtained from $\rho_{XY}(\xi, \eta)$ by marginalization, i.e., if $Q(d\eta) = \int_X Q(d\xi, d\eta)$, then

$$\rho_X(\xi) = \int_Y \rho(\xi, \eta) Q(d\eta). \tag{4.26}$$

Usually from the known stochastic model of the noise and (4.27) one is able to infer the conditional distribution of the data $Y = y$ given $X = x$, that is we known how to produce

$$\rho_Y(y\,|\,x) = P(Y = y\,|\,X = x) \tag{4.27}$$

and using the relationship

$$\rho_{XY}(x, y) = \rho_X(y\,|\,x)\rho_X(x) \tag{4.28}$$

and Bayes identity, we can produce a way of obtaining $\rho_X(x\,|\,y)$, i.e., the conditional probability of the unknown x (that is produced by the data y) given the datum y. This is achieved using (4.28) and (4.27) in the definitions

$$\rho_X(x\,|\,y) = \frac{\rho_{XY}(x,y)}{\rho_Y(y)} = \frac{\rho_X(y\,|\,x)\rho_X(x)}{\int \rho_X(y\,|\,\xi)\rho_X(\xi)Q(d\xi)}. \tag{4.29}$$

Here $Q(d\xi) = \int_Y Q(d\xi, d\eta)$. Once $\rho_X(x,y)$ is obtained, the *a posteriori* estimate of x is just

$$x^* = E[X\,|\,y] = \int \xi \rho_X(\xi\,|\,y) dQ(\xi).$$

We shall now briefly describe how to obtain (4.24) using the techniques described above.

Let us assume that n is a W-valued Gaussian random variable with variance R_{nn}, assume that X is a V-valued Gaussian random variable with variance R_{XX}. Then Y is a W-valued Gaussian random variable with variance $R_{YY} = AR_{XX}A^* + R_{nn}$, under the assumption of independence of n and X. From now on, to obtain (4.24) making use of (4.29) is down-wind sailing.

4.6 The method of maximum likelihood

In some sense this is the least probabilistic of the probabilistic methods. Actually it works in a set up much broader than the Hilbert space setup, and it is very simple to describe. Consider (4.25) again, i.e.,

$$y = Ax + n$$

but this time you do not want (or are forced not) to assume that x is a random variable. It is just an unknown vector in V. During the measurement of Ax noise comes in, in an additive way, and a mathematical model for it can be produced.

We shall assume that a positive measure $Q(d\eta)$ is a preassigned on $(W, B(W))$ and that $P(n \in d\eta) = \rho_n(\eta)Q(d\eta)$ describes the stochastic nature of the noise n. For each given x the distribution of $\eta = Ax + n$ has density $\rho_n(\eta - Ax)$ with respect to the translate $Q_{-Ax}(dy)$ of $Q(d\eta)$. The proof of this elementary fact is for the reader.

When, and only in this case, $Q(d\eta)$ is invariant under translations, i.e., $Q_a(d\eta) = Q(d\eta)$ for every $a \in W$, (or at least $a \in A(V)$) only then $\rho_n(\eta - Ax)$ is the probability density of $Ax + n$.

Assume for a moment that $Q(d\eta)$ is translation invariant, then the *method of maximum likelihood* consists of declaring $\hat{x}$ to be the *more* likely solution to $y = Ax + n$ if $\hat{x}$ is the value of (the parameter) vector x where the maximum of $\rho_n(\eta - Ax)$ is achieved.

A situation in which the extreme aspects of this method stand out is described in the following example.

Example 4.1. Assume that the noise in $Ax + n$ is modeled by a discrete random variable, taking value η_i with probability P_i. Then $y = Ax + n$ is a random variable taking values $y_i = Ax + \eta_i$ with probability P_i.

Assume that $(A^*A)^{-1}$ exists. Two extreme cases occur for the maximum likelihood method.

Case 1: There is only one index $\hat{\imath}$ such that $P_{\hat{\imath}} = \max\{P_i : i \geq 1\}$. This time the more likely solution to $Ax + n = y$ is

$$\hat{x} = (A^*A)^{-1}A^*(y_{\hat{\imath}} - \eta_{\hat{\imath}}).$$

Case 2: All P_i are equal. This time all solutions

$$x^{(i)} = (A^*A)^{-1}A^*(y_i - \eta_i)$$

are equally likely.

But in two of the most commonly used models: Poisson and Gaussian, the resulting distribution are unimodal, and we are in a situation like that of Case 1 above, i.e., a unique more likely solution exists. See for example: [1] and [4].

4.7 Bibliographical comments and references

For a very detailed treatment of regularization in Hilbert spaces, more detailed and covering much more material than what we did here, check with [1].

Two standard references for basics on functional analysis are Rudin's [5-2] or Reed and Simon's [6].

The books cited in Chapter 3 contain much more on regularization procedures than what we presented here, and Gaussian regularization plus some interesting applications are presented in Franklin's [3].

More on the Bayesian and maximum likelihood approaches can be seen in Richardson's [2] or Vardi et al. [5].

For some other variations on the theme check with Golan's review [4].

References

[1] Engel, H., Hanke, M. and Neubauer, A. *"Regularization of inverse problems"*. Kluwer Acad. Pub., Dordrecht, 1996.

[2] Richardson, W. H. *"Bayesian -Based Iterative Method of Image Restoration"*. J. Optical. Soc. Am., Vol. 62, No. 1. (1972). p. 55.

[3] Franklin , J. *"Well-posed stochastic extensions of ill-posed linear problems"*. J. Math. Anal. and Appl., Vol. 31 (1970). pp. 682-716.

[4] Golan, A. *"Maximum entropy, likelihood and uncertainty: A comparison"*. in Maximum Entropy and Bayesian Methods (Maxent'97 Workshop), Erickson, G. and Ryckert, J. (eds.) Kluwer Acad. Pub., Dordrecht (1998).

[5] Vardi, Y, Shepp, L. A. and Kaufman, L. *"A statistical model for position emission tomography"*. J. Am. Stat. Assoc., Vol. 80 (1985). pp. 8-37.

[6] Reed, M. and Simon, B. *"Methods of Mathematical Physics Vol. I"*. Acad. Press., N. Y., 1975.

Chapter 5

Maxentropic approach to linear inverse problems

This chapter is devoted to some basic aspects of the method of maximum entropy in the mean (MEM for short) as used to solve linear inverse problems with non-linear, convex constraints.

We shall describe what the routine is about and where the delicate issues lie.

5.1 Heuristic preliminaries

Consider a bounded, not necessarily injective linear operator

$$A : V \to W$$

where V and W are Banach spaces. We are interested in solving

$$Ax = y, \quad x \in C \tag{5.1}$$

where C is a closed convex subset of V which models the constraints, and $y \in W$. When there is noise in the measurement of the data vector y, we are led to consider two alternative approaches.

$$\begin{aligned}\textbf{(a)} & \quad Ax \in B_M(y,T), \quad x \in C \\ \textbf{(b)} & \quad y = Ax + n, \quad x \in C. \end{aligned} \tag{5.2}$$

In case (a), $B_M(y,T) = \{\eta \in W : d_M(y,\eta) < T\}$ denotes a ball in W determined by a metric M. This metric is related to the noise in the measurement, whereas the tolerance T measures how far off we are willing to be from the actual data. We could think of (5.2) differently. If $d_M(y,\eta)$ denotes the distance in the norm on W, and if $y \notin A(V)$, then we cannot hope to find a solution unless $B_M(y,T)$ intersects $A(V)$, thus T measures the distance from y to $A(V)$.

For case (b), n denotes a realization of a W-valued random variable η defined on a probability space $(\Omega, \mathcal{F}_n, Q_n)$ which is used to model the noise. In this case is an extension of (5.1), and MEM will provide us with an estimate x^* of the signal and an estimate n^* of the measurement noise.

The essence of the MEM consists of transforming (5.1) or (5.2) into a problem in convex optimization. The solution of (5.2) is built on the method developed to solve (5.1), and carried out in Sections 6 and 7 below. So let us begin with (5.1).

The first step in the MEM procedure consists of proposing a measure space $(\Omega, \mathcal{F}, Q)$, where the "sample space" Ω is usually the convex set C and $\mathcal{F}$ is the σ-algebra of subsets of C induced by the Borel subsets of W, i.e. $\mathcal{F} = C \cap B(W)$. In Chapter 8 we shall come across examples in which the natural sample space Ω differs from C, and for the sake of generality we distinguish them, but you should identify them in your mind.

The measure Q is usually a probability measure (but could be just a positive σ-finite measure on $(\Omega, \mathcal{F})$ which we shall call the "*a priori*" or "prior" measure (but not in the Bayesian sense of the term) for it reflects the information we have about the second mathematical object we need to introduce: the random variable

$$X : \Omega \to C \tag{5.3}$$

which when $\Omega = C$ is just the identity map, and again, you should think about it that way.

The only requisite placed upon the probability measure Q is that the closure of the convex hull generated by $supp(Q)$ is the constraint space C. The reason for this is that we shall want the probabilities $P << Q$ to have $supp(P) = C$ so that

$$E_P[X] = \int \xi dP \in C \tag{5.4}$$

and that the integral be defined, of course.

Actually, a simple consequence of the Hahn-Banach Theorem is that

Lemma 5.1. *Let $P << Q$ be a probability measure such that $supp(P) = C$ and that $E_P[\|X\|] < \infty$. Then*

$$\int \xi dQ(\xi) \in C.$$

Proof. If $x := \int \xi dP(\xi) \notin C$, $\exists \lambda \in V^*$ such that $\langle \lambda, (x - \xi) \rangle > 0 \quad \forall \xi \in C$. Integrating both sides with respect to P we obtain a contradiction. □

5.1.0.1 Restatement of the inverse problems

Instead of solving problem (5.1) or (5.2) we shall restate them as:

Problem P1: Find $P \ll Q$ such that

$$E_P[AX] = AE_P[X] = \int_\Omega A\xi dP(\xi) = y. \tag{5.5}$$

Problem P2: Find $P \ll Q$ such that

$$E_P[AX] = AE_P[X] = \int_\Omega A\xi dP(\xi) \in B_M(y, T). \tag{5.6}$$

Notice that if such P exists, $x = \int X dP$ solves (5.1) or (5.2), whichever the case, because of Lemma 5.1 $x \in C$ automatically. Actually, when the constraint set is compact, the following classical result is even stronger.

Theorem 5.1. *Let V be a Banach space such that V^* separates points. Let K be a compact subset of V such that the closed convex hull C of K is compact. Then $y \in C$ if and only if there exists a regular Borel probability measure P on C such that*

$$\int_C \xi P(d\xi) = y.$$

Proof. The idea of the proof consists of considering the dual $\mathbf{C}(C)^*$ of the Banach space $\mathbf{C}(C)$ of all continuous functions on C with the sup norm. $\mathbf{C}(C)^*$ consists of all Borel measures on C.

One then defines

$$\Psi : \mathbf{C}(C)^* \to V$$

by $\Psi(m) = \int_C \xi m(d\xi)$. The theorem asserts that $(\Psi(\mathbf{C}(C)^*)) = C$. The proof of the theorem reduces to proving that:

i.) The probability measures are a weak-*-compact subset of $\mathbf{C}(C)^*$.
ii.) The mapping Ψ is continuous when $\mathbf{C}(C)^*$ is given by its weak-*-topology.

See Rudin's book [2] for further details. □

Comment: In our case, besides the lack of compactness of C we are demanding $P \ll Q$ as well as fulfillment of (5.5) for the problem to have a solution as restated.

We also have

Proposition 5.1. *Let V be such that V^* separates points. Let the measure Q and the convex set C be such that $cl(convex\ hull(supp(Q))) = C$.*

Consider $\mathcal{M} = \{\rho \in L_1(C,Q)_+ : \rho > 0 \text{ on } C, \int \rho(\xi)dQ(\xi) = 1, \int_C \xi\rho(\xi)dQ(\xi) \text{ exists}\}$. If $int(C) \neq \emptyset$, then $x_0 = \int \xi\rho_0(\xi)dQ(\xi) \in int(C)$ for any $\rho_0 \in \mathcal{M}$.

Proof. Assume that $x_0 \in \partial C$. By the Hahn-Banach Theorem, then there exists a $\lambda_0 \in V^*$ such that $\langle\lambda,(x-x_0)\rangle \geq 0$ for all $x \in C$.

Define $\rho_1(\xi) = \begin{cases} \frac{\alpha}{\beta}\rho_0(\xi), & \text{if } \langle\lambda_0,\xi\rangle < 0 \\ \frac{1-\alpha}{1-\beta}\rho_0(\xi), & \text{if } \langle\lambda_0,\xi\rangle \geq 0 \end{cases}$

where $\alpha > \beta$ and $\beta = P_0(\{\langle\lambda_0,\xi\rangle < 0\}) > 0$. Then $\int \rho_1(\xi)dQ = \int \rho_0(\xi)dQ = 1$.

Note now that $\tilde{x} = E_{\rho_1}[X] \in C$ by Lemma 5.1, but

$\langle\lambda_0,\tilde{x}\rangle - \langle\lambda,x_0\rangle$
$= (\frac{\alpha}{\beta} - 1)\int_{\{\langle\lambda_0,\xi\rangle<0\}} \langle\lambda_0,\xi\rangle\, dP_0 + (\frac{1-\alpha}{1-\beta} - 1)\int_{\{\langle\lambda_0,\xi\rangle\geq 0\}} \langle\lambda_0,\xi\rangle\, dP_0 < 0$,

which contradicts our choice of λ_0. Thus x_0 must be in $int(C)$. □

The issue now consists of producing a $\rho(\xi)$ so that $P(d\xi) = \rho(\xi)dQ(\xi)$ satisfies (5.5) or (5.6). Notice that the class

$$\mathcal{P}(y) = \left\{\rho \in L_1(C,Q) : \rho \geq 0, \int \rho(\xi)dQ(\xi) = 1, A\int \xi\rho(\xi)dQ(\xi) = y\right\}$$

is convex, and what better way is there for picking a point in a convex set than with the aid a concave function. Enters the entropy functional.

Definition 5.1. The functional $S(\rho) = S(P,Q)$ defined on $\mathcal{M}$ by

$$S(\rho) = -\int \ln\left(\frac{dP}{dQ}\right)dP = -\int_C \rho(\xi)\ln(\rho(\xi))dQ(\xi) \qquad (5.7)$$

when $|\ln(\rho(\cdot))|$ is dP-integrable or $-\infty$ if not is a well-defined, concave functional.

The next lemma is a key ingredient in the heuristics of MEM.

Lemma 5.2. Let $Z : \Omega \to \mathbb{R}$ be a random variable such that $\int e^{-Z}dQ < \infty$, and $\int |Z|e^{-Z}dQ < \infty$. The maximum of $S(P,Q)$ over $\{P \ll Q : \int |Z|dP < \infty\}$ is reached at $P^* = Ne^{-Z}Q$, where $N^{-1} = \int e^{-Z}dQ$ is a normalization constant.

The proof is an easy consequence of the following

Proposition 5.2. Let $f(\xi), g(\xi)$ be two probability densities such that both $|\ln(f(\xi))|$ and $|\ln(g(\xi))|$ are dQ-integrable. Then

$$K(f,g) = \int f(\xi)\ln\left(\frac{f(\xi)}{g(\xi)}\right)dQ(\xi)$$

is convex in f, positive and vanishes only when $f = g$.

Proof. See Appendix D. □

Proof. [Of Lemma 5.2] Notice that

$$S(P, P^*) = -\int \ln\left(\frac{dP}{dP^*}\right) dQ = -\int dP \ln\left(\frac{dP}{dQ}\frac{dQ}{dP^*}\right)$$
$$= -\int \ln\left(\frac{dP}{dQ}\right) dQ - \int Z dP + \ln N$$
$$S(P, Q) - \int Z dP + \ln N \geq 0.$$

Where the last step is a consequence of the proposition. Therefore

$$S(P, Q) \leq \int Z dP + \ln N \geq 0$$

with equality holds whenever $P = P^*$. □

Comment: In statistics $K(f, g)$ is called the Kullback-Leibler's divergence between the densities f and g.

To relate to our problem, suppose that for $\lambda \in W^*$

$$Z(\lambda) = \int_C \exp(-\langle \lambda, A\xi \rangle) dQ(\xi) < \infty$$

and define the probability

$$dP_\lambda(\xi) = \rho_\lambda(\xi) dQ(\xi) = \frac{\exp(-\langle \lambda, A\xi \rangle)}{Z(\lambda)} dQ(\xi). \quad (5.8)$$

According to what we saw above

$$\Sigma(\lambda) := \ln(Z(\lambda)) + \langle \lambda, y \rangle \geq S(\rho). \quad (5.9)$$

Notice as well that when P_λ were in $\mathcal{P}(y)$, according to Lemma 5.2

$$S(P_\lambda, Q) = \Sigma(\lambda) = \ln(Z(\lambda)) + \langle \lambda, y \rangle. \quad (5.10)$$

The procedure now is clear: Try to find $\lambda^* \in W^*$ such that $P_{\lambda^*} \in \mathcal{P}(y)$, i.e. P_{λ^*} satisfies the constraints $E_{P_{\lambda^*}}(AX) = y$. According to (5.9) such P_{λ^*} maximizes $S(P_\lambda, Q)$ over $\mathcal{P}(\dagger)$ and therefore

$$x^* = E_{P_{\lambda^*}}[X] \quad (5.11)$$

is the maxentropic solution to (5.1) that we are after.

To find the λ^* we have to check whether

$$\inf \{\Sigma(\lambda) : \lambda \in W^*\} = \sup \{S(P, Q) : P \in \mathcal{P}(y)\} \quad (5.12)$$

and both are finite and coincide. After we examine some properties of the two functionals, we consider the two optimization problems.

5.2 Some properties of the entropy functionals

We have introduced two entropy functionals. One of them $S(\rho)$ actually defined on the cone $K = \{\rho \in L^1(C,Q) : \rho \geq 0, \int \rho dQ = 1\}$ by

$$S(\rho) := -\int_\Omega \rho(\xi) \ln(\rho(\xi)) dQ(\xi)$$

whenever the integral is finite or $-\infty$ otherwise.

We also defined on W^* the "dual entropy" functional

$$\Sigma(\lambda) := \ln(Z(\lambda)) + \langle \lambda, y \rangle$$

for given $y \in A(C)$ (or $y \in W$ if you prefer) and set $S(\lambda) = +\infty$ whenever

$$Z(\lambda) = \int_\Omega \exp(-\langle \lambda, AX \rangle) dQ = +\infty.$$

Note that if we transport the measure Q from Ω to C by means of X, we can regard both integrals above as integrals over C.

Lemma 5.3. *Both entropy functionals are respectively strongly concave and strongly convex, on the sets where they are finite.*

Proof.

a) For $S(\rho)$ the strong concavity follows from the strong convexity of $x \to x \ln(x)$ on $(0, \infty)$.

b) An application of the Hölder inequality yields the convexity of $\Sigma(\lambda)$ as follows: let $0 < \alpha < 1$ and $\beta = 1 - \alpha$, and let λ_1, λ_2 be such that $Z(\lambda_i) < \infty$. Then setting $\frac{1}{p} = \alpha$, $\frac{1}{q} = \beta$ we obtain

$$Z(\alpha\lambda_1 + \beta\lambda_2) = \int_\Omega \exp(-\alpha \langle \lambda_1, A\xi \rangle) \exp(-\beta \langle \lambda_2, A\xi \rangle) dQ(\xi)$$

and invoking Hölder's inequality

$$Z(\alpha\lambda_1 + \beta\lambda_2) < (Z(\lambda_1))^\alpha (Z(\lambda_2))^\beta$$

from which the strict convexity of $\Sigma(\lambda)$ drops out. $\square$

Comment: An implicit corollary of the lemma is that

$$\text{dom } Z = \{\lambda \in W^* : Z(\lambda) < \infty\}$$

is a convex set whenever it is not empty.

Lemma 5.4. $\ln(Z(\lambda))$ *is a lower semi-continuous on dom Z.*

Proof. We need to verify that for real α the set
$$\{\lambda : \ln(Z(\lambda)) \leq \alpha\}$$
is closed. For that let $\|\lambda_n - \lambda\|_{W^*} \to 0$. Then $\xi \in C, \langle \lambda_n, A\xi \rangle \to \langle \lambda, A\xi \rangle$ and since $Z(\lambda_n) \leq e^{\alpha}$ Fatou's lemma yields
$\int \exp(-\langle \lambda, A\xi \rangle) dQ(\xi) = \int \liminf \exp(-\langle \lambda_n, A\xi \rangle) dQ(\xi)$
$\qquad \leq \liminf \int \exp(-\langle \lambda_n, A\xi \rangle) dQ(\xi) \leq e^{\alpha}.$ □

But things are not as nice for $S(\rho)$. On one hand, we have

Lemma 5.5. *Suppose Q is a σ-finite measure. The sets*
$$\left\{ \rho > 0, \int \rho dQ = 1 : S(\rho) \geq \alpha \right\}$$
may not be compact in the strong, weak or weak-$*$ topologies.

Proof. Consider a sequence C_n of sets in $\mathcal{F}$ such that $Q(C_n) \to \infty$, and define $\rho_n = \frac{1}{Q(C_n)} I_{C_n}$. Then $S(\rho_n) = \ln Q(C_n) > \alpha$ for n large enough. But this sequence cannot have a limiting density. □

Comments: If we restrict ourselves to ρ such that $\int \rho dQ = 1$ we may put a bound on $\{\rho \in L_1(C, Q) : \rho > 0, \int \rho dQ = 1, \int \|\xi\| \rho(\xi) dQ(\xi) < \infty \}$, but regretfully $\{\rho : S(\rho) \geq \alpha\}$ may not be compact. Simple examples can be constructed when C is not compact. Try it. But if we consider a larger space of probabilities, something like $\{P << Q : P(C) = 1, E_P[\|X\|] < \infty\}$, we could perhaps obtain compactness.

Even worse, the simple minded use of Lagrange multipliers may not be justified because it also happens that

Lemma 5.6. *The class $K = \{\rho \in L^1(C, Q) : \rho \geq 0\}$ has an empty interior.*

Proof. For the reader. Notice that in any neighborhood of a density one can have non-positive functions. □

Comment: This fact makes the usual "calculus" argument used to find maxentropic densities wrong, or at least suspect.

5.3 The direct approach to the entropic maximization problem

The results in this section are a particular case of the results in the following section. What we do is basically the standard approach using maximum entropy method.

We saw at the end of Section 1, that all that was needed to provide a solution to (5.1) is to verify that there exists $\lambda^* \in W^*$ such that (5.12) holds, i.e.

$$\inf\{\Sigma(\lambda) : \lambda \in W^*\} = \sup\{S(P,Q) : P \in \mathcal{P}\}. \tag{5.13}$$

If these problems are to have solution, we better make

Assumption 1: Both sets, $\mathcal{P}(y)$ and $dom\, Z$ are not empty.

Comment: $\mathcal{P}(y)$ being not empty means that the infimum is finite and $dom\, Z$ not empty means that the supremum is finite.

Notice that when C is bounded and $\int_C dQ = 1$, then $dom\, Z = W^*$, for then

$$Z(\lambda) \leq \int_C e^{\|A^*\lambda\|\|\xi\|} dQ(\xi) \leq e^{\|A^*\lambda\|M}$$

where $M = \sup\{\|\xi\| : \xi \in C\}$. In this case we have

Proposition 5.3. *Under Assumption 1 if C is bounded and if the infimum is reached at $\lambda^* \in W^*$, then $P_{\lambda^*} \in \mathcal{P}(y)$, where $dP_{\lambda^*}(\xi) = \dfrac{e^{-\langle\lambda^*,A\xi\rangle}dQ(\xi)}{Z(\lambda^*)}.$*

Proof. We shall verify that $\Sigma(\lambda)$ is Frechet differentiable at λ^*. For this note to begin with that for any real a

$$e^{-a} - 1 = -a + a^2 \int_0^1 e^{-at}(1-t)dt.$$

Therefore for $h \in W^*$, $0 < \|h\| < \infty$ we have

$\int (e^{-\langle h, A\xi\rangle} - 1) dP_\lambda(\xi)$
$= -\int \langle h, A\xi\rangle\, dP_\lambda(\xi) \; + \int_0^1 (1-t) \left(\int \langle h, A\xi\rangle^2 e^{-t\langle h, A\xi\rangle} dP_\lambda(\xi)\right) dt.$

We want to compute the Frechet derivative of $\Sigma(\lambda)$ at λ^*. The candidate being

$$DS(\lambda^*) = -\int A\xi dP_{\lambda^*}(\xi) + y \tag{5.14}$$

thus we have to verify that

$$\frac{1}{\|h\|}|S(\lambda^* + h) - S(\lambda^*) - DS(\lambda^*)(h)| \longrightarrow 0 \text{ when } \|h\| \to 0.$$

But notice that

$$\frac{1}{\|h\|}|S(\lambda^* + h) - S(\lambda^*) - DS(\lambda^*)(h)|$$

$$= \frac{1}{\|h\|}\left|\ln\left(\int (e^{-\langle h, A\xi\rangle} - 1)dP_{\lambda^*}(\xi) + \int \langle h, A\xi\rangle\, dP_{\lambda^*}(\xi)\right)\right|$$

$$= \frac{1}{\|h\|}\left|\ln\left(\int_\Omega \langle h, A\xi\rangle^2 \left[\int_0^1 (1-t)e^{-\langle h, A\xi\rangle}dt\right] dP_{\lambda^*}(\xi)\right)\right|$$

$$\leq \frac{1}{2\|h\|}\int_\Omega e^{\|h\|\|A\xi\|} \langle h, A\xi\rangle^2\, dP_{\lambda^*}(\xi).$$

Clearly the last term tends to zero as $\|h\| \to 0$, and we are done. Thus if λ^* is a minimum, $DS(\lambda^*)$ is the zero operator and

$$\int A\xi dP_{\lambda^*}(\xi) = y$$

or $P_{\lambda^*} \in \mathcal{P}$. □

Actually, the same proof allows us to conclude

Proposition 5.4. *Under Assumption 1, if* $\inf\{\Sigma(\lambda) : \lambda \in W^*\}$ *is reached at* λ^* *belonging to* $\text{int}W^*$, *then* $P_{\lambda^*} \in \mathcal{P}$ *and (5.13) holds.*

Proposition 5.5. *Assume that there exists a subspace* W_1^* *of* W^*, *an* $\varepsilon > 0$ *such that*

$$\int_C e^{\|A^*\gamma\|\|\xi\|}dP_{\lambda^*}(\xi) < \infty$$

for all $\|\gamma\| \leq \varepsilon$ *and* $\gamma \in W_1^*$, *where* λ^* *is the point in* W^* *at which* $\Sigma(\lambda)$ *reaches its minimum. Then*

$$\langle \gamma, (y - Ax^*)\rangle = 0 \tag{5.15}$$

where $x^* = \int \xi dP_{\lambda^*}(\xi)$.

Proof. Fix $\gamma \in W_1^*$, $\|\gamma\| \leq \varepsilon$ and consider the convex function defined on an interval $(-t_0, t_0)$ by

$$t \to S(\lambda^* + t\gamma).$$

Under our assumptions, this is a differentiable function having a minimum at $t = 0$. Computing its derivative we have $0 = \langle \gamma, (y - Ax^*)\rangle$. □

This result is the input for

Proposition 5.6. *Assume that $W_1 \subset W$ is a closed subspace, that W_1^* considered as a subspace of W^* via the Hahn-Banach theorem satisfies (5.5), then either (a) or (b) holds.*

a) $R(A) = W$, or
b) $R(A)$ is strictly contained in W_1.

Then if $\lambda^* \in W^*$ minimizes $\Sigma(\lambda)$, $Ax^* = y$ and $P_{\lambda^*}(d\xi)$ is in $\mathcal{P}$.

Proof. We know from Proposition 5.5 that for any $\gamma \in W_1^*$,

$$\langle \gamma, (y - Ax^*) \rangle = 0.$$

Under assumption a) this is only possible if $Ax^* = y$, since $\ker(A^*) = 0$. Similarly, if $y \neq Ax^*$, there exists $\gamma \in W_1^*$ such that

$$\langle \gamma, (y - Ax) \rangle \neq 0,$$

but this contradicts (5.15). $\square$

Comments: Observe that from the representation in Proposition 5.6, the non-uniqueness of the solution is clear that if we add $\eta \in Ker(A^*)$ to λ^* the representation does not change. Of course, this does $\Sigma(\lambda)$. This explains it.

The assumption in Proposition 5.5 amounts to saying that $\lambda^* \in ri(dom\, Z)$. Recall that this means that the positive cone generated by $dom\, Z - \lambda^*$ is a subspace.

Before quitting this line of analysis we direct the reader to references [3]-[8]. These deal with the problem consisting of finding P^* that realizes

$$\sup\{S(P,Q) : P << Q, E_P[AX] = y\}.$$

It is addressed from the point of view of duality theory.

5.4 A more detailed analysis

In this section we shall see that under one key assumption the basic identity (5.11) holds, and that in this case the MEM can be applied as well. Let us introduce some notation. The results in this section follow Cherny and Maslov's [9] almost verbatim.

Put $Y = AX - y$ and observe that $\Sigma(\lambda) = \ln \int e^{-\langle \lambda, Y \rangle} dQ$ is finite on $dom\, Z$. Introduce as well $\Psi(\lambda) = \left(\ln \int e^{-\langle \lambda, Y \rangle} dQ \right)^{-1} \left(\int X e^{-\langle \lambda Y \rangle} dQ \right)$, which is finite on $\{\lambda \in W^* | \int \|X\| e^{-\langle \lambda, Y \rangle} dQ < \infty\}$. Consider the induced measure $Q \circ Y^{-1}$ on W and denote by $\mathcal{L}$ the affine hull generated by $supp(Q \circ Y^{-1})$.

We have already established that in the interior on $dom\, Z$ we have $y - A\Psi(\lambda) = \nabla \Sigma(\lambda)$.

Theorem 5.2. *Suppose that $y \in A(C)$ and 0 is in the relative interior of $\mathcal{L}$. Then*
(a) $sup\{S(P, Q) : P \in \mathcal{P}(y)\} = \inf\{\Sigma(\lambda) \lambda \in W^*\}$.
(b) If there exists a $\lambda^* \in W^*$ such that $\Sigma(\lambda^*) = \min\{\Sigma(\lambda) | \lambda \in W^*\}$ and $A\Psi(\lambda^*) = y$, then $\sup\{S(P,Q) | P \in \mathcal{P}(y)\}$ is attained at the unique P^*. Otherwise, $\sup S(P,Q)$ is not attained.

Proof. (a) We saw in Lemma 5.2 that if $dom\, Z \neq \emptyset$ then $S(P, Q) \leq \Sigma(\lambda)$. To prove the opposite inequality, we proceed as follows. Since Q is σ-finite, there exist sets $A_n \in \mathcal{F}$ such that $0 < Q(A_n) < \infty$, $A_n \subset A_{n+1}$ and $\cup A_n = \Omega$. Now set $B_n = A_n \cap \{\|Y\| \leq n\}$ and $Q_n(\cdot) = Q(\cdot \cap B_n)$. Now define $\Sigma_n(\lambda) = \ln \int e^{-\langle \lambda, Y \rangle} dQ_n$. Clearly $\Sigma_n(\lambda) \uparrow \Sigma(\lambda)$. Observe that for $\|\lambda\| > 1$

$$\Sigma_n(\lambda) \geq \ln Q(B_n) + \|\lambda\| \ln(\int e^{-\langle u, Y \rangle} \frac{dQ_n}{Q(B_n)})$$

where $u = \lambda/\|\lambda\|$. Thus, $\sigma_n(\lambda) \to \infty$ as $\|\lambda\| \to \infty$.

Thus, Σ_n being strictly convex, it attains an infimum at some λ_n^*. Consider the probability law P^* with density $e^{-\langle \lambda^*, AX \rangle}(Z(\lambda^*))^{-1}$ relative to Q_n. Then $\int Y dQ_n = -\nabla \Sigma_n(\lambda_n^*) = 0$. Notice as well that

$$\Sigma_n(\lambda_n^*) = S(P_n, Q_n) = S(P_n, Q) \leq \sup S(P, Q).$$

Keep the point λ_n fixed but vary the function to obtain $\Sigma(\lambda_n) \leq \sup S(P, Q)$, therefore

$$\inf \Sigma(\lambda) \leq S(P, Q)$$

thus concluding the proof of item (a). To prove the first assertion of (b), if λ^* exists such that $\Sigma(\lambda^*) = \min\{\Sigma(\lambda) | \lambda \in W^*\}$ and $A\Psi(\lambda^*) - y = 0$, then $\sup S(p, Q)$ over $\mathcal{P}(y)$ is attained at P^* described above (as guaranteed by Lemma 5.2).

The proof that if no such λ^* exists, then $S(P, Q)$ cannot attain a maximum is carried out by *reductio ad absurdum*: If no such λ^* exists and a maximum of $S(P, Q)$ is attained, a contradiction occurs. The proof is rather long and we skip it. The interested reader should consult [9]. □

5.5 Convergence of maxentropic estimates

We saw in Chapter 3 that both the data gathering process and the need for numerical computations, forced us to consider finite dimensional versions of the problem of solving $Ax = y$ under $x \in C$. The data gathering process forces us to consider $y \in W$ with $\dim W = M$ a finite number, and in order to describe x numerically we should consider $A_n x = y$ with $A_n = A|_{V_n}$, with $\dim V_n = n > M$.

If we denote by x_n^* the maxentropic solution to $A_n x = y$, and if $V_n \uparrow V$, the question is: Do the $x_n^* \to x^* \in V$? Here x^* is the maxentropic solution to $Ax = y$.

Let now $x_m^* \in V$ denote the maxentropic solution to $Ax = y_m$. If $y_m \in W_m \to y \in W$ (or if $y_m \to y$ in W) does $x_m^* \to x^* =$ the maxentropic solution to $Ax = y$?

The second problem has received considerable attention. See for example [13]-[18] as well as in [10]-[11]. Below we shall present a simple approach to these issues.

Let us begin with

Proposition 5.7. *Assume that either the constraint set C is bounded or that Q is such that for every positive α, $\int_C \exp(\alpha \|\xi\|) dQ(\xi) < \infty$.*

Denote by λ_n^* the point in W^* at which (5.13) holds for each given A_n, and assume that $\|A_n - A\| \to 0$ and $\|y_n - y\| \to 0$.

Then, in a more or less obvious notation: $P_n(\lambda_n^*) \to P(\lambda)$ weakly and $x_n^* \to x$ in V.

Proof. Under the stated assumptions $\|\lambda_n - \lambda\| \to 0$ (see Proposition 5.9 below) and

$$\|A_n^* \lambda_n - A\lambda\| \leq \|A_n - A\| \|\lambda_n\| + \|A^*\| \|\lambda_n - \lambda\| \to 0.$$

Therefore, for each $\xi \in V$ the sequence $e^{-\langle \lambda_n^*, A_n \xi \rangle}$ converges to $e^{-\langle \lambda^*, A\xi \rangle}$ pointwise (and boundedly) and therefore $Z_{A_n}(\lambda_n^*) \to Z_A(\lambda^*)$. Also for every continuous bounded function defined on V

$$\int g(\xi) dP_n(\lambda_n^*) = \int g(\xi) \frac{e^{-\langle \lambda_n^*, A_n \xi \rangle}}{Z_{A_n}(\lambda_n^*)} dQ \to \int g(\xi) \frac{e^{-\langle \lambda_n^*, A\xi \rangle}}{Z_A(\lambda)} dQ = \int g(\xi) dP(\lambda).$$

$\square$

In particular when C is bounded, take $g(\xi) = \langle \mu, \xi \rangle$ for arbitrary $\mu \in V^*$ to obtain $x_n^* \to x^*$. When C is not bounded, but $\int \exp(a \|\xi\|) dQ(\xi) < \infty$ for $a > 0$, the same holds true.

Comment: Clearly

$$S(\lambda_n^*) = \langle \lambda_n^*, y \rangle + \ln(Z_{A_n}(\lambda_n^*))$$
$$= S(P_n(\lambda_n^*), Q) \to S(\lambda^*)$$
$$= \langle \lambda^*, y \rangle + \ln(Z_A(\lambda)) = S(P(\lambda^*), Q).$$

Note that, from the point of view of numerical estimation, only a measure of the difference $\|\lambda_n^* - \lambda_m^*\|$ is to be expected. The following result relates $\|A_n^*\lambda_n - A_m^*\lambda_m\|$ to $\|x_n^* - x_m^*\|$.

Proposition 5.8. *Under the same assumptions as in Proposition 5.7,*

$$\|x_n^* - x_m^*\| \le M \|A_n^*\lambda_n - A_m^*\lambda_m\|. \tag{5.16}$$

Proof. Just follow the following estimates

$$\|x_n^* - x_m^*\| = \left\| \int \xi \frac{e^{-\langle \lambda_n^*, A_n \xi \rangle}}{Z_{A_n}(\lambda_n^*)} dQ(\xi) - \int \xi \frac{e^{-\langle \lambda_m^*, A_m \xi \rangle}}{Z_{A_m}(\lambda_m^*)} dQ(\xi) \right\|$$

$$= \left\| \int \xi \frac{e^{-\langle \lambda_n^*, A_n \xi \rangle}}{Z_{A_n}(\lambda_n^*)} \left(1 - e^{-\langle \mu, \xi \rangle} \frac{Z_n(\lambda_n^*)}{Z_m(\lambda_m^*)} \right) dQ(\xi) \right\|$$

$$\le \int \|\xi\| \frac{e^{-\langle \lambda_n^*, A_n \xi \rangle}}{Z_{A_n}(\lambda_n^*)} \left| 1 - e^{-\langle \mu, \xi \rangle} \frac{Z_n(\lambda_n^*)}{Z_m(\lambda_m^*)} \right| dQ(\xi)$$

where we have set $\mu = A_n^*\lambda_n - A_m^*\lambda_m$. We can further rewrite the last inequality as

$$\|x_n^* - x_m^*\| \le \int \|\xi\| \frac{e^{-\langle \lambda_n^*, A_n \xi \rangle}}{Z_n(\lambda_n^*)} \left| e^{\|\mu\|\|\xi\|} \frac{Z_n(\lambda_n^*)}{Z_m(\lambda_m^*)} - 1 \right| dQ(\xi)$$

$$= \int \|\xi\| \left| e^{\|\mu\|\|\xi\|} \frac{Z_n(\lambda_n^*)}{Z_m(\lambda_m^*)} - 1 \right| dP_n(\lambda_n^*).$$

Notice as well that

$$\frac{Z_n(\lambda_n^*)}{Z_m(\lambda_m^*)} = \int \frac{e^{-\langle \lambda_n^*, A_n \xi \rangle}}{Z_m(\lambda_m^*)} dQ = \int e^{\langle \mu, \xi \rangle} dP_m(\lambda_m^*)$$
$$\le \int e^{\|\mu\|\|\xi\|} dP_m(\lambda_m^*).$$

Let us now invoke the obvious inequality
$$e^{\|\mu\|\|\xi\|} \leq 1+\|\mu\|\left(e^{\|\xi\|}-1\right)$$
which implies that
$$\frac{Z_n(\lambda_n^*)}{Z_m(\lambda_m^*)} \leq 1+\|\mu\| G_m(\lambda_m^*) = 1+\|\mu\|\int\left(e^{\|\xi\|}-1\right)P_m(\lambda_m^*)$$
as well as
$$\|x_n^* - x_m^*\| \leq \int \|\xi\|\left|\left(1+\mu\left(e^{\|\xi\|}-1\right)\right)\left(1+\|\mu\| G_m(\lambda_m^*)\right)-1\right|dP_n(\lambda_n^*)$$
$$\leq \|\mu\| M$$
and we leave for the reader to obtain M. □

Comment: In the proof above, when λ^* is known, a similar estimate is obtained for $\|x_n^* - x_m^*\|$, namely
$$\|x_n^* - x_m^*\| \leq m \|A_n^* \lambda_n^* - A^* \lambda^*\| \qquad (5.17)$$
for some appropriate M.

To provide at least one answer to the second question consider

Proposition 5.9. *Assume that either the constraint set C is bounded or for any $\alpha > 0$, $\int \exp(a\|\xi\|)dQ(\xi) < \infty$. Assume that for some $0 < m < M < \infty$,*
$$m = \inf\{B(\eta,\eta) : \eta \in W^*, \|\eta\|=1\}, M = \sup\{B(\eta,\eta) : \eta \in W^*, \|\eta\|=1\},$$
where the bilinear form $B(\lambda_1, \lambda_2)$ on $W^* \times W^*$ is defined by
$$B(\lambda_1, \lambda_2) = \int \langle \lambda_1, (A\xi - y)\rangle \langle \lambda_2, (A\xi - y)\rangle dP(\lambda^*)$$
and where $y = A\int \xi dP(\lambda^*)$. Then a change δy in the datum y produces a change $\delta\lambda$ in minimizing $\lambda^* = \arg\inf\{\Sigma(\lambda) = \langle \lambda, y\rangle + \ln(Z(\lambda)) : \lambda \in W^*\}$ satisfying
$$\|\delta\lambda\| \leq \frac{1}{m}\|\delta y\|.$$

Proof. Think of
$$y = \int A\xi dP(\lambda)$$
as a function of λ. Its directional derivative in the direction of a $\delta\lambda$ in W^* is given by
$$\nabla y(\delta\lambda) = -\int A\xi \langle \delta\lambda, A\xi\rangle dP(\lambda) + y(\lambda)\langle \delta\lambda, y\rangle$$

as an element of W. (Note that $\nabla y : W^* \to W$, since W^* is the tangent space to W^* at λ.)

The left-hand side of the last identity coincides, up to first order in $\|\delta\lambda\|$ with the increment δy in y as λ changes by δy. Denote it by δy. Applying $\delta\lambda$ to both sides of the identity and rearranging we obtain

$$\langle \delta\lambda, \delta y \rangle = -\int \langle \delta\lambda, (A\xi - y) \rangle \langle \delta\lambda, (A\xi - y) \rangle \, dP(\lambda).$$

If we use the characterization $\|\delta y\| = \sup_{\|\delta\lambda\|=1} |\langle \delta y, \delta\lambda \rangle|$, we obtain from

$$m \|\delta\lambda\|^2 \leq |\langle \delta\lambda, \delta y \rangle| \leq M \|\delta\lambda\|^2$$

the inequality $m \|\delta\lambda\| \leq \|\delta y\| \leq M \|\delta\lambda\|$, which is what we are after. $\square$

Comment: If we invoke the obvious extension of (5.16) to this situation, since both A_n and A_m equal A, we conclude that

$$\|x^*(y + \delta y) - x^*\| \leq M_1 \|\delta y\| \tag{5.18}$$

for some appropriate constant M_1.

This last problem is dealt with in different guises and context in references [12]-[17]. There, different, but somewhat related results are obtained. See in particular the survey [16].

5.6 Maxentropic reconstruction in the presence of noise

Now instead of (5.1) the problem consists of solving

$$Ax + n = y \quad x \in C \tag{5.19}$$

where n describes an additive noise, i.e., a W-valued random variable, which appears when we measure Ax.

The situation is similar, but a bit more general than that treated in Section 4.6 of Chapter 4. We must assume that we have a model for the noise, i.e., some probability space $(\Omega', \mathcal{F}', P')$ and a random variable $n : \Omega' \to W$.

We shall assume the statistical properties if the noise leads to some positive definite variance operator Σ, so that solving (5.19) can be achieved when we find $x \in V$ satisfying

$$Ax \in B_{\Sigma^{-1}}(y, T), \quad x \in C \tag{5.20}$$

where
$$B_{\Sigma^{-1}}(y,T) = \{\eta \in W : \|\Sigma^{-1}(\eta - y)\| \leq T\}$$
where $\|...\|$ is the norm on W. This choice is motivated by the procedure used in the final dimensional situation in which
$$n = y - Ax$$
and usually $n = \Sigma\varepsilon$ where ε is some "white noise" (whatever that may mean) with standard statistical properties. Then
$$\varepsilon = \Sigma^{-1}(y - Ax)$$
and the reconstruction criterion is that we admit a reconstruction $\hat{\eta}$ as "good" or "typical" when $\|\Sigma^{-1}(y - A\hat{\eta})\| \leq T$ for this corresponds to acceptable values $E\|\varepsilon\|^2$ of the "white noise" ε. The gist of what is behind the translation of (5.19) into (5.20), consists of getting rid of the randomness in (5.19) by replacing searching for exact solutions by a quest for solutions within some tolerance.

To proceed as above we shall assume we have constructed $(\Omega, \mathcal{F}, Q)$ such that $X : \Omega \to C$ is a given random variable, or just a measurable function if Q is not a probability but only some σ-finite measure. The next step consists of assuming that the class
$$\mathcal{P}(y,T) = \{P << Q : E_P[AX] \in B_{\Sigma^{-1}}(y,T)\}$$
is nonempty, i.e., that our problem has a solution.

On $\mathcal{P}(y,T)$ one defines the visual entropy functional
$$S(P,Q) = -\int_\Omega \rho(\xi) \ln(\rho(\xi)) dQ(\xi)$$
where $\rho(\xi) = \frac{dP}{dQ}(\xi)$ is the Radon-Nikodym derivative of P with respect to Q. $S(P,Q) = -\infty$ when $E_P[|\ln(\rho(\xi))|] = +\infty$.

To see that the whole game is pretty much as above note that
$$\mathcal{P}(y,T) = \cup_{\eta \in B_{\Sigma^{-1}}(y,T)} \mathcal{P}(\eta)$$
where $\mathcal{P}(\eta) = \{P << Q : E_P[AX] = \eta\}$. Therefore
$$\sup\{S(P,Q) : P \in \mathcal{P}(y,T)\}$$
$$= \sup\{\sup\{S(P,Q) : P \in \mathcal{P}(\eta)\} : \eta \in B_{\Sigma^{-1}}(y,T)\}.$$

We already know that whenever $\sup\{S(P,Q) : P \in \mathcal{P}(\eta)\}$ is achieved, it must equal $\inf_\lambda \{S(\lambda, \eta) : \lambda \in W^*\}$, where, recall that $S(\lambda, \eta) = \ln(Z(\lambda)) +$

$\langle \lambda, \eta \rangle$ for $\lambda \in W^*$ (we are adding η to emphasize). Therefore

$$\sup\{S(P,Q): P \in \mathcal{P}(y,T)\} = \sup_{\eta}\left\{\inf_{\lambda}\{S(\lambda,\eta)\}\right\}$$

$$= \inf_{\lambda}\left\{\sup_{\eta}\{\ln(Z(\lambda)) + \langle \lambda, \eta \rangle\}\right\}$$

$$= \inf_{\lambda}\left\{\ln(Z(\lambda)) + \sup_{\eta}\{\langle \lambda, \eta \rangle : \eta \in B_{\Sigma^{-1}}(y,T)\}\right\}.$$

To compute the supremum at the last step note that

$$\eta \in B_{\Sigma^{-1}}(y,T) \Leftrightarrow \eta = y + T\Sigma\rho, \quad \rho \in B(0,1)$$

where of course $B(0,1) = \{\zeta \in W : \|\zeta\| = 1\}$. Now by definition, for $\lambda \in W^*$, $\|\lambda\| = \sup_{\rho \in B(0,1)} \langle \lambda, \rho \rangle$, therefore

$$\sup\{\langle \lambda, \eta \rangle : \eta \in B_{\Sigma^{-1}}(y,T)\} = \langle \lambda, y \rangle + T\sup_{\rho}\{\langle \Sigma^*\lambda, \rho \rangle\}$$

$$= \langle \lambda, y \rangle + T\|\Sigma^*\lambda\|.$$

To sum up,

$$\sup\{S(P,Q): P \in \mathcal{P}(y,T)\} = \inf_{\lambda}\{S_T(\lambda) = S_0(\lambda) + T\|\Sigma\lambda\|\} \quad (5.21)$$

where $S_0(\lambda) = \ln(Z(\lambda)) + \langle \lambda, y \rangle$, and we collect everything under

Theorem 5.3. *Assume that*

$$\mathcal{P}(y,T) = \{P << Q : E_P[AX] \in B_{\Sigma^{-1}}(y,T)\}$$

is not empty, that $\inf S_T(\lambda)$ *is reached in* $\text{int}\mathcal{D}(A,Q)$. *Then (5.21) holds and*

$$dP^* = \frac{e^{-\langle \lambda^*, AX \rangle}}{Z(\lambda)}(\xi)dQ(\xi)$$

is such that $A(E_{P^*}[X]) = Ax^* \in B_{\Sigma^{-1}}(y,T)$.

Comments:

1) The fact that we defined $B_{\Sigma^{-1}}(y,T)$ using the given norm on W was use to invoke the fact that $\|\lambda\| = \sup\{\langle \lambda, \eta \rangle : \eta \in B(0,1)\}$ for $\lambda \in W^*$.
2) The exchange of infimum and supremum in

$$\sup_{\eta}\left\{\inf_{\lambda}\{S(\lambda,\eta)\}\right\} = \inf_{\lambda}\left\{\sup_{\eta}\{S(\lambda,\eta)\}\right\}$$

made a few lines above, asserts that for $S(\lambda,\eta)$, which is convex in λ and concave in η, the value at a saddle point can be obtained by either of the two ways.

5.7 Maxentropic reconstruction of signal and noise

A geometric interpretation of the method proposed in the previous section to deal with the presence of noise is the following: When there is noise in the measurement of Ax, we end up with a datum $y \notin A(V)$. To find x^* satisfying (5.19) we put a ball $B_{\Sigma^{-1}}(y,T)$ around the datum, with the radius T large enough for it to intersect $A(V)$, and then search for x^* lying in that intersection.

Here we shall present a totally different way of looking at

$$y = Ax + n, \quad x \in C_s$$

which was originally proposed in [37], and made fully explicit for the finite dimensional context in [41]. We just extend it to the infinite dimensional setup.

In the present approach it is also assumed that n is a random variable, defined in a probability space $(\Omega_n, \mathcal{F}_n, Q_n)$ and taking values in a closed, convex subset C_n of W. Actually, as above, we may as well assume that $\Omega_n = C_n$, $\mathcal{F}_n = \mathcal{B}(C_n)$ and that the a priori measure Q_n is such that the closed convex hull generated by $\mathrm{supp}Q_n$ equals C_n.

Now, instead of (5.19) we shall consider

$$y = A\widehat{x}, \quad \widehat{x} \in C \tag{5.22}$$

where $\widehat{A} : V \times W \to W$ is defined on $\widehat{x} = (x, \eta) \in V \times W$ by $\widehat{A}\widehat{x} = Ax + \eta$. we put $C = C_s \times C_n$.

To bring the maxentropic approach, we want to think of (5.22) as a quest for a measure P on $C = C_s \times C_n$, $\mathcal{F} = \mathcal{B}(C_s) \otimes \mathcal{B}(C_n)$ such that $P << Q := Q_s \otimes Q_n$ and

$$\widehat{A}E_P[\widehat{X}] = y. \tag{5.23}$$

As above, the maxentropic P^* will be of the form

$$dP^* = \frac{\exp\left(-\left\langle \lambda, \widehat{A}\widehat{X} \right\rangle\right)}{Z(\lambda)} dQ \tag{5.24}$$

where $Z(\lambda)$ is the usual

$$Z(\lambda) = \int_C \exp\left(-\left\langle \lambda, \widehat{A}\widehat{X} \right\rangle\right) dQ$$

where $\widehat{X} : C \to C$ is the identity mapping $\widehat{X}(\widehat{\xi}) = \widehat{\xi}$ for any $\widehat{\xi} = (\xi, \eta) \in C$. The special structure of (5.22) allows us to conclude

Lemma 5.7. *With the notations introduced above*

$$Z(\lambda) = Z_s(\lambda) Z_n(\lambda)$$

where

$$Z_s(\lambda) = \int_{C_s} \exp\left(-\langle \lambda, A\xi \rangle\right) dQ_s(\xi), \quad Z_n(\lambda) = \int_{C_n} \exp\left(-\langle \lambda, \eta \rangle\right) dQ_n(\eta).$$

The Q-entropy of any member of the exponential family is

$$S_Q(P(\lambda)) \equiv \Sigma(\lambda) = \ln Z_s(\lambda) + \ln Z_n(\lambda) + \langle \lambda, y \rangle \qquad (5.25)$$

and if all goes well and we set

$$\lambda^* = \arg\inf \Sigma(\lambda) = \arg\inf \{\ln Z_s(\lambda) + \ln Z_n(\lambda) + \langle \lambda, y \rangle\}$$

and substitute in (5.24) we obtain the important

Proposition 5.10. *With the notations introduced above, if a maxentropic measure P^* satisfying (5.23) exists, it must factor as $dP^*(\xi) = dP_s^*(\xi) dP_n^*(\eta)$ with*

$$dP_s^*(\xi) = \frac{\exp\left(-\langle \lambda^*, A\xi \rangle\right)}{Z_s(\lambda^*)} dQ_s(\xi), \quad dP_n^*(\eta) = \frac{\exp\left(-\langle \lambda^*, \eta \rangle\right)}{Z_n(\lambda^*)} dQ_n(\eta). \qquad (5.26)$$

Proof. It is easy. Just combine Lemma 5.7 with (5.24). □

Comment: It is through (5.25) that the data determines how noise and state variables become entangled in (5.26).

Notice that, even though the *a priori* mean of the noise, given by $E_{Q_n}[\eta]$, was zero, it may well happen that

$$n^* = E_{P_n^*}[\eta] \neq 0$$

and therefore the maxentropic reconstruction x^* satisfies

$$Ax^* = y - \eta^*$$

which can be interpreted as follows: first correct the noisy data by subtracting an average noise from it, then solve the resulting equation using the method of maximum entropy in mean, of course!

5.8 Maximum entropy according to Dacunha-Castelle and Gamboa. Comparison with Jaynes' classical approach

5.8.1 Basic results

In this section we present a slightly modified version of the results obtained by Dacunha-Castelle and Gamboa (D&G) in [21], in order to compare their version with the approach that we developed above. Also we shall in a later section see how some classical approaches, namely that of Jaynes and Burg follow from their approach.

We shall carry out this program in two versions of the problem consisting of finding a probability distribution $(p) = (p_1, p_2, ..., p_n, ...)$ satisfying

$$\sum A_{ij} p_j = y_i, \quad 1 \leq i \leq k \tag{5.27}$$

where we shall assume that the first row of A is $A_{1j} \equiv 1$ for all j and $y_1 = 1$, so that the constraint on (p) is automatically part of the data. We think of the A_{ij} as the values of functions $A_i : S \to \mathbb{R}$, where S is some countable set.

In the physicist's jargon S describes microscopic states of some system and the A_i describe observables (statistics) of the system. The problem thus consists of finding a probability distribution with respect to which the observables (statistics) have prescribed values.

We shall think of (p) as an element of ℓ^1_+, to bring our approach and D&G's closer we consider $\Omega = [0, \infty)^N$, $\mathcal{F} = \mathcal{B}(\Omega)$ and $Q = \otimes_{j \geq 1} Q_j$, and later on we will consider specific choices for the Q_j's. We consider the mapping

$$X : \Omega \to \ell^1_+ \quad X(\xi) = (X_1(\xi), ..., X_n(\xi), ...) = (\xi_1, ..., \xi_n, ...)$$

where $(\xi) = (\xi_1, ..., \xi_n, ...)$ is a typical element in Ω.

The only restriction we impose on the Q_j's is such that

$$E_Q [\|X\|_1] = E_Q \left[\sum X_j(\xi) \right] = \sum E_{Q_j} [X_j(\xi)] < \infty$$

which would imply that $X(\xi) \in \ell^1_+$ a.s.-Q.

So far we differ slightly from D&G. Next they define, first the Laplace transform of Q (or partition function of Q in the physicist's jargon)

$$Z_Q(t) = E_{Q_j} \left[e^{-\langle t, X \rangle} \right]$$

and then the moment generating function, or log-Laplace transform

$$\psi(t) = \ln \left(Z_Q(t) \right) \tag{5.28}$$

for $t \in \ell_\infty$. Since we have taken $Q = \otimes_{j \geq 1} Q_j$, the components of X are independent and

$$\ln(Z_Q(t)) = \sum_j \ln\left(E_Q\left[e^{-t_j X_j}\right]\right). \tag{5.29}$$

This will be true for any $Q = \otimes_{j\geq 1} Q_j$, even when the resulting Q is not a probability measure. Also our choice of sign follows the traditional convention for Laplace transforms, but that is non-essential. It is nevertheless important to note that $Z(t)$ has a domain of finiteness defined by

$$\mathcal{D}(Q) = \{t \in \ell_\infty : Z(t) < \infty\}. \tag{5.30}$$

When, for example, Q_j is a Poisson law $Q_j(X_j = k) = \dfrac{e^{-q_j} q_j^k}{k!}$, then

$$\ln(Z_Q(t)) = \sum_j \ln\left(E_Q\left[e^{-t_j X_j}\right]\right) = \sum_j q_j(e^{-t_j} - 1) \tag{5.31}$$

which is finite for $t \in \ell_+^\infty$ as long as $\sum_j q_j < \infty$.

Now we come to the main (substantial) difference between the approach of D&G and ours: it lies in the way the entropy functionals are brought in and in the class of objects on which they are defined. They define, for every $\eta \in \ell_+^1$

$$S_Q(\eta) = \inf\{\psi(t) + \langle t, \eta\rangle : t \in \ell_\infty\} \tag{5.32}$$

which, when independence is brought in, amounts to

$$S_Q(\eta) = \sum_{j\geq 1} s_j(\eta_j)$$

where obviously

$$s_j(\eta_j) = \inf\{\ln(z_j(t_j)) + t_j \eta_j : t_j \in \mathbb{R}\}$$

where the more or less obvious and annoying subscripts are to emphasize that we are dealing with sequence valued functions.

In the particular example $Q = \otimes_{j\geq 1} Q_j$ with each Q_j being a Poisson distribution with parameter q_j we obtain from (5.31)

$$S_Q(\eta) = \eta_j - q_j - \eta_j \ln\left(\frac{\eta_j}{q_j}\right) \tag{5.33}$$

and for any $\eta \in \ell_+^1$

$$S_Q(\eta) = \sum_{j\geq 1}\left\{\eta_j - q_j - \eta_j \ln\left(\frac{\eta_j}{q_j}\right)\right\} \tag{5.34}$$

which, when both q and η satisfy $\sum q_j = \sum \eta_j = 1$ reduces to

$$S_Q(\eta) = -\sum_{j\geq 1} \eta_j \ln\left(\frac{\eta_j}{q_j}\right). \tag{5.35}$$

The objects defined in (5.28) and (5.32) are the starting point for D&G's approach to the maximum entropy method. We shall see how it winds down further below. An interesting feature of this approach is that it has a direct counterpart within the standard optimization methods as exemplified by Borwein and Lewis.

Also, D&G's approach reduces to the standard Jaynes' maximum entropy method and goes beyond it. In Jaynes' approach, a fruitful method for characterizing probability distributions is played by (5.34) and/or (5.35).

In D&G's approach the central role of the functional based on $-x\lg x$ is lost, and it is replaced by $S_Q(\eta)$ based on $Z_Q(t)$.

To further exemplify the issue, let us choose $dQ_j(\xi_j) = q_j e^{-q_j \xi_j} d\xi_j$. Then $z_j(t_j) = \dfrac{q_j}{(t_j + q_j)}$ (defined for $t_j > q_j$) and therefore $s_j(\eta_j) = \ln(q_j\eta_j) + (1 - \eta_j q_j)$. Notice that $s_j(\eta_j) \leq q_j\eta_j(e-1)$ and therefore

$$S_Q(\eta) = \sum_{j\geq 1}\{\ln(q_j\eta_j) + (1 - \eta_j q_j)\} \tag{5.36}$$

converges for $\eta \in \ell^1_+$ and $q \in \ell^\infty_+$.

The $S_Q(\eta)$ in (5.36) is known as Burg's entropy and appears in many problems in signal and time series analysis.

To get a feeling for what goes on, the reader is urged to go ahead and take a shot at

Exercise 5.1. Consider $\Omega = [0,1]^N$, $\mathcal{F} = \mathcal{B}(\Omega)$ and $Q = \otimes_{j\geq 1} Q_j$, with each Q_j being either the uniform measure on $[0,1]$ or a Bernoulli law $dQ_j(\xi) = q_j \varepsilon_0(d\xi) + p_j \varepsilon_1(d\xi)$, and compute both $Z_Q(t)$ and $S_Q(\eta)$ for each.

The moral will be obvious: the entropy functional is dependent on the framework.

To continue with D&G's approach, we have to find $p^* \in \ell^1_+$ such that

$$p^* = \arg\max\left\{S(\eta) : \eta \in \ell^1_+, \sum A_{ij}\eta_j = m_i \text{ for } 1 \leq i \leq k\right\} \tag{5.37}$$

which will be the maxentropic solution to (5.27).

Let us write, for $m \in \mathbb{R}^k$, with components m_i

$$\mathcal{P}(m) = \left\{\eta \in \ell^1_+ : A\eta = m\right\}. \tag{5.38}$$

Note that since $A_{1j} \equiv 1$ and $m_1 = 1$, then if $\eta \in \ell_+^1$ is in $\mathcal{P}(m)$, then η is a probability distribution. The following is proved in a somewhat different setting in [21].

Theorem 5.4. *Assume that*

i) *The closure of the convex envelope of supp(Q) is Ω.*
ii) *$\mathcal{D}(Q)$ (defined in (5.30)) has a nonempty interior. Actually $\mathcal{D}(Q)$ is convex when nonempty.*
   Note that $A^* : \mathbb{R}^k \to \ell^\infty$ when $\sup_{i,j} |A_{ij}| < \infty$.
iii) *The set $\mathcal{D}^* = \{\lambda \in \mathbb{R}^k : A^*\lambda \in \mathcal{D}(Q)\}$ is not empty.*

Then the following are equivalent

a) *Problem (5.27) has a solution.*
b) *There exists a measure Q, which together with its log-Laplace transform $\psi(t)$ satisfies i)-iii) spelt out above and*

$$p^* = -\frac{\partial}{\partial t_j} \ln(z_j(t_j))\Big|_{t_j = (A^*\lambda^*)_j} \quad (5.39)$$

satisfies (5.27) where λ^* is obtained as

$$\lambda^* = \arg\inf \{S_Q(\lambda) = \ln(Z_Q(\lambda)) + \langle \lambda, m \rangle : \lambda \in \mathbb{R}^k\} \quad (5.40)$$

where $Z_Q(\lambda) = Z_Q(A^*\lambda)$ for $\lambda \in \mathcal{D}^*$.
c) *For a pair (Q, Z_Q) satisfying assumptions i)-iii) listed above, problem (5.27) has a solution given by (5.39) where λ^* is defined by (5.40).*

Comments: The proof in [21] uses probabilistic techniques related to large deviation theory. We stated their result some what differently to make comparison with our and Jaynes frameworks apparent.

To compare with our scheme, once $(\Omega, \mathcal{F}, Q)$ and the coordinate maps $X : \Omega \to C$, we would be searching for $P \ll Q$ such that

$$E_P[AX] = m.$$

For us the entropy functional will be defined on

$$\mathcal{P}(m) = \{P \ll Q : E_P[AX] = m\}$$

and would be given by the usual Boltzmann-Gibbs $-\rho \ln \rho$ as

$$S_Q(P) = -\int_\Omega \rho(\xi) \ln(\rho(\xi)) dQ(\xi).$$

Here we do as Jaynes did, but in this approach the entropy functional is defined directly on the class of objects we are searching. So you have three approaches to choose from to solve some problems. As above, our maxentropic distribution (assuming it exists) will be given by

$$dP^*(\xi) = \frac{1}{Z_Q(\lambda)} e^{-\langle \lambda^*, AX(\xi) \rangle} dQ(\xi)$$

where λ^* has to minimize $S_Q(\lambda) = \ln(Z_Q(\lambda)) + \langle \lambda, m \rangle$.

Note that $Z_Q(\lambda)$ is the same for us and for D&G. Not only that, when λ^* is an interior point of $\mathcal{D}^*$

$$p_j^* = E_Q[X_j] = -\frac{\partial}{\partial t_j} \ln(z_j(t_j)) \bigg|_{t_j = (A^*\lambda^*)_j}$$

as for D&G. Actually, had we followed D&G step by step the results will be somewhat different, but now the reader will be able to catch the differences readily.

In passing we note that when Q_j is Poisson with intensity q_j, (5.39) yields

$$p_j^* = \frac{1}{q_j + (A^*\lambda^*)_j} \tag{5.41}$$

as maxentropic solution to (5.27).

If we wanted to use Jaynes' approach, we would have defined $S_Q(\eta)$ given by (5.35) on $\mathcal{P}(m)$ specified in (5.38) rewritten as

$$\mathcal{P}(\widehat{m}) = \left\{ \eta \in \ell_+^1 : \hat{A}\eta = \widehat{m}, \|\eta\|_1 = 1 \right\}$$

where $\hat{A} : \ell_1 \to \mathbb{R}^{k-1}$ is the "matrix obtained from A by deleting the first row and $\widehat{m}$ is obtained from m by deleting the first element.

The next step in Jaynes' approach would be as in the beginning of this chapter: to find $\lambda^* \in \mathbb{R}^{k-1}$ such that

$$p_j^* = \frac{e^{-(\hat{A}^*\lambda^*)_j}}{\zeta_j(\lambda^*)}$$

maximizes $S_Q(\eta) = -\sum \eta_j \ln \eta_j$, where now for $\lambda \in \mathbb{R}^{k-1}$

$$\zeta(\lambda^*) = \sum_{j \geq 1} e^{-(\hat{A}^*\lambda^*)_j}$$

and again λ^* is to be found minimizing $\ln(\zeta(\lambda)) + \langle \lambda, \widehat{m} \rangle$. What you will definitely not obtain is a representation like (5.41).

5.8.2 Jaynes' and Dacunha and Gamboa's approaches

Let us state, without further ado, the approaches of Jaynes' (who started all this) and D&G's ways to use a maxentropic procedure to find a positive, continuous, probability density $\rho(x)$ on [0,1] such that

$$\int_0^1 \Phi(x)\rho(x)P_a(dx) = m \qquad (5.42)$$

where $\Phi : [0,1] \to \mathbb{R}^k$ is a given continuous function and $m \in \mathbb{R}^k$. $P_a(dx)$ is some preassigned or *a priori* measure and $P(dx) = \rho(x)P_a(dx)$ is what we look for. Usually $P_a(dx) = q(x)dx$ for some positive, continuous $q(x)$.

Jaynes' scheme consists of defining an entropy functional on

$$\mathcal{P}_J(m) = \left\{ P << P_a : \int_0^1 \Phi(x) dP(x) = m \right\} \qquad (5.43)$$

by

$$S_{P_a}(\rho) = -\int_0^1 \rho(x)\ln(\rho(x))P_a(dx) \qquad (5.44)$$

and then finding

$$p^* = \arg\max\{S_{P_a}(\rho) : \rho \in \mathcal{P}_j\}. \qquad (5.45)$$

As you see, the approach presented from the beginning of this chapter just translates Jaynes' approach to a different setting.

If we set

$$\mathcal{D}(\Phi) = \left\{ \lambda \in \mathbb{R}^k : E_{P_a}\left[e^{-\langle \lambda, \Phi(x) \rangle}\right] < \infty \right\} \qquad (5.46)$$

then the output of Jaynes' approach is described in

Theorem 5.5. *Let $Q(dx)$ be a probability on $([0,1], \mathcal{B}([0,1]))$. If $\mathcal{P}_J$ is not empty and $\mathcal{D}(\Phi)$ is open, there is a unique $\lambda^* \in \mathcal{D}(\Phi)$ such that*

$$P^*(dx) = \frac{e^{-\langle \lambda^*, \Phi(x) \rangle}}{Z_Q(\lambda^*)} P_a(dx) \qquad (5.47)$$

is in $\mathcal{P}_J$ and $\lambda^* = \arg\min\{\ln(Z_{P_a}(\lambda)) + \langle \lambda, m \rangle : \lambda \in \mathbb{R}^k\}$.

This is the way the maximum entropy is used to characterize exponential families of probability distributions in physics, statistics and a variety of other applications.

To describe D&G's approach, to (5.42). Here **we do not** require that $\rho(x)$ be a probability. Only that it is a continuous function satisfying some constraints. Neither do we require that $P_a(dx)$ is a probability measure anymore. Since the solution scheme depends on the constraints, we assume that we want to solve (5.42) for $\rho(x)$ continuous and satisfying some convex constraint like $a \leq \rho(x) \leq b$, with $-\infty \leq a < b \leq \infty$. To proceed according to the simplest version of D&G, we choose any interesting measure Q on $[a, b]$. We require that:

H1) The convex envelope of $supp(Q) = [a, b]$
H2) $\mathcal{D}(Q) = \{\tau \in \mathbb{R} | \int_a^b e^{\tau y} Q(dy) < \infty\}$ is a non-empty open interval. (If non-empty it would be convex, hence an interval.)
H3) $\mathcal{D}^* = \{\lambda \in \mathbb{R}^k : \forall x; \langle \lambda, \Phi(x) \rangle \in \mathcal{D}(Q)\}$ is not empty and coincides with the set $\{\lambda \in \mathbb{R}^k : \int_a^b |\ln(\zeta_Q(\langle \lambda, \Phi(x) \rangle)) P_a(dx)| < \infty\}$.

Here $\zeta_Q(\tau) = \int_a^b e^{\tau y} Q(dy)$. Observe that $\zeta_Q(\langle \lambda, \Phi(x) \rangle)$ is quite different from the $Z_{P_a}(\lambda)$ that came up in Jaynes' approach. They are both Laplace transforms, but of different measures.

To continue, denote by $\sigma(\lambda)$ the function

$$\Sigma(\lambda) = \int_0^1 \ln\left(Z_Q(\langle \lambda, \Phi(x) \rangle)\right) dP(x) - \langle \lambda, m \rangle$$

and, once more, but $\psi(t) = \ln(Z_Q(t))$, then D&G prove

Theorem 5.6. *Let $\Phi : [0, 1] \to \mathbb{R}^k$ be a continuous function. The problem consists in finding a $\rho(x)$ satisfying (5.42) and such that $a \leq \rho(x) \leq x$ for $x \in [0, 1]$.*

The following are equivalent

A) *Our problem has a solution $\rho^*(x)$.*
B) *There exists a measure Q on $[a, b]$ such that conditions H1)-H3) spelt out above hold and $\rho^*(x)$ is given by*

$$\rho^*(x) = \psi'\left(\langle \lambda^*, \Phi(x) \rangle\right) \qquad (5.48)$$

where λ^* minimizes $\Sigma(\lambda)$, i.e., solving

$$\int_0^1 \psi'\left(\langle \lambda^*, \Phi(x) \rangle\right) \Phi(x) dP_a(x) = m.$$

C) *For any Q such that $Z_Q(t)$ satisfies H1)-H3), our problem has a solution given by (5.47) with λ^* as in B).*

5.9 MEM under translation

We want to solve
$$Ax = y, \quad x \in C_1 \qquad (5.49)$$
where C_1 is a closed convex set in V. It may sometimes happen that for some $\xi_0 \in C_1$, $C_1 = \xi_0 + C_0$, where C_0 is again a closed, convex set.

A typical instance of this situation arises when V is a vector lattice (usually a space of functions defined on some domain) and $C_1 = \{x \in V : \xi_0 \leq x \leq \xi_1\}$ for some $\xi_0, \xi_1 \in V$, and for some reason we prefer to solve (5.49) for $x \in C_0 = \{x \in V : 0 \leq x \leq \xi_2\}$, where now $\xi_2 = \xi_1 - \xi_0$.

Thus, after translation we want to solve for x in
$$Ax = y - Ax_0 = y_0, \quad x \in C_0. \qquad (5.50)$$

We expect that the formalism introduced above is such that, if x_1^* and x_0^*, respectively denote the solutions to (5.49) and (5.50), provided by the maxentropic methods developed above, then $x_1^* = x_0^* + \xi_0$. Let us verify it.

Let $Q_1(d\xi)$ be the *a priori* measure on $(C_1, \mathcal{B}(C_1))$ and let $Q_0(d\xi)$ be its translate to $(C_0, \mathcal{B}(C_0))$. That is, for any integrable F on C_1, we have, see Appendix B,
$$\int_{C_1} F(\xi) dQ_1(\xi) = \int_{C_0} F(\xi_0 + \xi) dQ_0(\xi). \qquad (5.51)$$

Now, let us put
$$Z_0(\lambda) = \int_{C_0} \exp(-\langle \lambda, A\xi \rangle) dQ_0(\xi), \quad Z_1(\lambda) = \int_{C_1} \exp(-\langle \lambda, A\xi \rangle) dQ_1(\xi)$$
and use (5.51) to verify that
$$Z_1(\lambda) = \exp\left(-\langle \lambda, A\xi_0 \rangle\right) Z_0(\lambda). \qquad (5.52)$$

If we denote by $\Sigma_1(\lambda)$ and $\Sigma_0(\lambda)$ the Σ_Q-dual entropies of the exponential families associated with Q_1 and Q_0, then
$$\Sigma_1(\lambda) = \ln Z_1(\lambda) + \langle \lambda, y \rangle = -\langle \lambda, A\xi_0 \rangle + \ln Z_0(\lambda) + \langle \lambda, y \rangle$$
$$= \ln Z_0(\lambda) + \langle \lambda, y - A\xi_0 \rangle = \ln Z_0(\lambda) + \langle \lambda, y \rangle = \Sigma_0(\lambda)$$
where we brought in (5.52) at the second step. Therefore
$$\lambda^* = \arg\inf \{S_0(\lambda) : \lambda \in W^*\} = \arg\inf \{S_1(\lambda) : \lambda \in W^*\}$$
and the corresponding maxentropic maximizers are
$$dP_0^*(\xi) = \frac{\exp\left(-\langle \lambda^*, A\xi \rangle\right)}{Z_0(\lambda^*)} dQ_0(\xi), \quad dP_1^*(\xi) = \frac{\exp\left(-\langle \lambda^*, A\xi \rangle\right)}{Z_1(\lambda^*)} dQ_1(\xi)$$

which are supported by C_0 and C_1 respectively, and the λ^* appearing in their definition is the same.

To complete, use (5.51) to verify that

$$x_1^* = E_{P_1^*}[X] = \int_{C_1} \xi \frac{\exp(-\langle \lambda^*, A\xi \rangle)}{Z_1(\lambda^*)} dQ_1(\xi)$$

$$= \int_{C_1} (\xi_0 + \xi) \frac{\exp(-\langle \lambda^*, A\xi \rangle)}{Z_0(\lambda^*)} dQ_0(\xi) = \xi_0 + x_0^*.$$

5.10 Maxent reconstructions under increase of data

Some generic situations covered by the title of this section can be described in terms of the generalized moment problem consisting of finding $x(t)$ from

$$\int_D \Phi_1(t) x(t) dm(t) = y \tag{5.53}$$

where $\Phi_1(t) : D \to \mathbb{R}^{m_1}$ is given, $(D, \mathcal{B}(D), m)$ are part of the set up of the problem, and then compare with the solution of

$$\int_D \Psi(t) x(t) dm(t) = y^1 \tag{5.54}$$

where now $\Psi(t) : D \to \mathbb{R}^{m_1+m_2}$ and $\Psi(t) = \begin{pmatrix} \Phi_1(t) \\ \Phi_2(t) \end{pmatrix}$, where $\Phi_2(t) : D \to \mathbb{R}^{m_2}$.

Two examples exemplify the situation: Consider $D = [0,1]$, and $m(dt) = dt$ the Lebesgue measure, and let $\Phi_1(t)$ be the vector with components t^i, $i = 1, ..., m_1$ and $\Phi_2(t)$ be the vector with components t^{j+m_1}, $j = 1, ..., m_2$. Or let $D = [0,1]^2$ and let $\Phi_1(t)$ correspond to projections on m_1 given planes and $\Phi_2(t)$ to further projections on other m_2 planes.

Two obvious questions are: How do the corresponding reconstructions compare with each other? and more important: When to stop gathering information?

To go abstract consider solving for x in

$$A_1 x = y_1, x \in C \tag{5.55}$$

where $A_1 : V \to W_1$, $y_1 \in W_1$ and C is a closed, convex subset of V, and consider as well solving for x in

$$Ax = \begin{pmatrix} A_1 \\ A_2 \end{pmatrix} x = \begin{pmatrix} y_1 \\ y_2 \end{pmatrix} = y \tag{5.56}$$

where $A_2 : V \to W_2$, $y_2 \in W_2$.

Let us put
$$\zeta(t) = \int_C \exp\left(-\langle t, \xi \rangle\right) dQ(\xi)$$

which is the Laplace transform of $Q(d\xi)$ and suppose, for not to be stopped by mathematical formalism, that $\zeta(t)$ is defined for all $t \in V^*$.

Assume that both (5.55) and (5.56) have a maxentropic solution, which are to be obtained integrating, respectively, with respect to

$$dP^*_{\lambda_0} = \frac{\exp\left(-\langle \lambda_0, AX \rangle\right)}{\zeta(A_1^* \lambda_0)} dQ \tag{5.57}$$

or

$$dP^*_{(\lambda_1, \lambda_2)} = \frac{\exp\left(-\langle A_1^* \lambda_1 + A_2^* \lambda_2, X \rangle\right)}{\zeta(A_1^* \lambda_1 + A_2^* \lambda_2)} dQ. \tag{5.58}$$

Notice as well that $P^*_{(\lambda_1, \lambda_2)}$ belongs to the class

$$\mathcal{P}_1(A_1, y_1) = \{P << Q : E_P[A_1 X] = y_1\}$$

and that the $\lambda_0 \in W_1^*$ in (5.57) need not be equal to $\lambda_1 \in W_1^*$ in (5.58).

The first is to be obtained minimizing $\Sigma_1(\eta) = \ln \zeta(A_1^* \eta) + \langle \eta, y_1 \rangle$ over W_1, whereas (λ_1, λ_2) is obtained minimizing $\Sigma(\mu, v) = \ln \zeta(A_1^* \mu + A_2^* v) + \langle \mu, y_1 \rangle + \langle v, y_2 \rangle$ over $W_1 \times W_2 = W$.

It is then clear that

$$\Sigma_1(\lambda_0) = S_q(P^*_{\lambda_1}) \geq \Sigma(\lambda_1, \lambda_2) = S_Q\left(P^*_{(\lambda_1, \lambda_2)}\right) \tag{5.59}$$

because $P^*_{\lambda_1}$ is obtained maximizing $S_Q(\cdot)$ over $\mathcal{P}(A_1, y_1)$.

Thus, as physicists (enjoy to) say: increasing information (or available data) decreases the entropy.

To describe the practical consequence derived from (5.59) we need some more notation. Let A and y be as the beginning of this section. Let W_n be a nested sequence of finite dimensional subspaces of W. Suppose for definiteness that $dim W_n = n$ and assume there exist subspaces $\widehat{W}_n$, of W

$$W = W_n \oplus \widehat{W}, \quad \|y - y_n\| \to 0 \quad \text{as} \quad n \to \infty$$

and write $y = y_n + \widehat{y}_n$, and let $\Pi_n : W \to W_n$ denote the projection determined by this decomposition, and

$$A_n : V \to W_n, \quad \text{with} \quad A_n = \Pi_n A.$$

Now let $\lambda_n^* \in W_n^*$ be such that $P(\lambda_n^*) \in \mathcal{P}(A_n, y_n)$ and let $x_n^* = E_{P(\lambda_n^*)}[X]$. Then $A_n x_n^* = y_n$.

If λ_∞^* is such that $P(\lambda_\infty^*) \in \mathcal{P}(A, y)$, then $P(\lambda_n^*)$ and $P(\lambda_\infty^*)$ are both in $\mathcal{P}(A_n, y_n)$ and therefore (5.59) extends to

$$S_Q(P(\lambda_n^*)) \geq S_Q(P(\lambda_{n+1}^*)) \geq S_q(P(\lambda_\infty^*))$$

for all n. To sum up, the practical recommendation derived from this inequality goes as follows: solve your maxentropic problem for each datum y_n and stop increasing the dimension of the data when further increase of data does not bring "significant" decrease of entropy. The pending issue is how to quantify this rule.

In a few cases we tried, this works all right. Some examples are provided in Chapter 10.

Exercise 5.2. (Another way of looking at the standard maxent method) Consider the problem of finding $\{p_j : j = 1, ..., n\}$ such that

$$\sum_{j=1}^n A_{ij} p_j = a_i, \quad i = 1, ..., m$$

$$\sum_{j=1}^n p_j = 1.$$

Define $\Omega = \{x \in \mathbb{R}^n : 0 \leq x_j \leq 1, \sum x_j = 1\}$. Set $\mathcal{F} = \mathcal{B}(\Omega)$ and $X : \Omega \to \Omega$ is the identity mapping. Let $\{q_j : j = 1, ..., n\}$ be such that $\sum q_j = 1$. (But this is not really necessary) and define $dQ = \sum_{j=1}^n q_j \varepsilon_{e_j}(d\xi)$, where $\varepsilon_{e_j}(d\xi)$ is the Dirac point mass at the unit vector e_j in $\mathbb{R}^n$. Note that for $\lambda \in \mathbb{R}^m$

$$Z(\lambda) = \int_\Omega e^{-\langle \lambda, AX \rangle} dQ = \sum_{j=1}^n q_j e^{-\langle \lambda, Ae_j \rangle} = \sum_{j=1}^n q_j e^{-\sum_{i=1}^n \lambda_i A_{ij}}.$$

If dP is a measure on $(\Omega, \mathcal{F})$ which is $dP << dQ$, then it must be true that $dP = \sum_{j=1}^n \rho_j q_j \varepsilon_{e_j}(d\xi)$. Compute $S_Q(P)$? Verify that this coincides with the usual maxent approach (in the finite dimensional case). Redo for the infinite dimensional but denumerable case.

We shall make use of this setup in Section 7.8.

5.11 Bibliographical comments and references

We have been able to trace the origin of our version of maximum entropy in the mean down to [1]. A basic reference of a functional analysis from which we borrowed some stuff is [4-2]. A collection, which is a must for reader interested in the approach to convex optimization in general is, in approximate chronological order, [2], [3], [4], [5], [6] and [7]. A few items about large deviations were taken from [8]. And some applications of these ideas in the formulation of the maximum entropy approach to inverse problems can be found in [9]-[18].

For minimum principles see [6-3]. See also [19] by Sion.

Dacunha-Castelle and Gamboa's paper contains much more than the missing details. It contains the basics by following [10]-[11].

Two quite different proofs of Theorem 5.6 are contained in [22] and [23].

Jaynes seminal paper [24] is the basis for all instances in [25]-[36], which is basically a collection of case studies in which the maximum entropy method works, plus some more or less speculative philosophical essays. Take a look at [37] for applications in econometrics.

You can take a look at [38]-[39] for full blown approaches to the problem of solving integral equations for maxentropic methods along the lines of [21]. A way of connecting the maxentropic formulation to the standard variational methods is described in [40]. This is rather neat paper.

To close, consider the yet unpublished manuscript [41] by Leitner and Temnov in which they deal with the related problem of estimating a density by maximum entropy methods when the information available consists of moments obtained from experimental data.

References

[1] Rietch, E. *"A maximum entropy approach to inverse problems"*. J. Geophys. 42 (1977), 489-506.

[2] Rudin, W. *"Functional Analysis"*. McGraw Hill, New York, 1973.

[3] Rockafellar, R. T. *"Convex Analysis"*. Princeton Univ. Press, New Jersey, 1970.

[4] Rockafellar, R. T. *"Conjugate Duality and Optimization"*. SIAM, Philadelphia, 1974.

[5] Borwein, J. M. and Lewis, A. S. *"Duality relationships for entropy-like minimization problems"*. SIAM, J. Control and Optimization. Vol. 29 (1991), pp. 325-338.

[6] Borwein, J. M., Lewis, A. S. *"Partially finite convex programming: I and II"*. Mathematical Programming, Vol. 57 (1992), pp. 15-48 and 49-83 resp.

[7] Borwein, J. M., Lewis, A. S. and Limber, M. *"Entropy minimization with lattice bounds"*. J. Approximation Theory, Vol. 79 (1994), pp. 1-16.

[8] Decarreau, A., Hilhorst, D., Demarechal, C. and Navaza, J. *"Dual methods in entropy maximization. Application to some problems in crystallography"*. SIAM. J. Optimization. Vol. 2 (1992). pp. 173-197.

[9] Cherny, A. S. and Maslov, V. P. (2003) "On maximization and minimization of entropy functionals in various disciplines". Theory of Probability and its Applications, Vol. 3 (2003), pp. 447-464.

[10] Gamboa, F. and Gassiat, E. *"Bayesian methods and maximum entropy for ill posed problems"*. Annals of Statistics, Vol. 25 (1997), pp. 328-350.

[11] Gamboa, F. *"New Bayesian methods for ill posed problems"*. Statistics and Decisions, Vol. 17 (1999), pp. 315-337.

[12] Robert, C. *"An entropy concentration theory: applications in artificial intelligence and descriptive statistics"*. J. Apply. Prob. Vol. 27 (1990), pp. 303-313.

[13] Mead, L. R. and Papanicolau, N. *"Maximum entropy in the problem of moments"*. J. Math. Phys. Vol. 25 (1989), pp. 2404-2417.

[14] Forte, B., Hughes, W. and Pales, Z. *"Maximum entropy estimators and the problem of moments"*. Rendiconti di Matematica, serie VII. Vol. 9 (1989), pp. 689-699.

[15] Borwein, J. M. and Lewis, A. S. *"The convergence of moment problems"*. Transaction Amer. Math. Soc. Vol. 325 (1991), pp. 249-271.

[16] ibid. *"Convergence of best entropy estimates"*. SIAM J. Optimization, Vol. 1 (1991), pp. 191-205.

[17] ibid. *"A survey of convergence results for maximum entropy methods"*. In "Maximum Entropy and Bayesian methods", Mohammed-Djafari and G. Demoment (eds). Kluwer Academic Publishers, Dordrecht 1993.

[18] Lewis, A. S. *"The convergence of entropy -based in approximations for moment systems"*. Optimization, Vol. 28 (1194), pp. 383-395.

[19] Sion, M. *"On general minimax theorems"*. Pacific J. Math. Vol. 8 (1958), pp. 171-175.

[20] Aubin, J. P. *"Optima and Equilibria: an Introduction to Nonlinear Analysis"*. Springer-Verlag, New York, 1993.

[21] Dacunha-Castelle, D. and Gamboa, F. *"Maximum d'entropie et probleme des moments"*, Am. Inst. Henri Poincaré. Vol. 26 (1990), pp. 567-596.

[22] Kullback, S. *"Information Theory and Statistics"*, Dover Pubs, New York, 1968.
[23] Csiszar, I. *" I-divergence geometry of probability distributions and minimization problems"*. Am. Prob. Vol. 3 (1975), pp. 148-158.
[24] Jaynes, E. *"Information theory and statistical physics"*. Phys. Rev. Vol. 106 (1957), pp. 620-630.
[25] Erickson, G. and Rychert, J. (eds.) *"Maximum Entropy and Bayesian Methods"*. Kluwer Acad. Pubs., Dordrecht, 1998.
[26] Hanson, K. M. and Silver, R. (eds.) *"Maximum Entropy and Bayesian Methods"*. Kluwer Acad. Pubs., Dordrecht, 1996.
[27] Skilling, J. and Sibusiso, S. (eds.) *"Maximum Entropy and Bayesian Methods"*. Kluwer Acad. Pubs., Dordrecht, 1995.
[28] Heidbreder, G. R. (ed.) *"Maximum Entropy and Bayesian Methods"*. Kluwer Acad. Pubs., Dordrecht, 1994.
[29] Mohamad-Djafari, A. and Demoment, G. (eds.) *"Maximum Entropy and Bayesian Methods"*. Kluwer Acad. Pubs., Dordrecht, 1993.
[30] Ray Smith, G., Erickson, G. and Neudorfer, P. O. (eds.) *"Maximum Entropy and Bayesian Methods"*. Kluwer Acad. Pubs., Dordrecht, 1992.
[31] Grandy, W. T. and Schick, L. H.(eds.) *"Maximum Entropy and Bayesian Methods"*. Kluwer Acad. Pubs., Dordrecht, 1991.
[32] Fougere, P. (ed.) *"Maximum Entropy and Bayesian Methods"*. Kluwer Acad. Pubs., Dordrecht, 1990.
[33] Skilling, J. (ed.) *"Maximum Entropy and Bayesian Methods"*. Kluwer Acad. Pubs., Dordrecht, 1989.
[34] Erickson, G. and Ray Smith, G. (eds.) *"Maximum Entropy and Bayesian Methods"*. Vols. I and II, Kluwer Acad. Pubs., Dordrecht, 1988.
[35] Kapur, J. N. *" Maximum Entropy Models in Science and Engineering"*. J. Wiley, New Delhi, 1989.
[36] Bevense, R. M. *"Maximum Entropy Solutions to Scientific Problems"*. Prentice Hall, New York, 1993.
[37] Golan, A., Judge, G. and Miller, D. *"Maximum Entropy Econometrics"*. J. Wiley, New York, 1996.
[38] Gamboa, F. and Gzyl, H. *"Maxentropic solutions of linear Fredholm equations"*. Math. & Comp. Modeling. Vol. 25 (1997), pp. 23-32.
[39] Csiszar, I., Gamboa, F. and Gassiat, E. *"MEM pixel correlated solutions for generalized moment problems"*. Submitted to IEEE. T.I.T.

[40] Marechal, P. and Lannes, A. *"Unification of deterministic and probabilistic methods for the solution of linear inverse problems via the principle of maximum entropy on the mean"*. Inv. Problems, Vol. 13 (1997), pp. 135-151.

[41] Leitner, J. and Temnov, G. *"Estimation by fitting generalized moments"*. Preprint.

Chapter 6

Finite dimensional problems

Most of the time the data gathering process, or the need of numerical calculations, naturally lead to finite dimensional problems. In this chapter we are going to examine some methods of solving them. Some of the methods we call classical, as opposed to maxentropic methods, as well as the maxentropic methods.

6.1 Two classical methods of solution

The following material could have been presented in Chapters 2 and/or, but we might as well use it as our starting point. Consider the usual

$$Ax = y \tag{6.1}$$

where A is an $m \times n$ matrix, x is an unknown vector in $\mathbb{R}^n$ and y is a given vector in $\mathbb{R}^m$. The results in this section will also be valid when both or one of the intervening spaces in (6.1) is an infinitely dimensional Hilbert space.

Suppose that $n > m$ and we decide to go around the issue of non-unique solutions by finding

$$\min\left\{\frac{1}{2}\|x\|^2 : Ax = y\right\}. \tag{6.2}$$

To eliminate the constraint set note that

$$\sup\left\{\langle \lambda, Ax - y \rangle : \lambda \in \mathbb{R}^m\right\} = 0 \quad \text{or} \quad +\infty \tag{6.3}$$

depending on whether $Ax = y$ or not. Therefore, (6.2) reads

$$\inf\left\{\frac{1}{2}\|x\|^2 : Ax = y\right\} = \inf_x \sup_\lambda \left\{\frac{1}{2}\|x\|^2 - \langle \lambda, Ax - y \rangle\right\} \tag{6.4}$$

on account of (6.3). Notice that the constraint set has disappeared in (6.4). To find the optimal value indicated, first exchange the operations indicated and obtain

$$\inf\left\{\frac{1}{2}\|x\|^2 : Ax = y\right\} = \sup_\lambda \inf_x \left\{\frac{1}{2}\|x\|^2 - \langle \lambda, Ax - y\rangle\right\}.$$

To obtain the minimum of $\frac{1}{2}\|x\|^2 - \langle \lambda, Ax - y\rangle$ for each fixed λ is a piece of cake: either use calculus or complete the squares. Either way one sees that the minimizing value is $x_{op}(\lambda) = A^*\lambda$ and the function value is $\langle \lambda, y\rangle - \frac{1}{2}\langle \lambda, AA^*\lambda\rangle$. To complete we must find

$$\sup\left\{\langle \lambda, y\rangle - \frac{1}{2}\langle \lambda, AA^*\lambda\rangle : \lambda \in \mathbb{R}^m\right\}$$

which can again be obtained by completing squares (do it!). The result is

$$\lambda_{op} = (AA^*)^{-1}y$$

provided $(AA^*)^{-1}$ exists. Inserting this in the expression for $x_{op}(\lambda) = A^*\lambda$ found above we obtain

$$\inf\left\{\frac{1}{2}\|x\|^2 : Ax = y\right\} = \frac{1}{2}\langle y, AA^*y\rangle$$

which reached at

$$x_{op} = A^*(AA^*)^{-1}y \qquad (6.5)$$

(again, only if $(AA^*)^{-1}$ exists). We leave it as an exercise for the reader to verify that the usual Lagrange multipliers technique applied to (6.2) also yields (6.5).

A very easy alternative method of solution exists when $(A^*A)^{-1}$ exists. Just multiply both sides of (6.1) by A^* to obtain $(A^*A)x = A^*y$, which can be solved yielding

$$x^* = (A^*A)^{-1}A^*y. \qquad (6.6)$$

Still another way of understanding how (6.6) comes about is to consider, instead of (6.2), the problem of finding

$$\inf\left\{\frac{1}{2}\|Ax - y\|^2 : x \in \mathbb{R}^n\right\}. \qquad (6.7)$$

Since the gradient (with respect to x) of the quadratic form $\frac{1}{2}\|Ax - y\|^2$ is $A^*Ax - A^*y$, which equals zero at x^* given by (6.6) when $(A^*A)^{-1}$ exists.

Observe also that when any of $(A^*A)^{-1}A^*$ and $A^*(AA^*)^{-1}$ is defined, it is a generalized inverse of A, i.e., they satisfy

$$AA_g^{-1}A = A.$$

See reference [2] for more about generalized inverses. What is of importance for us here is to decide when to use (6.2) or (6.7) as starting point, besides the fact that either (or both) $A^*(AA^*)^{-1}A^*$ and $(AA^*)^{-1}A^*$ have to be defined in each case.

To use (6.2) it is necessary to know for sure that the data vector y is in Ran(A), whereas to use (6.7) we do not really need that. Hence the name quasi solution to (6.6) introduced in Chapter 3.

To understand (6.6) from the point of view of the regularization methods described in Chapter 4, notice that since $\mathbb{R}^n = \ker(A) \oplus \text{Ran}(A^*)$, then the equation $A^*Ax = A^*y$ can be regarded as a projection of $Ax = y$ onto Ran(A^*) on which A^*A is invertible.

To understand (6.5) from different point of view think of (6.1) as the collection

$$\langle A_i, x \rangle = y_i, \quad i = 1, ..., m \tag{6.8}$$

where $A_i \in \mathbb{R}^n$ stand for the rows of the matrix A. To make life easy, assume these to be independent vectors in $\mathbb{R}^n$. Thus to use the only information available, we should consider writing

$$x = \sum_{i=1}^{m} \beta_i A_i^* = A^*\beta$$

where β is a column vector in $\mathbb{R}^m$, and the issue now consists in determining the coefficients β_i, $1 \le i \le m$. Substituting in (6.8) we obtain

$$\Gamma\beta = AA^*\beta = y$$

therefore

$$x = A^*(AA^*)^{-1}y$$

whenever (AA^*) is invertible.

A totally different approach to solving (6.8) was developed by Kaczmarz in the 1930's. See [5] for a review of related ideas and applications. Think of (6.8) as a set of conditions for x to be in a collection of hyperplanes

$$H_i = \{\xi \in \mathbb{R}^n : \langle A_i, \xi \rangle = y_i\}, \quad i = 1, ..., m.$$

If we define

$$P_i\xi = \xi - \{\langle A_i, \xi \rangle - y_i\} \frac{A_i}{\langle A_i, A_i \rangle},$$

the projection operator on the i-th hyperplane, and for any starting vector x_0, define x_{n+1} by

$$x_{n+1} = (P_1 P_2 ... P_m) x_n \quad n \ge 0$$

the $x_n \to \hat{x}$ satisfying $A\hat{x} = y$. When $x_0 = 0$, $x_n \to x_*$ satisfying (6.1) and realizing (6.2)!

When A^*A is not invertible, there is still another approach to try when we want to solve (6.7) combined with (6.2). Such approach is related to the SVD-regularization procedure described in Chapter 4. In this case we have to find x^* which realizes

$$\inf \left\{ \frac{1}{2} \|Ax - y\|^2 + \frac{\alpha}{2} \|x\|^2 : x \in \mathbb{R}^n \right\}. \qquad (6.9)$$

This time we are led to solve

$$(A^*A + \alpha I)x_* = A^*y$$

and here it does not matter whether A^*A has zero eigenvalues or not because $A^*A + \alpha I$ is invertible for every positive α. The x^* realizing (6.9) is

$$x^* = (A^*A + \alpha I)^{-1} A^* y. \qquad (6.10)$$

Had we chosen to minimize $\|x\|$ or $\|Ax - y\|$ instead of $\|x\|^2$ or $\|Ax - y\|^2$ life would have been slightly harder. See [1] or [3]-[4] for a very general approach to these problems using the full machinery of convex analysis. From [4] we shall take the basic result in Fenchel duality to deal with constrained problems.

6.2 Continuous time iteration schemes

An obvious question rises when considering (6.2) or (6.7) (or (6.9) if we are lazy to invert matrices): What happens when A^*A or AA^* are not invertible? Well, not much. The three problems consist of minimizing a quadratic function with or without constraints. Let us examine an alternative way of approaching these problems.

To solve (6.2) using the Lagrange multipliers method we define the Lagrangian function

$$\mathcal{L}(x, \lambda) = \frac{1}{2} \|x\|^2 + \langle \lambda, y - Ax \rangle \qquad (6.11)$$

and searches for the zeros of

$$\nabla_x \mathcal{L} = x - A^* \lambda$$

$$\nabla_\lambda \mathcal{L} = y - Ax.$$

When $(AA^*)^{-1}$ is defined, this set leads to (6.6). When not one, defines the following dynamical system of the gradient type

$$\frac{dx}{dt} = -\nabla_x \mathcal{L} = A^*\lambda - x, \quad x(0) = x_0$$

$$\frac{d\lambda}{dt} = -\nabla_\lambda \mathcal{L} = Ax - y, \quad \lambda(0) = \lambda_0 \quad (6.12)$$

for any (x_0, λ_0) in $\mathbb{R}^n \times \mathbb{R}^m$. If $A : \mathbb{R}^n \to \mathbb{R}^m$ it is very easy to see that if (x^*, λ^*) is a zero of the right-hand side of (6.12), it is unique and $V(x, \lambda) = \mathcal{L}(x, \lambda) - \mathcal{L}(x^*, \lambda^*)$ decreases along the trajectories of (6.12). This happens since

$$\frac{d}{dt}V(x(t), \lambda(t)) = -\langle \nabla_x \mathcal{L}, \nabla_\lambda \mathcal{L}\rangle = -\left\{ \|\nabla_x \mathcal{L}\|^2 + \|\nabla_\lambda \mathcal{L}\|^2 \right\}.$$

To sum up V is positive and decreases to 0 along the solution to the differential equation, and achieves its minimum value at (x^*, λ^*), the equilibrium solution of the system.

To minimize (6.9) we proceed similarly. This time things are easier for we do not need Lagrange multipliers. We set

$$\mathcal{L}(x) = \frac{1}{2}\|Ax - y\|^2 + \frac{\alpha}{2}\|x\|^2 \quad (6.13)$$

and the corresponding dynamical system is

$$\frac{dx}{dt} = -\nabla_x \mathcal{L} = -(AA^* + \alpha AI)x + A^*y, \quad x(0) = x_0. \quad (6.14)$$

Certainly, if x^* is an equilibrium point for (6.14) it satisfies $(AA^* + \alpha I)x^* = A^*y$ and it is unique for $\alpha > 0$.

As a Lyapunov function for (6.14) we consider the obvious $V(x) = \mathcal{L}(x) - \mathcal{L}(x^*)$. Obviously

$$\frac{dV}{dt}(x(t)) = -\langle \nabla_x \mathcal{L}, \nabla_x \mathcal{L}\rangle < 0$$

and $V(x(t))$ decreases to its unique 0, which is attained at x^*.

6.3 Incorporation of convex constraints

6.3.1 Basics and comments

In a variety of applications the actual presentation of (6.1) is the following: Solve

$$Ax = y, \quad x \in C \quad (6.15)$$

where C denotes a given convex set in $\mathbb{R}^n$ and every thing else is as above. Usually C is given by an intersection of half spaces, for example

$$C = \{x \in \mathbb{R}^n : -\infty \leq a_i \leq x_i \leq b_i \leq +\infty\}, \tag{6.16}$$

where $a_i < b_i$ for $1 \leq i \leq n$ and the possibility of any of them being infinite is admitted.

Up to this point, with C as in (6.16) the problem has been extensively studied within the context of linear programming, where besides satisfying (6.15) and (6.16), the solution has to minimize some linear functional $\ell(x) = \sum w_i x_i$. In other words, we shall only be concerned with producing feasible solutions to the linear programming problem.

A different, but frequent class of constraints, consists of requiring C to be a sphere of radius R

$$C = \left\{x \in \mathbb{R}^n : \sum x_i^2 \leq R^2\right\}.$$

Note that the case $C = \left\{x \in \mathbb{R}^n : \sum (x_i - x_i^0) M_{ij}(x_j - x_j^0) \leq R^2\right\}$ can be reduced to the previous case by linear changes of variables.

Let us comment on some of the usual techniques for solving (6.15). We have already mentioned that (6.15) with C given by (6.16) amounts to finding feasible solutions to a linear programming problem.

To solve the constrained variant of (6.2), i.e., to find x^* realizing

$$\inf \left\{\frac{1}{2} \|x\|^2 : Ax = y, \ x \geq 0\right\} \tag{6.17}$$

where now $C = \{x \in \mathbb{R}^n : x_i \geq 0\}$ is known as the constrained quadratic programming problem, and prepackaged algorithms for solving this problem exist. It can be proved that solving (6.17) is equivalent to solving the linear problem

$$\begin{aligned} Ax &= y \\ x + A^* u - v &= 0 \\ x \geq 0, \ v \geq 0 \ x.v &= 0 \end{aligned} \tag{6.18}$$

with u unconstrained. (The non-linear condition $x.v = 0$ amounts to require that $x_i^* > 0$ if and only if $v_i^* = 0$ and vice versa.)

Although the full theorem asserts that (6.18) has a solution x^* if and only if x^* realizes (6.17), let us just prove the sufficiency

Theorem 6.1. *Suppose x^*, v^*, u^* solve (6.18). Then x^* realizes (6.17). Actually, x^* is the unique solution to (6.17).*

Proof. Put $q(x) = \frac{1}{2}\|x\|^2$ and consider for $y \in \Re^n$

$$q(x^* + y) = q(x^0) + \langle x^*, y \rangle + \frac{1}{2}\|y\|^2.$$

If x is feasible for (6.17), i.e., $Ax = y$ and $x \in C$, let $\xi = x - x^*$. Then ξ must satisfy

$$A\xi = 0, \; \xi_j \geq 0 \text{ if } x_i^* = 0.$$

Since x^*, v^*, u^* satisfy (6.18)

$$x^* = v^* - A^*u^*.$$

Then

$$\langle \xi, x^* \rangle = \langle \xi, v^* \rangle - \langle \xi, A^*u^* \rangle = \langle \xi, v^* \rangle$$

since $\langle \xi, A^*u^* \rangle = \langle A\xi, u^* \rangle = 0$.

But $v^* \geq 0$ and $v_j^* > 0$ implies $x_j^* = 0$ which implies $\xi_j > 0$. Thus $\langle v^*, \xi \rangle \geq 0$, and

$$q(x^* + \xi) = q(x) \geq q(x^*) + \frac{1}{2}\|\xi\|^2 \geq q(x^*)$$

or x^* minimizes $q(x)$. And uniqueness follows similarly. $\square$

To verify that problem (6.17) has a solution is easy, once we know there is at least one solution.

Lemma 6.1. *If $\{x : Ax = y, \; x \in C\} \neq \emptyset$, then (6.17) has a solution.*

Proof. If $\{x : Ax = y, \; x \in C\} \neq \emptyset$, there exist an $R > 0$ such that the closed bounded set

$$\{x : Ax = y, \; x \geq 0, \; \|x\| \leq R\}$$

is not empty. Thus $\inf \left\{ \frac{1}{2}\|x\|^2 : Ax = y, \; x \geq 0, \; \|x\| \leq R \right\}$ is achieved at some x^*. $\square$

This lemma brings up a few remarks to the relationship between (6.2) and the problem of finding x_0 such that $x_0 = \arg\inf \left\{ \frac{1}{2}\|x\|^2 : Ax = y, \; y \in C \right\}$.

The issue may be described graphically in figures 6.1, 6.2 and 6.3.

Certainly the situations depicted in figure 6.1 or 6.3 are the easiest to deal with. The first one corresponds to the case of nonexistence of solutions: either we remodel our problem or we conclude that no solutions exist and that is it.

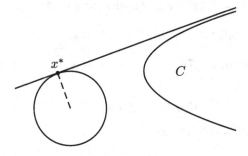

Fig. 6.1 $C \cap A^{-1}y = \emptyset$.

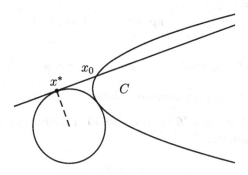

Fig. 6.2 $C \cap A^{-1}y = \emptyset$. $x^* \neq x_0$.

The case described by figure 6.3 is nice for the opposite reason: if we are able to ascertain at the outset the solution to the unconstrained case is in the constraint set.

If the solution to the unconstrained case happened to lie in the interior of C, then the continuous iteration scheme provides us with a method to pin it down.

But if the solution to the unconstrained case lies on the boundary of C, or if we are in the case depicted in figure 6.2, in which the solution x^* to the unconstrained problem is in $\bar{C}$, but there is an x_0 in $A^{-1}\{y\} \cap C$ closest to x^*, we can still apply some standard calculus to find it.

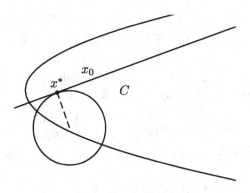

Fig. 6.3 $C \cap A^{-1}y \ni x^* = x_0$.

6.3.2 Optimization with differentiable non-degenerate equality constraints

Let us consider a general case, and then comment on our particular case.

First we want to consider the problem of finding x^* realizing

$$\inf \{F(x) : x \in \mathbb{R}^n, g(x) = 0\} \tag{6.19}$$

where both $F(x) : \mathbb{R}^n \to \mathbb{R}$ and $g(x) : \mathbb{R}^n \to \mathbb{R}^k$ are twice continuously differentiable functions.

The basic results characterizing x^* is

Theorem 6.2. *Assume that x^* realizes (6.19). Assume that x^* is a regular point for the constraints (i.e., the $\nabla g_i(x^*)$ are linearly independent vectors). Then there exists a $\lambda^* \in \mathbb{R}^*$ such that*

$$\nabla F(x^*) + \langle \lambda^*, \nabla g(x^*) \rangle = 0$$

and for every $\xi \in \mathbb{R}^n$ such that $\langle \nabla g(x^*), \xi \rangle = 0$, we have

$$\langle \xi, D^2 F(x)\xi \rangle \geq 0$$

where $\langle \lambda^*, \nabla g(x^*) \rangle = \sum_{j=1}^{k} \lambda_j^* \nabla g_j(x^*)$ *(is a vector in $\mathbb{R}^n$), and* $D^2 F(x)_{ij} = \frac{\partial^2 F(x^*)}{\partial x_i \partial x_j}$.

The systematic way to approach (6.19) consists in putting together a Lagrangian function

$$\mathcal{L}(x, \lambda) = F(x) + \langle \lambda, g(x) \rangle$$

defined on $\mathbb{R}^n \times \mathbb{R}^k$. To find the extreme points of $\mathcal{L}(x, \lambda)$ one solves
$$\nabla_x \mathcal{L}(x, \lambda) = \nabla F(x) + \langle \lambda, \nabla_x g(x) \rangle = 0$$
$$\nabla_\lambda \mathcal{L}(x, \lambda) = g(x). \tag{6.20}$$

The condition for (x^*, λ^*) to be a local-minimum is that $D^2 \mathcal{L}(x^*, \lambda^*)$ be positive definite, i.e., for any $(\xi, \eta) \in \mathbb{R}^n \times \mathbb{R}^k$

$$(\xi, \eta) \begin{pmatrix} D_{xx}^2 F(x) & \nabla_x g(x^*) \\ (\nabla_x g)^t(x^*) & 0 \end{pmatrix} \begin{pmatrix} \xi \\ \eta \end{pmatrix} \geq 0 \tag{6.21}$$

where $(\nabla_x g)_l(x) = \frac{\partial g_l}{\partial x_i}$ for $i = 1, ..., n$ and $l = 1, ..., k$. This means that for any $(\xi, \eta) \in \mathbb{R}^{n+k}$

$$\langle \xi, D^2 F \xi \rangle + 2 \langle \xi, \langle \eta, \nabla_x g \rangle \rangle \geq 0$$

or

$$\langle \xi, D^2 F \xi \rangle + 2 \langle \eta, \langle \xi, \nabla_x g \rangle \rangle \geq 0$$

from which condition (6.21) is clear. Again, to get at a minimum we form the dynamical system of the gradient type

$$\frac{dx}{dt} = -\nabla_x \mathcal{L}, \quad x(0) = x_0$$
$$\frac{d\lambda}{dt} = -\nabla_\lambda \mathcal{L}, \quad \lambda(0) = \lambda_0. \tag{6.22}$$

Again, form $V(x, \lambda) = \mathcal{L}(x, \lambda) - \mathcal{L}(x^*, \lambda^*)$, where (x^*, λ^*) is a minimum of $\mathcal{L}(x, \lambda)$, i.e., satisfies (6.22). Then (x^*, λ^*) is a zero of $V(x^*, \lambda^*)$ and

$$\frac{dV}{dt}(x(t), \lambda(t)) = -\langle \nabla_x \mathcal{L}, \nabla_\lambda \mathcal{L} \rangle = -\left\{ \|\nabla_x \mathcal{L}\|^2 + \|\nabla_\lambda \mathcal{L}\|^2 \right\}.$$

When (x_0, λ_0) is a zero of the right-hand side of (6.22) and $\mathcal{L}(x, \lambda)$ reaches a minimum there, then there is a trajectory of the system (6.22) converging to (x_0, λ_0).

Theorem 6.3. *Let $(x^*, \lambda^*) \in \mathbb{R}^{n+k}$ be an equilibrium point for (6.22), which is a local minimum of $\mathcal{L}(x, \lambda)$. Then (x^*, λ^*) is a stable point for (6.22).*

Proof. Define $V(x, \lambda) = \mathcal{L}(x, \lambda) - \mathcal{L}(x^*, \lambda^*)$ in an appropriate neighborhood U of (x^*, λ^*). U is chosen so that $V(x, \lambda)$ is positive in $U - \{(x^*, \lambda^*)\}$, and (x^*, λ^*) is the only extremum point of $\mathcal{L}(x, \lambda)$ there. Since $\frac{dV}{dt}(x(t), \lambda(t)) < 0$ in $U - \{(x^*, \lambda^*)\}$, (x^*, λ^*) is a stable equilibrium point for (6.22). That is for any starting point $(x_0, \lambda_0) \in U$, $(x(t), \lambda(t)) \to (x^*, \lambda^*)$. □

Comment: In our applications (x^*, λ^*) will be unique, therefore we won't have to worry about where we start at in our iterative scheme.

See [11] for more about Lyapunov functions and gradient systems.

6.3.3 Optimization with differentiable, non-degenerate inequality constraints

The problem we want to solve is to find x^* realizing

$$\inf \{F(x) : x \in \mathbb{R}^n, \; g(x) \geq 0, \; h(x) = 0\} \tag{6.23}$$

where the objective function $F(x) : \mathbb{R}^n \to \mathbb{R}$, and the constraint functions $g(x) : \mathbb{R}^n \to \mathbb{R}^M$, and $h(x) : \mathbb{R}^n \to \mathbb{R}^N$ are all $C^2(\mathbb{R}^n)$. The characterization of x^* is split into a necessity part and a sufficiency part. Both can be checked out in [8]-[10].

Let us put $E = \{i : g_i(x^*) = 0\}$, $I = \{i : g_j(x^*) > 0\}$ and *assume* we have chosen labels so that $E = \{1, ..., k\}$, $I = \{k+1, ..., M\}$. With these conventions, we have the following theorem:

Theorem 6.4. *Assume x^* is a local minimum of $F(x)$ on*

$$C = \{x \in \mathbb{R}^n : g(x) \geq 0, \; h(x) = 0\}.$$

Assume that $\nabla_x[g_E, h](x^0)$ has maximal rank. Then there exists $\lambda_1, ..., \lambda_M$, $\mu_1, ..., \mu_N$ such that

a) $\nabla_x[F + \langle \lambda, g \rangle + \langle \mu, h \rangle](x^*) = 0.$
b) $\lambda_j \geq 0, \; 1 \leq j \leq M.$
c) $\lambda_j g_j(x^*) = 0, \; 1 \leq j \leq M$ *(then $\langle \lambda, g(x^*) \rangle = 0$).*
   Furthermore
d) $\langle \xi, D^2 F(x)\xi \rangle = 0$ *for any ξ such that*

$$\langle \xi, \nabla_x g_E(x^*) \rangle = 0, \; \langle \xi, \nabla_x h(x^*) \rangle = 0.$$

Comment: $\nabla_x[g_E, h]$ denotes an $n \times (k + N)$ matrix whose columns are the functions determining the effective constraints.

Theorem 6.5. *Let F, g, h be as above. Assume that $g(x^*) \geq 0$, $h(x^*) = 0$, and that these exist $\lambda_0, \lambda_1, ..., \lambda_M, \mu_1, ..., \mu_N$ so that*

a) $\lambda_i \geq 0, \quad \forall i \geq 0$
b) $\lambda_j g_j(x) = 0, \quad 1 \leq j \leq M$
c) $\nabla_x[\lambda_0 F(x^*) + \langle \lambda, g(x^*) \rangle + \langle \mu, h(x^*) \rangle] = 0$ *and the condition of minimality*
d) $\langle \xi, D^2\{\lambda_0 F(x^*) + \langle \lambda, g(x^*) \rangle + \langle \mu, h(x^*) \rangle\}\xi \rangle > 0$ *for all nonzero vectors ξ satisfying $\lambda_0 \langle \xi, \nabla F(x^*) \rangle \equiv 0$, $\{\lambda_j \langle \xi, \nabla_x g_j(x^*) \rangle\} = 0$ for $1 \leq j \leq M$ and $\langle \xi, \nabla_x h(x^*) \rangle = 0$.*

Then there is a neighborhood U of x^0 such that $F(x) > F(x^*)$ for all $x \neq x^*$ in

$$U \cap \{x : g(x) \geq 0,\ h(x) = 0\}.$$

6.4 The method of projections in continuous time

Here we review the method for obtaining

$$x^* = \arg\inf\{F(x) : x \in C\} \tag{6.24}$$

when C is defined by

$$C = \{x \in \mathbb{R}^n : g(x) \geq 0\}.$$

When $x_{op} = \arg\inf\{F(x) : x \in \mathbb{R}^n\}$ is in C^c, it is clear that $x^* \in \partial C$. Actually $x^* \in \{x : F(x) \leq t_{op}\}$ where t_{op} is defined as

$$\inf\{t \in \mathbb{R} : \{x : F(x) \leq t\} \cap C \neq \emptyset\} = \sup\{t \in \mathbb{R} : \{x : F(x) \leq t\} \cap C = \emptyset\}.$$

We shall assume that both $F : \mathbb{R}^n \to \mathbb{R}$ and $g : \mathbb{R}^n \to \mathbb{R}^m$ have at least two continuous derivatives, so that $\partial C = \{x : g(x) = 0\}$ has a continuously differentiable set of normals at every point.

Define the projection onto the normal space to $\xi \in \partial C$ as follows

$$Q(\xi) = \sum_{k,j}^{m} \hat{\eta}_k(\xi) G_{ij}(\xi) \hat{\eta}_j(\xi)$$

where the $\hat{\eta}_j(\xi) = \frac{\nabla g_i(\xi)}{\|\nabla g_i(\xi)\|}$ are the normal vectors to $\{g(x) = 0\}$ and $G_{ij}(\xi)$ is the (i,j)-th element of the inverse of the matrix with (i,j)-th entry given by $\langle \hat{\eta}_i(\xi), \hat{\eta}_j(\xi) \rangle$. We shall assume that the components of g are functionally independent, which means that the m normals $\hat{\eta}_j(\xi)$ are linearly independent.

It is thus clear that $Q^2 = Q$ and that $\Pi(\xi) = I - Q(\xi)$ projects onto the tangent space to ∂C at ξ (which we denote by the customary TC_ξ).

Consider the dynamical system in $\mathbb{R}^n$ defined by

$$\frac{d\xi}{dt} = -\Pi(\xi).\nabla F(\xi), \quad \xi(0) = x_0 \in \partial C. \tag{6.25}$$

It is a simple exercise to verify that $\xi(t) \in \partial C$ for all t, and if we assume that $x^* \in \partial C$ is a global minimum for (6.24), then $V(\xi) = F(\xi) - F(x^*)$ is a Lyapunov function for (6.25) since

$$\frac{dV}{dt} = -\langle \nabla F(\xi), \Pi(\xi).\nabla F(\xi) \rangle \leq 0$$

since $\Pi(\xi)$ is a self-adjoint linear operator on $\mathbb{R}^n$. This is trivial to verify from its definition. Just keep in mind that $\Pi(\xi) : \mathbb{R}^n \to TC_\xi$ linearly at every $\xi \in \partial C$.

To compare the approach described by (6.25) with the usual method of projection see [12] for example, let us recall it. Define $P : \mathbb{R}^n \to C$ by

$$P(x) := \arg\inf \{\|x - y\| : y \in C\}. \tag{6.26}$$

Such P is certainly a projection, for $P(P(x)) = P(x)$ and $P(x) = x$ if $x \in C$. P is usually nonlinear, but it is nevertheless a contracting map, that is $\|P(x_1) - P(x_2)\| \leq k \|x_1 - x_2\|$ for some positive k.

The usual numerical procedure to get at x^* consists of carrying out the following iteration scheme

$$\begin{aligned} x_{n+1} &= P(x_n - \lambda_n \nabla F(x_n)) \\ x_0 &= P(x_0) \text{ by definition} \end{aligned} \tag{6.27}$$

and the λ_n are either fixed or chosen according to some appropriate description.

To see what happens at steps of λ-time units a part, consider an $x_0 \in \partial C$, then moving along the tangent plane we would end up at

$$\xi_{ap}(\lambda) = x_0 - \lambda \Pi(x_0) \nabla F(x_0) \tag{6.28}$$

whereas moving according to (6.27) we would end up at

$$\xi_{pr}(\lambda) = P(x_0 - \lambda \nabla F(x_0)). \tag{6.29}$$

Even though the separation between $\xi_{ap}(\lambda)$ and the exact $\xi(\lambda)$ given by solving (6.25) is $\mathcal{O}(\lambda^2)$, the difference between the $\xi_{pr}(\lambda)$ and $\xi(\lambda)$ or $\xi_{ap}(\lambda)$ does not seem easy to estimate. But (6.28) and (6.29) suggest that an alternative iterative scheme could be

$$\xi_n = P(\xi_{n-1} - \lambda_n \Pi(\xi_n - 1) \nabla F(\xi_{n-1})) \tag{6.30}$$

which seems to be only slightly more computationally demanding than (6.27), the comparison and relative merits of the different iteration schemes may be an unresolved problem.

6.5 Maxentropic approaches

In the various subsections of the present section we shall consider the equation (6.1) with various typical constraints imposed on the unknown vector x.

6.5.1 Linear systems with band constraints

We will look for solutions to (6.1) when the constraint set C is of type, (6.16), i.e.,
$$C = \{x \in \mathbb{R}^n : a_i \leq x_i \leq b_i\}$$
with $-\infty < a_i < b_i < +\infty$. Such band or box constraints appear, for example, when discretizing a Fredholm equation
$$y(t) = \int A(t,s)x(s)ds$$
and requiring that the solution $x(s)$ lies in the band $[a(s), b(s)]$.

To proceed as specified in Chapter 5, we must produce a probability space $(\Omega, \mathcal{F}, Q)$ such that the coordinate maps satisfy the constraints.

The easiest way out is to consider
$$\Omega = \prod_{j=1}^{n}[a_j, b_j], \quad \mathcal{F} = \mathcal{B}(\Omega) \tag{6.31}$$
and note that this time the coordinate maps
$$X_j : \Omega \to \mathbb{R}, \; \xi = (\xi_1, ..., \xi_n) \to X_n(\xi) = \xi_n \tag{6.32}$$
automatically satisfy the constraints.

The two usual choices we shall make for the *a priori* measure dQ on Ω consist of setting $dQ(\xi) = \otimes dQ_i(\xi_i)$, i.e., the coordinate maps X_n are assumed independent to begin with, and then we take either the uniform measures
$$dQ_j(\xi_j) = \frac{d\xi_j}{(b_j - a_j)} \tag{6.33}$$
on each $[a_j, b_j]$, or we take the Bernoulli measures
$$dQ_j(\xi_j) = p_j \varepsilon_{a_j}(d\xi_j) + q_j \varepsilon_{b_j}(d\xi_j) \tag{6.34}$$
on each $[a_j, b_j]$.

We now want to look for $P(d\xi) = \rho(\xi)dQ(\xi)$ such that
$$AE_P[X] = y \tag{6.35}$$
where $X : \Omega \to \Omega \subset \mathbb{R}^n$ is an $\mathbb{R}^n$-valued random variable whose components are the coordinate maps introduced above in (6.32).

We saw in Chapter 5 that $\rho(\xi)$ would come out being a member of the exponential family
$$\rho(\xi, \lambda) = \frac{e^{-\langle \lambda, AX \rangle(\xi)}}{Z(\lambda)}$$

where $Z(\lambda)$ is produced as follows

$$Z(\lambda) = \int_\Omega e^{-\langle \lambda, AX\rangle(\xi)} dQ(\xi).$$

Denote by $z_j(t_j)$ the integral

$$z_j(t_j) = \int_{a_j}^{b_j} e^{-t\xi_j} dQ_j(\xi).$$

If we denote by $(A^*\lambda)_j$ the j-th component of $A^*\lambda$, then

$$Z(\lambda) = \prod_{j=1}^n z_j\left((A^*\lambda)_j\right). \tag{6.36}$$

When $dQ_j(\xi)$ is of type (6.33)

$$z_j(t_j) = \frac{1}{t_j(b_j - a_j)}[e^{-t_j a_j} - e^{-t_j b_j}] \tag{6.37}$$

and when $dQ_j(\xi)$ is of type (6.34)

$$z_j(t_j) = p_j e^{-t_j a_j} + q_j e^{-t_j b_j} \tag{6.38}$$

from which the corresponding expression for $Z(\lambda)$ can be obtained via (6.36).

Now to find λ^* one has to form

$$\Sigma(\lambda) = \ln(Z(\lambda)) + \langle \lambda, y\rangle$$

and find the λ^* that makes $\Sigma(\lambda^*) = \min\{S(\lambda) : \lambda \in \mathbb{R}^n\}$.

Once such λ^* has been found we have

Lemma 6.2. *The maxentropic solution of $Ax = y$ when*

$$C = \{x \in \mathbb{R}^n : a_j \leq x_j \leq b_j\}$$

is given by

$$x_j^* = -\frac{\partial}{\partial t_j} \ln(z_j(t_j))\big|_{t_j=(A^*\lambda^*)_j}. \tag{6.39}$$

Proof. It involves an easy computation and it is the same proof throughout. The condition for λ^* to be a minimum of $S(\lambda)$ is

$$\frac{\partial}{\partial \lambda_i}\Sigma(\lambda^*) = \sum_{j=1}^n \frac{\partial}{\partial t_j} \ln(z_j(t_j))\big|_{t_j=(A^*\lambda^*)_j} \frac{\partial t_j}{\partial \lambda_i} + y_i = 0.$$

Since $\frac{\partial t_j}{\partial \lambda_i} = A_{ij}$, we can rewrite the previous identity as

$$\sum A_{ij}\left(-\frac{\partial}{\partial t_j} \ln(z_j(t_j))\big|_{t_j=(A^*\lambda^*)_j}\right) = y_i$$

which is what we wanted to obtain. □

Comment: The fact that $Q = \otimes Q_j$ and then (6.36) is of product type is essential for (6.39) to hold.

When we use (6.37), the reconstruction is

$$x_j^* = \frac{a_j \exp(-a_j t_j^*) - b_j \exp(-b_j t_j^*)}{\exp(-a_j t_j^*) - \exp(-b_j t_j^*)} - \frac{1}{t_j^*} \tag{6.40}$$

and when we use (6.38) the reconstruction is

$$x_j^* = \frac{p_j a_j \exp(-a_j t_j^*) + q_j b_j \exp(-b_j t_j^*)}{p_j \exp(-a_j t_j^*) + q_j \exp(-b_j t_j^*)} \tag{6.41}$$

where in each case $t_j^* = (A^* \lambda^*)_j$ and λ^* is the minimizer of the corresponding energy.

Comment: We only insist on the fact that (6.40) and (6.41) provide an explicit representation of the solution to the linear problem in which the constraints and the data appear explicitly. Only notice that *the independence on the data comes through λ^* and is nonlinear.*

Exercise 6.1. Consider a triangular distribution and a beta distribution on each (or some) $[a_j, b_j]$. Work out the corresponding analogues to (6.40) and (6.41).

Exercise 6.2. Replace the Bernoulli measure (6.34) by something like

$$dQ_j(\xi) = \sum_{i=1}^{k} P_{j,i} \varepsilon_{c_{j,i}}(d\xi),$$

where $c_{j,1} = a_j < c_{j,2} < ... < c_{j,k} = b_j$ and $\sum_{i=1}^{k} P_{j,i} = 1$. (The number k of point masses may depend on j.) Work out the expression for the x_j^*.

Exercise 6.3. (In lieu of comment). In all these situations it was not really necessary to have $dQ_j(\xi)$ to be a probability measure. Obtain the x_j^* for a few examples of your choice.

6.5.2 Linear system with Euclidean norm constraints

This time we shall consider (6.1) but how the constraint set will be

$$C = \{x \in \mathbb{R}^n : \|x\| \leq R\}.$$

Again we take $\Omega = C$, $\mathcal{F} = \mathcal{B}(\Omega)$ to be the Borel subsets of Ω. The coordinate map

$$X : \Omega \to \mathbb{R}^n, \ X(\xi) = \xi$$

automatically satisfies the constraints.

As in the previous case, we shall consider two *a priori* measures on $(\Omega, \mathcal{F})$. One of them will be the uniform measure on Ω, of total mass one, i.e.,

$$dQ(\xi) = \frac{d\xi}{V_n(R)} \qquad (6.42)$$

where $d\xi$ is just Lebesgue measure on $\mathbb{R}^n$ and $V_n(R)$ is just the volume of the ball of radius R in $\mathbb{R}^n$. The other alternative we will play with consists of

$$dQ(\xi) = \frac{d\sigma(\xi)}{\Sigma_n(R)} \qquad (6.43)$$

where $d\sigma(\xi)$ is the (induced) Lebesgue measure on the surface

$$\{x \in \mathbb{R}^n : \|x\| = R\} = S_n(R)$$

of the sphere and $\Sigma_n(R)$ is the area of the sphere of radius R. This is an analogue of the Bernoulli *a priori* given by (6.34).

The computation of $Z(\lambda)$ in each case involves the computation of the following integral

$$\zeta_n(t) = \int_{S_n(1)} e^{-\langle t,\xi\rangle} \frac{d\sigma(\xi)}{\Sigma_n(1)}$$

where $t \in \mathbb{R}^n$ is a fixed vector. By orienting the coordinate system with its n-th axis along t, we can write

$$\zeta_n(t) = \frac{\Sigma_{n-1}(1)}{\Sigma_n(1)} \int_0^\pi e^{-\|t\|\cos(\theta)} (\sin(\theta))^{n-2} d\theta$$

after introducing spherical coordinates in $\mathbb{R}^n$. The integral is just a representation of the modified spherical Bessel function of the first kind $I_{\frac{n-2}{2}}(|t|)$ and therefore

$$\zeta_n(t) = \frac{\Sigma_{n-1}(1)}{\Sigma_n(1)} I_{\frac{n-2}{2}}(|t|)^1. \qquad (6.44)$$

When $dQ(\xi)$ is given by (6.42) we shall write for $\lambda \in \mathbb{R}^m$, $Z(\lambda) = z(A^*\lambda)$, for $t \in \mathbb{R}^n$

$$z(t) = \int_\Omega e^{-\langle t,x\rangle(\xi)} dQ(\xi)$$

$$= \frac{\Sigma_{n-1}(1)}{v_n(R)} \int_0^R \gamma^{n-1} \left(\int_0^\pi e^{-\|t\|\cos(\theta)} (\sin(\theta))^{n-2} d\theta \right) d\gamma$$

$$= \frac{\Sigma_{n-1}(1)}{v_n(R)} \int_0^R \gamma^{n-1} \zeta_n(\|t\|\gamma) d\gamma$$

---

[1] for $\rho > -1$ and real s, $I_\rho(s) = \sum_{k=0}^\infty \frac{1}{k!} \frac{1}{\Gamma(k+\rho+1)} \left(\frac{s}{2}\right)^{2k+\rho}$.

from which we obtain

$$Z(\lambda) = z(A^*\lambda) = \frac{\Sigma_{n-1}(1)}{v_n(R)} \int_0^R \gamma^{n-1} \zeta_n\left(\|A^*\lambda\|\gamma\right) d\gamma \qquad (6.45)$$

after rescaling.

When $dQ(\xi)$ is given by (6.43), a similar computation yields

$$Z(\lambda) = z(A^*\lambda) = \zeta_n\left(\|A^*\lambda\| R\right). \qquad (6.46)$$

In each case, proceeding according to the routine, we have to minimize

$$S(\lambda) = \ln(Z(\lambda)) + \langle \lambda, y \rangle$$

where in one case $Z(\lambda)$ is given by (6.45) and in the other by (6.46). Once the point λ^* at which the minimum is achieved.

Now, the appropriate variant of Lemma 6.2 yields, when dQ is given by (6.43) and $Z(\lambda)$ by (6.5), the maxentropic solution to (6.1) is given by

$$x_j^* = -\left.\frac{\partial \ln z(t)}{\partial t_j}\right|_{t=A^*\lambda} = -\frac{RF_n'\left(\|A^*\lambda\| R\right)}{\zeta_n\left(\|A^*\lambda\| R\right)} \frac{(A^*\lambda)_j}{\|A^*\lambda\|}. \qquad (6.47)$$

We leave to the reader to work out a similar representation for the case when dQ is given by (6.42) and $Z(\lambda)$ by (6.45).

Comment: Notice that there may be values of R for which a λ^* minimizing $S(\lambda)$ may not exist.

6.5.3 Linear systems with non-Euclidean norm constraints

Consider now solving (6.1) and demanding that the solutions lie in the convex set

$$C = \{x \in \mathbb{R}^n : \|x\|_1 = \sum_{j=1}^n |x_j| \leq R\}.$$

This time for not to drag the R around assuming $R = 1$.

Let us start with the standard choice $(\Omega, \mathcal{F}) = (C, \mathcal{B}(C))$. Now for any given dQ on $(\Omega, \mathcal{F})$ we would have to know how to compute

$$z(t) = \int_\Omega e^{-\langle t, \xi \rangle} dQ(\xi)$$

but this time the geometry is slightly more complicated, and to compute the above integral when $dQ(\xi) = \frac{d\xi}{V(C)}$ seems to be hard. Since we use an analogue of (6.34), let e_j denote the unit vector along the j-th axis, and put

$$dQ(\xi) = \sum_{j=1}^n \alpha_j (p_j \varepsilon_{e_j}(d\xi) + q_j \varepsilon_{-e_j}(d\xi))$$

where for each j, $p_j + q_j = 1$ and $\sum_{j=1}^{n} \alpha_j = 1$. In this case

$$z(t) = \sum_{j=1}^{n} \alpha_j (p_j e^{-t_j} + q_j e^{t_j}) \qquad (6.48)$$

from which $Z(\lambda) = z(A^*\lambda)$ is obtained. Once the λ^* that minimizes $S(\lambda)$ is obtained (if it exists), then the solution to the problem is given by

$$x_j^* = -\left.\frac{\partial \ln z(t)}{\partial t_j}\right|_{t=A^*\lambda^*} = \frac{1}{Z(A^*\lambda^*)} \alpha_j \left(p_j e^{-(A^*\lambda^*)_j} - q_j e^{-(A^*\lambda^*)_j} \right). \qquad (6.49)$$

To wind down this section, let us propose for the reader to compute

$$z(t) = \int_{\Omega} e^{-\langle t, \xi \rangle} dQ(\xi)$$

when dQ is either normalized Lebesgue measure on $\Omega = \{x \in \mathbb{R}^n : \|x\|_1 \leq 1\}$ or dQ is the (normalized) induced surface measure on $\partial\Omega = \{x \in \mathbb{R}^n : \|x\|_1 = 1\}$.

By means of appropriate reflections, the problem reduces to computing the integrals when Ω is replaced by $\Omega_0 = \{x \in \mathbb{R}^n : x_j \geq 0, \sum x_j \leq 1\}$ and $\partial\Omega$ by $\partial\Omega_0 = \{x \in \mathbb{R}^n : x_j \geq 0, \sum x_j = 1\}$.

As a final **exercise** for the reader, we propose her (him) to determine the set

$$\mathcal{D}(A, Q) = \{\lambda \in \mathbb{R}^m : Z(\lambda) < \infty\}$$

for each of the examples described above.

6.5.4 Linear systems with solutions in unbounded convex sets

This time we shall look for solutions to $Ax = y$ with

$$x \in C = \{x \in \mathbb{R}^n : -\infty \leq a_i \leq x_i \leq b_i \leq +\infty\}$$

in which either a_i or b_i or both are allowed to be either $-\infty$ or $+\infty$ respectively.

Again the routine will be pretty much as above on $(\Omega, \mathcal{F}) = (C, \mathcal{B}(C))$ we shall define some appropriate *a priori* measure and compute the corresponding Laplace transforms (partition function) and write the reconstruction that would be if the minimization problem could be solved.

Consider to begin with the semi-bounded case

$$C = \{x \in \mathbb{R}^n : -\infty \le a_i \le x_i < +\infty\} = \prod_{j=1}^{n} [a_j, \infty).$$

An *a priori* measure on $(\Omega, \mathcal{F})$ with which it is easy to compute consists of setting $dQ = \otimes dQ_j(\xi_j)$ with each $dQ_j(\xi_j)$ being of the Γ-type

$$dQ_j(\xi_j) = \frac{\mu_j^{\rho_j+1}}{\Gamma(\rho+1)} (\xi_j - a_j)^{\rho_j} e^{-\mu_j(\xi_j - a_j)} d\xi_j. \qquad (6.50)$$

The corresponding $z(t) = E_Q \left[e^{-\langle t, x \rangle} \right]$ factors nicely, each factor being

$$z_j(t_j) = \frac{\mu_j^{\rho_j+1}}{\Gamma(\rho+1)} \int_0^\infty (\xi_j - a_j)^{\rho_j} e^{-\mu_j(\xi_j - a_j)} e^{-t_j \xi_j} d\xi_j$$

$$= \left(\frac{\mu_j}{\mu_j + t_j} \right)^{\rho_j+1} e^{-t_j a_j}$$

and therefore, $\ln(z(t)) = \sum \ln(z_j(t_j))$, from which recurring to Lemma 6.2 we obtain

$$-\frac{\partial \ln z(t)}{\partial t_j} = -\frac{\partial \ln z_j(t_j)}{\partial t_j} = a_j + (\rho_j + 1)\frac{1}{\mu_j + t_j}$$

which after the evaluation at $t_j = (A^*\lambda^*)_j$, when the λ^* minimizing $S(\lambda)$ is found, yields

$$x_j^* = a_j + (\rho_j + 1)\frac{1}{\mu_j + (A^*\lambda^*)_j}. \qquad (6.51)$$

In the present example it is clear that

$$\mathcal{D}(A, Q) = \cap_{j=1}^{n} \{\lambda \in \mathbb{R}^n : \mu_j + (A^*\lambda)_j > 0\}.$$

It is certainly a nice open convex set. Its boundary is determined by hyperplanes $\mu_j + (A^*\lambda)_j = 0$, and $\ln(Z(\lambda))$ becomes undefined there.

If instead of (6.50) had we considered $dQ(\xi) = d\xi_j$ (a non finite measure on $[a_j, \infty)$), instead of (6.51) we would have obtained $x_j^* = a_j + \frac{1}{t_j}\big|_{t_j = (A^*\lambda^*)_j}$ and this time

$$\mathcal{D}(A, Q) = \cap_{j=1}^{n} \{\lambda \in \mathbb{R}^n : (A^*\lambda)_j > 0\}.$$

Verify it!

Consider now a simple example in which the components X_j of X are not independent with respect to dQ. Set

$$dQ(\xi) = \frac{N(k)}{\Sigma_n(1)} \|\xi\|^k e^{-\langle \mu, \xi \rangle} d\xi \qquad (6.52)$$

where $k > 0$ is an integer and μ is a vector in $\mathbb{R}^n_+$. $N(k)$ is an appropriate normalization constant. This time

$$z(t) = N(k) \int_0^\infty \rho^{k+n+1} \zeta_n(\|t+\mu\|\rho) d\rho.$$

The integral can be readily computed and

$$z(t) = N(k) \left(-\frac{\partial}{\partial s}\right)^{k+n-1} \int_0^\infty e^{-s\rho} \frac{\Sigma_{n-1}(1)}{\Sigma_n(1)} I_{\frac{n-2}{2}}(\|t+\mu\|\rho) d\rho$$

$$= N(k) \frac{\Sigma_{n-1}(1)}{\Sigma_n(1)} \|t+\mu\|^{\frac{n-2}{2}} \left(-\frac{\partial}{\partial s}\right)^{k+n-1} \frac{\left(s-\sqrt{s^2-\|t+\mu\|^2}\right)^{\frac{n-2}{2}}}{\sqrt{s^2-\|t+\mu\|^2}} \Bigg|_{s=0}.$$

To obtain the maxentropic solution in this case, we begin by finding the minimum λ^* of

$$\Sigma(\lambda) = \ln(Z(A^*\lambda)) + \langle \lambda, y \rangle$$

over $\mathcal{D}(A, Q)$. After that is done, we compute

$$x_j^* = -\frac{\partial \ln z(t)}{\partial t_j}\Bigg|_{t_j = (A^*\lambda^*)_j}.$$

We leave the details for interested reader.

We also leave it to the reader to work out the following example: $dQ(\xi) = \otimes dQ_j(\xi_j)$ where, for each $1 \leq j \leq n$, $dQ_j(\xi_j)$ is a (Poisson) randomized sum of gamma densities (like those considered at the beginning of the section).

$$dQ_j(\xi_j) = e^{-s-\xi} \sqrt{\left(\frac{\xi_j}{s}\right)^\rho} I_\rho\left(2\sqrt{s\xi_j}\right) d\xi_j$$

where $s > 0$, $\rho > 0$ are free parameters. See Vol. II of Feller's books for more about this distribution.

Let us now consider two examples in which each $dQ_j(\xi_j)$ is a singular measure.

Consider first $dQ_j(\xi_j)$ to be a sum of point masses at each integer $1, ...$, i.e.,

$$dQ_j(\xi_j) = \sum_{k=0}^\infty \varepsilon_k(d\xi_j) \tag{6.53}$$

and of course $dQ = \otimes dQ_j$. Now

$$z_j(t_j) = \int_0^\infty e^{-t_j \xi_j} dQ_j(\xi_j) = \sum_{k=0}^\infty e^{-t_j k} = \frac{1}{1 - e^{-t_j}}$$

and once λ^* minimizing $S(\lambda) = \ln(Z(\lambda)) + \langle \lambda, y \rangle$ is found

$$x_j^* = -\left.\frac{e^{-t_j}}{1-e^{-t_j}}\right|_{t_j=(A^*\lambda^*)_j} = \frac{1}{e^{(A^*\lambda^*)_j}-1}$$

provide us with a solution to

$$Ax = y, \quad x_j \geq 0, \quad j = 1, 2, ..., n.$$

But instead of (6.53) we could have taken each dQ_j to be a probability measure, for example

$$dQ_j(\xi_j) = \sum_{k\geq 0} p_j^k(1-p_j)\varepsilon_k(d\xi_j)$$

so that $dQ = \otimes dQ_j$ and again the X_j come out being independent (relative to dQ) and

$$z_j(t_j) = \int_0^\infty e^{-t_j\xi_j}dQ_j(\xi_j) = \frac{(1-p_j)}{(1-p_je^{-t_j})}.$$

Again, for $z(t) = \prod_{j=1}^n z_j(t_j)$ and $Z(\lambda) = z(A^*\lambda)$ and minimize $\Sigma(\lambda)$ to find λ^*. We leave it to the reader to determine $\mathcal{D}(A,Q)$ for this example, and then to write down the resulting expression for x_j^*.

To finish

Exercise 6.4. Consider each $dQ_j(\xi_j)$ to be of Poisson type

$$dQ_j(\xi_j) = \sum e^{-\mu_j}\frac{\mu_j^k}{k!}\varepsilon_{h_jk}(d\xi_j)$$

which assigns probability $e^{-\mu_j}\frac{\mu_j^k}{k!}$ to the points $h_j k$, $k = 0, 1, ...,$ and the h_j are positive real numbers.

Verify as well that the maxentropic solution is provided by

$$x_j^* = \mu_j e^{-(A^*\lambda^*)_j}.$$

Comment: The moral so far is that what you output depends on what you input. Since the constraints on the solution are (conceivably) dictated by the nature of the problem. The only input which we have to be careful about is the *a priori* measure dQ. The only mathematical requirement on it is that $\sup(Q) = C$, but as far as we know, "physical" constraints (i.e., constraints dictated by the specific application) seem to be anybody's guess.

6.5.5 Linear equations without constraints

In this section we shall be considering the problem of solving (6.1) without any constraints imposed on the solution x to be found. There will be some overlap with the material in Maxentropic methods in Hilbert spaces, but the finite dimensionality of the set up makes things easier.

The list of examples below is put together with two criteria in mind: first, easy computability and second, the measures dQ are chosen so that $\int e^{-\langle t,\xi\rangle} dQ(\xi)$ is finite in a large enough set. For example, if $dQ(\xi) = d\xi$ on $(\Omega, \mathcal{F}) = (\mathbb{R}^n, \mathcal{B}(\mathbb{R}^n))$, then the integral is not defined for any $t \in \mathbb{R}^n$.

6.5.5.1 Gaussian reference measure

Consider to begin with

$$dQ(\xi) = \frac{1}{(2\pi)^{\frac{n}{2}}} e^{-\frac{1}{2}(\xi - x^0)^2} d\xi \qquad (6.54)$$

where x^0 is a given vector in $\mathbb{R}^n$. Certainly dQ is a product measure and $X(\xi) = \xi$ has independent components with respect to dQ. Notice that $E_Q[X] = x^0$, so if $Ax^0 = y$, our problem would already be solved. Now $z(t)$ is easy to compute

$$z(t) = \int e^{-\langle t,\xi\rangle - \frac{1}{2}(\xi - x^0)^2} \frac{d\xi}{(2\pi)^{\frac{n}{2}}} = e^{-\langle t,\xi^0\rangle + \frac{1}{2}\|t\|^2}$$

and

$$\ln(z(t)) = \frac{1}{2} \|t - x^0\|^2 - \frac{1}{2} \|x^0\|^2$$

therefore, for $Z(\lambda) = z(A^*\lambda)$ we obtain

$$\Sigma(\lambda) = \ln(Z(\lambda)) + \langle \lambda, y\rangle = \frac{1}{2} \|A^*\lambda - x^0\|^2 - \frac{1}{2} \|x^0\|^2 + \langle \lambda, y\rangle.$$

Here there is no doubt that a minimizing λ^* exist. But since a nice representation exists when $(AA^*)^{-1}$ is defined, let us consider this case first. Now

$$\lambda^* = (AA^*)^{-1}(Ax^0 - y)$$

and since $-\nabla_t \ln(z(t)) = x^0 - t$ we obtain that

$$x^* = x^0 + A^*(AA^*)^{-1}(y - Ax^0) \qquad (6.55)$$

and the role of the *a priori* measure in providing us with "*a priori* guess" and the maxentropic method to provide the correction cannot be more obvious. Notice also that when A is invertible, the correct solution is obtained.

When AA^* is not invertible, λ^* still exists but all we can conclude is that
$$x^* = x^0 - A^*\lambda^*. \tag{6.56}$$

When the reference measure is Gaussian, we can proceed to Section 5.6 of Chapter 5, and find the solution to
$$Ax \in B(y,r)$$
where $B(y,r) = \{\eta \in \mathbb{R}^m : \|\eta - y\| \leq r\}$ is given by $x^* = E_{P^*}[X]$, where $dP^* = e^{-\langle \lambda^*, AX\rangle} dQ$ and λ^* minimizes
$$\Sigma_r(\lambda) = \Sigma_0(\lambda) + r\,\|\lambda\| = \ln(Z(\lambda)) + \langle y, \lambda\rangle + r\,\|\lambda\|.$$

When dQ is given by (6.55) we end up having to minimize
$$\Sigma_r(\lambda) = \frac{1}{2}\left\|A^*\lambda - x^0\right\|^2 + \langle y, \lambda\rangle + r\,\|\lambda\|$$
which leads to having to solve
$$A(A^*\lambda - x^0) + y + r\frac{\lambda}{\|\lambda\|} = 0$$
or equivalently
$$AA^*\lambda + r\frac{\lambda}{\|\lambda\|} = Ax^0 - y.$$

The result is given by

Lemma 6.3. *If $(AA^*)^{-1}$ exists, the last equation above is solved by*
$$(AA^* + \alpha I)^{-1}(Ax^0 - y) \quad \text{if } r = \alpha\,\|(AA^* + \alpha I)^{-1}b\|$$
where $b = Ax^0 - y$.

Comment: The function $\alpha\,\|(AA^* + \alpha I)^{-1}b\|$ increases between 0 and $+\infty$ as α varies from 0 to $+\infty$. This is easy to see using the singular value decomposition of AA^*. Thus for each $r > 0$, there exists an α such that $r = \alpha\,\|(AA^* + \alpha I)^{-1}b\|$. This relates tolerances to penalization!

Exercise 6.5. Repeat the computations carried out above for the case
$$dQ(\xi) = \frac{(\det(C))^{\frac{1}{2}}}{(2\pi)^{\frac{1}{2}}} e^{-\frac{1}{2}\langle (\xi - x^0), C(\xi - x^0)\rangle} d\xi \tag{6.57}$$
where C is a symmetric positive definite matrix and x^0 is given. Now the components of ξ are not independent with respect to dQ.

6.5.5.2 The Laplace reference measure

Consider now the following *a priori* measure

$$dQ = \prod \left(\frac{\mu_j}{2}\right) \exp(-\sum \mu_j |\xi_j - x^0|) d\xi$$

consisting of a product of bilateral exponential probability laws.
Now

$$z(t) = e^{-\langle t, x^0 \rangle} \prod_{j=1}^{n} \frac{\mu_j^2}{\mu_j^2 - t_j^2}$$

(verify it!) and $\mathcal{D}(A, Q) = \{\lambda \in \mathbb{R}^n : -\mu_j < (A^*\lambda)_j < \mu_j\}$.
This time the λ^* minimizing

$$\Sigma(\lambda) = \sum \ln\left(\frac{1}{\mu_j^2 - (A^*\lambda)_j^2}\right) + \langle (y - Ax^0), \lambda \rangle$$

has to satisfy, for each $j = 1, ..., n$

$$2 \sum_j \frac{A_{ij}(A^*\lambda)_j}{\mu_j^2 - (A^*\lambda)_j^2} = (Ax^0 - y)_i.$$

Now, **assume** that A^*A is invertible. Then the last identity implies that the minimizing λ^* satisfies

$$\frac{2(A^*\lambda)_j}{\mu_j^2 - (A^*\lambda)_j^2} = (A^*A)^{-1} A^*(Ax^0 - y)_j$$

for each j. Note that the right-hand side equals $x^0 - (A^*A)^{-1} Ay$.
Since $-\frac{\partial \ln(z(t))}{\partial t_j} = x^0 - \frac{2t_j}{\mu_j^2 - t_j^2}$, and when we evaluate in $t = A^*\lambda^*$ we obtain

$$x^* = (A^*A)^{-1} Ay.$$

Unexpected huh?. Thus the important fact here is that

$$x_j^* = x_j^0 - \frac{2(A^*\lambda^*)_j}{\mu_j^2 - (A^*\lambda^*)_j^2}$$

still provides a solution to $Ax = y$ even when A^*A is not invertible, provided λ^* minimizing $S(\lambda)$ exists. Again the role of x^0 is apparent.

6.6 Linear systems with measurement noise

Consider the following variation in our opening theme
$$y = Ax + n, \qquad x \in [a,b]^n \qquad (6.58)$$
where as at the beginning A is an $m \times n$ matrix, x is an unknown vector in $\mathbb{R}^n$ which we want to satisfy the constraint $x \in [a,b]^n$. The observed or measured vector y is in $\mathbb{R}^m$. The vector $n \in \mathbb{R}^m$ denotes a particular realization of the measurement noise, which we want to estimate as well. The statistical nature of the noise is known, and we shall suppose that it is modeled by random variable η distributed like an $N(0, \Sigma)$, where Σ is a diagonal matrix with entries σ_j^2, for $j = 1, ..., m$.

To apply the extended version of MEM as developed in Section 5.7, and estimate both the signal and the noise, we first recast the problem as: Solve
$$y = \hat{A}z = [A\ I]\begin{pmatrix} x \\ n \end{pmatrix}; \quad z \in [a,b] \times \mathbb{R}^n. \qquad (6.59)$$
We know that to solve this constrained problem using MEM, we have to begin by defining $\Omega = [a,b]^n \times \mathbb{R}^m$, with $\mathcal{F}$ being the Borel subsets of Ω. As reference measure we shall consider
$$dQ(\xi, \eta) = \prod_{j=1}^{n} [p\delta_a(d\xi_j) + q\delta_b(d\xi_j)] \prod_{i=1}^{m} \frac{e^{-\eta^2/2\sigma_i^2} d\eta_i}{2\sqrt{\pi}\sigma_i}.$$
Now we have to search for a measure $dP(\xi, \eta) = \rho(\xi, \eta)dQ(\xi, \eta)$ on Ω such that
$$y = E_P[\hat{A}\zeta] = AE_P[\xi] + E_P[\eta]$$
where $\zeta = (\xi, \eta)^t$ denotes the (column) vector of coordinates on Ω. The maxentropic P^* has density
$$\rho(\zeta) = Z(\lambda^*)^{-1} e^{(-<\lambda^*, \hat{A}\zeta>)} = Z(\lambda^*)^{-1} e^{-<\lambda^*, A\xi> - <\lambda^*, \eta>}$$
with respect to dQ. To repeat ourselves, $Z(\lambda)$ is given by
$$Z(\lambda) = \int_\Omega e^{-<\lambda^*, \hat{A}\zeta>} dQ(\zeta) = \prod_{j=1}^{n} [pe^{-(A^t\lambda)_j a} + qe^{-(A^t\lambda)_j b}] \prod_{i=1}^{m} e^{\lambda_j^2 \sigma_j/2}.$$

To find λ^* we have to minimize $\Sigma(\lambda) = \ln Z(\lambda) + <\lambda, y>$. Once the λ^* is at hand, the reconstructed signal and noise are given by
$$\begin{aligned} x_j^* &= \frac{e^{-(A^t\lambda^*)_j a} + bqe^{-(A^t\lambda^*)_j b}}{pe^{-(A^t\lambda^*)_j a} + qe^{-(A^t\lambda^*)_j b}} \\ n_i^* &= -\lambda_i^* \sigma_i^*. \end{aligned} \qquad (6.60)$$

This is a good place to end this chapter. Numerical comparisons between the different reconstructions are described in a separate chapter below.

6.7 Bibliographical comments and references

The nice duality approach in Section 6.1 is borrowed almost verbatim from [1]-[6]. For issues related to quadratic and nonlinear, constrained an unconstrained optimization, consult with [7]-[11]. After these notes were completed, [12] was published. A worthwhile reference complementing the work by the authors cited in Chapter 5.

References

[1] Strang, G. *"Duality in the classroom"*. Am. Math. Monthly, Vol. 91 (1984) pp. 250-254.

[2] Metha, M. I. *"Matrix Theory"*. H.P.C., Delhi, 1977.

[3] Aubin, J. P. *"Optima and Equilibria: an Introduction to Nonlinear Analysis"*. Springer-Verlag, New York, 1993.

[4] Brezis, H. *"Analyse Fonctionelle"*, Mason Editeur, Paris, 1973.

[5] Deutsch, F. *"The method of alternating orthogonal projections"*. in Singh, S.P. (ed.) *"Approximation Theory Spline Functions and Applications"*, Kluwer Acad. Pubs., Dordrecht (1992).

[6] Franklin, J. *"Methods of Mathematical Economics"*, Springer-Verlag, Berlin, 1980.

[7] Dax, A. *"The relationship between theorems of the alternative, least norm problems, steepest descent directions and degeneracy: A review"*. Annals of Operations research. Vol. 46 (1993) pp. 11-66.

[8] Simon, C. *"Scalar and vector maximization: Calculus Techniques Economics"*, Reiter, S (ed.), MAA Studies in Mathematics, Vol. 25, MAA, 1986.

[9] Hadley, G. *"Nonlinear and Dynamic Programming"*. Addison-Wesley, Reading, Massachusetts, 1964.

[10] Bertzekas, D.P. *"Nonlinear Programming"*. Athena Scientific, Belmont, M.A., 1995.

[11] Smale, S and Hirsch, M. *"Differential Equations, Dynamical Systems and Linear Algebra."* Acad. Press, N. Y., 1979.

[12] Boerwein, J. M. and Lewis, A.S. *"Convex Analysis and Optimization,* CMS Books in Mathematics, Spriger Verlag, New York, 2000.

Chapter 7

Some simple numerical examples and moment problems

In this section we exemplify some of the issues discussed in Chapter 6 in the context of a specific problem: that of reconstructing a continuous function from its moments.

This is a rather well studied problem, but there are a few things we may add to it.

7.1 The density of the Earth

This was the situation considered by Rietsch in [5-1], where the setup was established. From astronomical measurements both the mass and the moment of inertia of the Earth about its axis of rotation are known.

If we assume the Earth to be spherically symmetric then the relation between the data and the unknown is contained in

$$\tfrac{4\pi}{3} \int_0^R \rho(r) r^2 dr = M$$

$$\tfrac{8\pi}{3} \int_0^R \rho(r) r^4 dr = I.$$

If we rescale $\frac{r}{R} = t$ and set $\frac{3M}{4\pi R^3} = m_1$ and $\frac{3I}{8\pi R^5} = m_2$, we shall reconstruct

$$\int_0^1 t^{2i} \hat{\rho}(t) dt = m_i$$

with $m_1 = 1835.2$ and $m_2 = 909.8$ (in appropriate units), and where $\hat{\rho}(t) = \rho(tR)$.

Since we are going to deal with exponentials, it is convenient to rescale further and set $y_i = \dfrac{m_i}{10^4}$ and $f(t) = \dfrac{\hat{\rho}(t)}{10^4}$. When we come to plot $\hat{\rho}(t)$ the factor 10^4 will be part of the scale.

7.1.1 Solution by the standard $L_2[0,1]$ techniques

If we were to apply the standard L_2 techniques, we would set $f(t) = x_1 t^2 + x_2 t^4$ and search for x_1, x_2 such that

$$\int_0^1 t^{2i} f(t) dt = y_i \tag{7.1}$$

holds. This leads to a system

$$\hat{A}x = \begin{pmatrix} \frac{1}{5} & \frac{1}{7} \\ \frac{1}{7} & \frac{1}{9} \end{pmatrix} \begin{pmatrix} x_1 \\ x_2 \end{pmatrix} = \begin{pmatrix} 0.18322 \\ 0.09098 \end{pmatrix} = y \tag{7.2}$$

and when we solve for x, plug in $f(t) = x_1 t^2 + x_2 t^4$ and plot we obtain figure 7.1 The solution happens to be negative near the surface of the Earth. Since the density of the Earth has to be positive, we have to search for a different method.

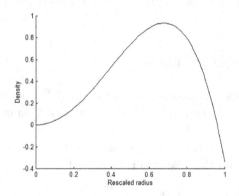

Fig. 7.1 Standard L_2 solution to (7.1).

If we say, well let us not demand exact agreement with (7.2) but penalize large solutions a bit, i.e., let us minimize $F(x) = \frac{1}{2} \left\| \hat{A}x - y \right\|^2 + \frac{\lambda}{2} \|x\|^2$ for several values of λ.

The results are displayed in figure 7.2 for various values of λ.

Note that for λ large, a positive solution is obtained, but for $\lambda < 0.0000133$, the solution becomes negative. This is natural for (7.2) has an exact solution and a small perturbation cannot destroy its features.

It should also be clear that searching for reconstructions with $x_1 \geq 0$ and $x_2 \geq 0$ will produce featureless solutions.

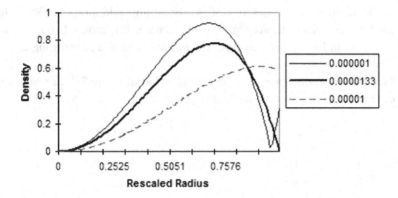

Fig. 7.2 Standard L_2 with penalization.

7.1.2 Piecewise approximations in $L_2([0,1])$

Let us now attempt a finer reconstruction: Approximate the continuous function in (7.1) by the simple function

$$f(t) = \sum_{j=1}^{n} x_j I_{[\frac{j-1}{n}, \frac{j}{n}]}(t)$$

where $n = 100$, for example. When such approximation is inserted in (7.1) we obtain

$$\sum_{j=1}^{n} A_{ij} x_j = y_i, \quad i = 1, 2, \tag{7.3}$$

where $A_{ij} = \int_0^1 t^{2i} I_{[\frac{j-1}{n}, \frac{j}{n}]}(t) dt = \frac{1}{2i+1}\left[\left(\frac{j}{n}\right)^{2i+1} - \left(\frac{j-1}{n}\right)^{2i+1}\right]$. In this version we can translate physical bounds for $\rho(r)$ onto $f(t)$. Certainly, the density of the Earth cannot be less than zero and it cannot be larger than that of the heaviest element. Thus it makes sense to require that $a \leq x_j \leq b$ for appropriate a and b.

In (7.3) we have (much) more unknowns than equations. To solve it we proceed to minimize penalized error given by $F(x) = \frac{1}{2}\|Ax - y\|^2 + \frac{\lambda}{2}\|x\|^2$. Here the first norm is in $\mathbb{R}^2$ and the second is in $\mathbb{R}^n$.

As described in Chapter 6, the best way to solve (7.3) is to use an iterative or gradient method. This is justified since $F(x)$ is convex. Denoting by x^n the n-th iterate in the rule

$$x^{n+1} = x^n - s\left\{A^*(Ax^n - y) + \lambda x^n\right\} \tag{7.4}$$

where A^* denotes the transpose of A, x^0 is any odd point in $\mathbb{R}^n$. The parameter s describes the step size in this Newton-Raphson scheme and we choose it to diminish as we get near the minimum of $F(x)$ according to the Barzilai-Borwein suggestion.

This would produce a reconstruction like that of figure 7.3, in which the reconstruction error is given by $\|Ax - y\|_2$.

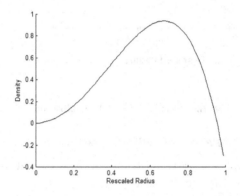

Fig. 7.3 Reconstruction with $\lambda = 10^{-7}$, $\varepsilon = 6.0063 \times 10^{-5}$.

When we decide to project onto the set of constraints, we proceed as described in Section 6.4 of Chapter 6 and iterate

$$x^{n+1} = P\{(1 - \lambda s)x^n + sA^*(y - Ax^n)\} \tag{7.5}$$

where the projection operator P is given by

$$(P(x))_j = \begin{cases} a & \text{if } x_j < a \\ x_j & \text{if } a \leq x_j \leq b \\ b & \text{if } b < x_j \end{cases}$$

In figures 7.4 and 7.5 we list some reconstructions. Different answers are obtained for different λ's and different pairs of constraints a, b.

7.1.3 Linear programming approach

Consider (7.3) as the following linear programming problem: Find $x^* \in \mathbb{R}^n$ such that

$$x^* = \arg\min \left\{ \sum_{i=1}^{n} x_i : Ax = y,\ 0 \leq a \leq x_j \leq b,\ \text{for all } j \right\}.$$

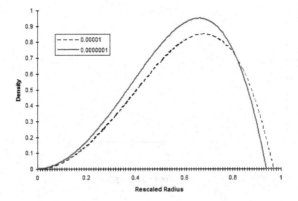

Fig. 7.4 Penalized least square reconstruction. Constrained $a = 0$, $b = 1$; parameters $\lambda = 10^{-5}$, 10^{-7}. Reconstruction errors $\varepsilon = 5.23 \times 10^{-3}$, 6.39×10^{-5}.

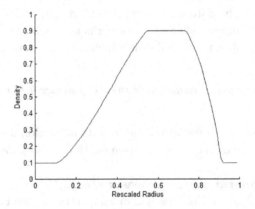

Fig. 7.5 Penalized least square reconstruction. Constrained $a = 0.1$, $b = 0.9$; parameters $\lambda = 10^{-5}$, 10^{-7}, $\varepsilon = 7.13 \times 10^{-5}$.

In other words, find an x^* satisfying (7.3), the constraints and which minimizes the "cost" objective $\hat{c}(x) = \sum_{j=1}^{n} x_j$.

The results are displayed in figure 7.6.

We shall reconsider this example at the end of Chapter 9 from the point of view of MEM. At this point the choice or arbitrariness of set up is apparent.

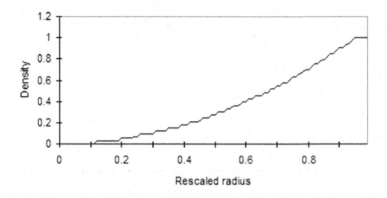

Fig. 7.6 Reconstructing using linear programming.

This example also points out another important issue. Since the actual density producing the data is unknown, we need "super criteria" (physical criteria) to select among different frameworks in which to set up and solve a given inverse problem and to decide which solutions are reasonable.

7.1.4 *Maxentropic reconstructions: Influence of* a priori *data*

In this section we are going to use the results of Section 6.5.1, and examine the dependence of the reconstruction on the *a priori* measure and the bounds.

Case 1. Uniform *a priori* measure on $[a, b]$.

As mentioned above this case was developed in Section 6.5.1. There we saw in (6.40) that

$$x_j^* = \left. \frac{ae^{-at_j^*} - be^{-bt_j^*}}{e^{-at_j^*} - e^{-bt_j^*}} - \frac{1}{t_j^*} \right|_{t^* = (A^*\lambda^*)_j}$$

where λ^* is to be found minimizing

$$S(\lambda) = \sum_j \ln \left\{ \frac{e^{-a(A^*\lambda)_j} - e^{-b(A^*\lambda)_j}}{(b-a)(A^*\lambda)_j} \right\} + \lambda_1 y_1 + \lambda_2 y_2$$

with λ ranging over $\mathbb{R}^2$. This minimization is done by walking against the gradient, with a self-adjusting step.

The important facts to note are

i) There are values of (a,b) for which there are no reconstructions, i.e., there does not exist λ^* minimizing $S(\lambda)$.
ii) The results of the reconstruction can depend drastically on the values of (a,b).

In figures 7.7 and 7.8 we show several reconstructions for different set of values of (a,b). In each case we list the error of the reconstruction as $\sum_{i=1}^{2} |y_i^* - y_i|$, where $y_i^* = \sum_{j=1}^{n} A_{ij} x_j^*$.

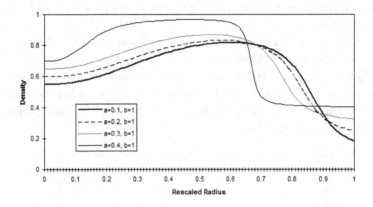

Fig. 7.7 Maxentropic reconstruction with uniform *a priori* on different $[a,b]$.

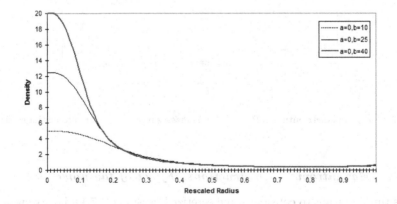

Fig. 7.8 Maxentropic reconstruction with uniform *a priori* on different $[a,b]$.

Case 2. Bernoulli *a priori* measure on $[a,b]$.

This case was worked out in Section 6.5.1 as well. There we saw for Bernoulli's *a priori* (6.34), the corresponding reconstruction is given by

$$x_j^* = \left.\frac{ape^{-at_j^*} - bqe^{-bt_j^*}}{pe^{-at_j^*} - qe^{-bt_j^*}}\right|_{t^*=(A^*\lambda^*)_j}$$

where now the λ^* are obtained by minimizing the function

$$S(\lambda) = \sum_j \ln\left\{pe^{-a(A^*\lambda)_j} - qe^{-b(A^*\lambda)_j}\right\} + \lambda_1 y_1 + \lambda_2 y_2.$$

Now besides the coordinates $a < b$ for possible values of the reconstructed function, we can vary p and q, the *a priori* probabilities for x_j to assume these values.

The moral is again the same. The choice of *a priori* data influences the reconstructions that we obtain.

In figures 7.9 and 7.10 we show several reconstructions in which the parameters a, b, p, q are varied. Again you see that when no detailed information about what to expect is at hand, we must at least explore as many alternatives as possible.

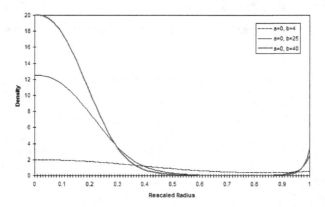

Fig. 7.9 *A priori* distribution of Bernoulli type masses $p = q = \frac{1}{2}$ and different intervals.

7.1.5 Maxentropic reconstructions: Effect of the noise

This time we want to consider the discretized version of (7.1), i.e., we want to consider solving (7.3) but we assume that the measurements of y_i are

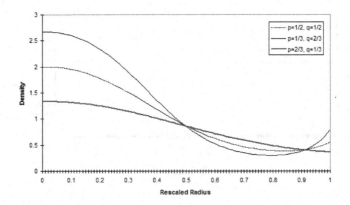

Fig. 7.10 A priori distribution of Bernoulli type on a fixed interval but different masses at the ends.

contaminated with noise. That is

$$y_i = \sum A_{ij} x_j + n_i \tag{7.6}$$

and we assume that it is reasonable to propose that noise is Gaussian, mean free with variance σ_i. We assume for simplicity that the variances are equal, i.e. $\sigma_1 = \sigma_2 = \sigma$.

As we did in Section 5.6, we shall call a reconstruction x admissible if

$$Ax \in B_\sigma(y, T) \tag{7.7}$$

where $B_\sigma(y,T) = \left\{ \eta \in \mathbb{R}^2 : \dfrac{1}{\sigma} \|\eta - y\| \leq T \right\}$. Also, do not forget that we want $a \leq x_j \leq b$ for all j. We shall again consider the cases of uniform and Bernoulli *a priori* measures on $[a, b]$.

Case 1. Uniform *a priori* measure on $[a, b]$.

As in the case treated under the same heading in Section 7.1, the reconstruction is once more given by

$$x_j^* = \left. \frac{ae^{-at_j^*} - be^{-bt_j^*}}{e^{-at_j^*} - e^{-bt_j^*}} - \frac{1}{t_j^*} \right|_{t_j^* = (A^*\lambda^*)_j}$$

where this time λ^* is obtained by minimizing

$$S(\lambda) = \sum_j \ln \left\{ \frac{ae^{-a(A^*\lambda)_j} - be^{-b(A^*\lambda)_j}}{(b-a)(A^*\lambda)_j} \right\} + \lambda_1 y_1 + \lambda_2 y_2 + \sigma T \|\lambda\|$$

with respect to $\lambda \in \mathbb{R}^2$. Similar conclusions are obtained, and different reconstructions are obtained varying a, b and σ, T.

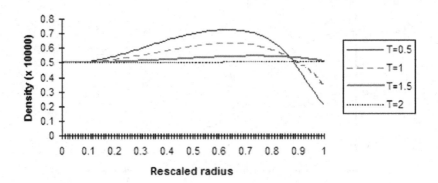

Fig. 7.11 Reconstructions with uniform *a priori* distribution on $[0,1]$, fixed $\sigma_1 = \sigma_2 = 0.01$ and varying T.

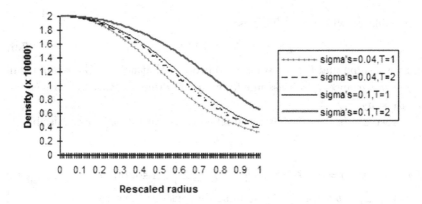

Fig. 7.12 Reconstructions with uniform *a priori* distribution on $[0,4]$, varying σs and Ts.

Some results are presented in figures 7.11 and 7.12.

Case 2. Bernoulli *a priori* measure on $[a,b]$.

Again, this time the corresponding reconstruction is given by

$$x_j^* = \left.\frac{ape^{-at_j^*} - bqe^{-bt_j^*}}{pe^{-at_j^*} - qe^{-bt_j^*}}\right|_{t_j^* = (A^*\lambda^*)_j}$$

where we must minimize

$$S(\lambda) = \sum_j \ln\left\{pe^{-a(A^*\lambda)_j} - qe^{-b(A^*\lambda)_j}\right\} + \lambda_1 y_1 + \lambda_2 y_2 + \sigma T \|\lambda\|$$

to obtain λ^*. The conclusions follow from some reconstructions obtained in figures 7.13 and 7.14 for different a, b, p, q, σ and T.

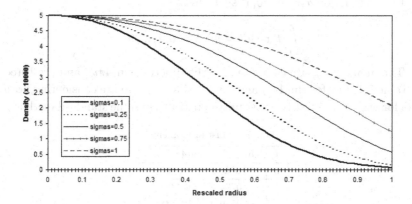

Fig. 7.13   Reconstructions with fixed Bernoulli distribution on $[0, 10]$ with $p = q = \frac{1}{2}$, $T = 1$ and $\sigma_1 = \sigma_2$ varying.

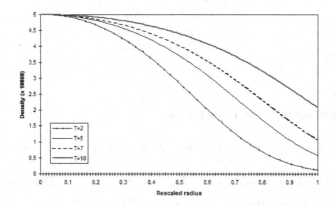

Fig. 7.14   Reconstructions under Bernoulli *a priori* distribution on $[0, 10]$ with $p = 1 = \frac{1}{2}$, fixed $\sigma_1 = \sigma_2 = 0.1$, but T varying as indicated.

7.2   A test case

In Section 7.1 we consider a reconstruction problem, with actual data, in which we have no way of knowing how "far off" the reconstruction is from the actual function that produces the data.

This section is devoted to a complementary problem: that of testing the tools. For that we consider the following moment problem.

Find a function $x(t) \in \mathcal{C}[0,1]$ such that

$$\int_0^1 t^i x(t) dt = y_i \quad i = 0, 1, ..., 9. \tag{7.8}$$

The moments y_i are computed for the function $\sin(4\pi t)$, and are listed in Table 7.1, but for $\sin(4\pi t)$ and for $1 + \sin(4\pi t)$, We are considering both lists for we want to test the positivity constraint in some standard methods.

Table 7.1 List of moments.

i	$\sin(4\pi t)$	$\sin(4\pi t) + 1$
0	0.0000	1.0000
1	−0.0796	0.4204
2	−0.0796	0.2538
3	−0.0766	0.1734
4	−0.0735	0.1265
5	−0.0699	0.0968
6	−0.0656	0.0772
7	−0.0610	0.0640
8	−0.0563	0.0548
9	−0.0518	0.0482

7.2.1 Standard $L_2[0,1]$ technique

As in Section 7.1.1, we may assume that $x(t) = \sum_{j=0}^{9} \xi_i t^i$, substitute in (7.8) to obtain the system $\hat{A}_{ij} = \int_0^1 t^{i+j} dt = \dfrac{1}{i+j+1}$. This amounts to projecting $x(t)$ onto the space spanned by $\{t^i : i = 0, 1, ..., 9\}$ and solving the corresponding problem there. Again, and depending on whether A is invertible and well conditioned or not we can proceed by either of the following routes

a) If A is invertible and well conditioned, i.e., $\left\|\hat{A}\right\| \cdot \left\|\hat{A}^{-1}\right\|$ is not too large (compared to 1) where $\left\|\hat{A}\right\| = \sqrt{tr(\hat{A}\hat{A})}$, then

$$\xi = \hat{A}^{-1} y.$$

b) When A is invertible, but has small eigenvalues we may regularize and obtain
$$\xi = (\lambda + \hat{A})^{-1} y.$$

c) When A is not invertible, we proceed as above and minimize
$$\frac{1}{2}\left\|\hat{A}\xi - y\right\|^2 + \frac{\lambda}{2}\|\xi\|^2.$$

Now since $\hat{A}$ is a square 10×10 matrix, both norms are defined on the space $\mathbb{R}^{10}$ and case (c) comprises both previous cases. As we have seen, we do not have to invert matrices but iterate:
$$\xi^0 = \xi_0, \quad \xi^{n+1} = \xi^n - s\left\{\hat{A}^*(A\xi^n - y) + \lambda \xi^n\right\}$$

and $\hat{A}^* = \hat{A}$ this time.

In figures 7.15 and 7.16 we present the reconstructions obtained, i.e., the plots of $\sum_{j=0}^{9} \xi_j t^j$, when the coefficients ξ are obtained for each list of moments and for different values of λ in each case.

The reconstruction error $\varepsilon = \sqrt{\sum (y_i - y_i^*)^2}$ is listed.

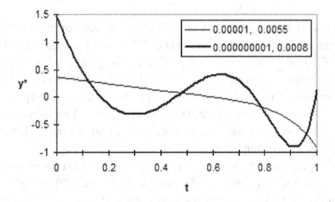

Fig. 7.15 Reconstructions from first list of moments. Different λs and its respectively error.

7.2.2 Discretized $L_2[0,1]$ approach

Here we proceed as in Section 7.1.2, but instead of projecting onto the space generated by $\{1, t, ..., t^9\}$ we decide to approximate $x(t)$ by $\sum_{j=1}^{N} x_j I_{[\frac{j-1}{N}, \frac{j}{N}]}(t)$

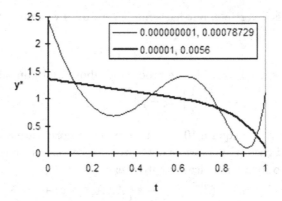

Fig. 7.16 Reconstructions from second list of moments. Different λs and its respectively error.

where $N = 100$ say. Substituting in (7.8) we end up with the system

$$\sum_{j=1}^{N} A_{ij} x_j = y_i \quad i = 0, 1, ..., 9 \qquad (7.9)$$

where A_{ij} is as below (7.3) and now there is no hope of inverting A. We therefore proceed as in case (c) of Section 7.2.1, and minimize the objective function $F(x) = \frac{1}{2} \|Ax - y\|_k^2 + \frac{\lambda}{2} \|x\|_N^2$, where $\|...\|_k$ and $\|...\|_N$ are the Euclidean norms on corresponding to $k = 10$, $N = 100$ respectively.

Furthermore, the projection technique allows us to incorporate constraints when information about them is available. The minimization is carried out by iteration, and the projection is obtained by projecting iterates as indicated in Chapter 6.

In figures 7.17 and 7.18 we present the reconstruction for several values of λ for each list of moments. In figures 7.19 and 7.20 we present the reconstruction with the constraints incorporated by means of projections.

7.2.3 Maxentropic reconstructions: Influence of a priori data

We shall now consider problem (7.8) in its discretized form (7.9), and we shall repeat the routine carried out in Section 7.1.4 Compared to the problem treated in Section 7.1, the difference lies in the fact that this time we know in advance what the functions producing the data listed as Table 7.1 look like.

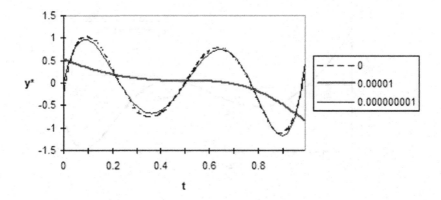

Fig. 7.17 Reconstructions for first list moments and several λ.

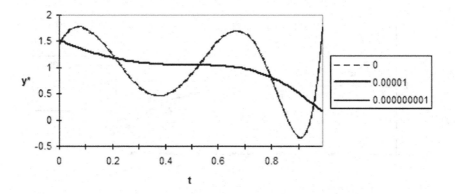

Fig. 7.18 Reconstructions for second list of moments and several λ.

Case 1. Uniform *a priori* measure on $[a, b]$.

We shall consider the interval $[a, b]$ to be either of $[-0.8, 0.8]$, $[-1, 1]$ or $[-3, 3]$, and carry out the reconstruction procedure outlined in Section 6.5.1 There we saw that the maxentropic solution to (7.9) is provided by

$$x_j^* = \left. \frac{ae^{-at_j^*} - be^{-bt_j^*}}{e^{-at_j^*} - e^{-bt_j^*}} - \frac{1}{t_j^*} \right|_{t^* = (A^*\lambda^*)_j}$$

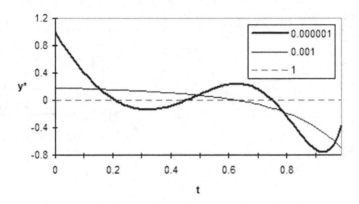

Fig. 7.19 Reconstructions for first list of moments and several λ, projected onto constraint space.

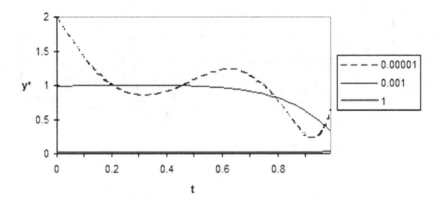

Fig. 7.20 Reconstructions for second list of moments and several λ, projected onto constrained space.

where λ^* is to be found minimizing

$$S(\lambda) = \sum_j \ln\left\{\frac{e^{-a(A^*\lambda)_j} - e^{-b(A^*\lambda)_j}}{(b-a)(A^*\lambda)_j}\right\} + \sum_{i=0}^{9} \lambda_i y_i$$

for λ ranging over $\mathbb{R}^{10}$.

Once λ^* is found, we obtain figure 7.21 in which we plot the reconstruction for the different intervals. We compute the reconstruction error as $\varepsilon = \sqrt{\sum (y_i - y_i^*)^2}$ and is listed.

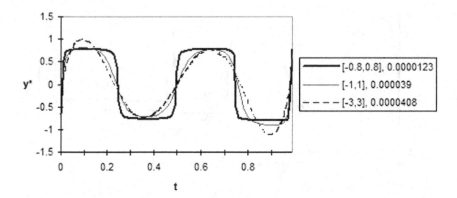

Fig. 7.21 Maxentropic reconstructions for different constraints and uniform *a priori* measure.

Case 2. Bernoulli *a priori* distribution on $[a, b]$.

Here we implement the same procedure as in Section 7.1.4, but now for 10 moments. The basics were spelt out in Section 6.5.1. For the Bernoulli *a priori* the maxentropic reconstruction is

$$x_j^* = \left. \frac{ape^{-at_j^*} - bqe^{-bt_j^*}}{pe^{-at_j^*} - qe^{-bt_j^*}} \right|_{t_j^* = (A^*\lambda^*)_j}$$

and λ^* is to be obtained minimizing

$$S(\lambda) = \sum_{j=1}^{100} \ln\left\{pe^{-a(A^*\lambda)_j} - qe^{-b(A^*\lambda)_j}\right\} + \sum_{i=0}^{9} \lambda_i y_i.$$

The reconstructions are displayed in figures 7.22 and 7.23, where the dependence on $[a, b]$ and p, q is illustrated. The reconstruction errors are computed by $\varepsilon = \sqrt{\sum (y_i - y_i^*)^2}$.

7.2.4 Reconstruction by means of cubic splines

Here we transform the standard moment problem: Find $x(t)$ continuous in $[0,1]$ such that

$$\int_0^1 t^i x(t) dt = y_i + \varepsilon_i$$

into an algebraic problem: Find a vector $x \in \mathbb{R}^n$ such that

$$\sum A_{ij} x_j = y_i + \varepsilon_i$$

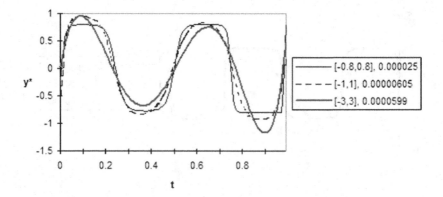

Fig. 7.22 Maxentropic reconstruction for different constrained and Bernoulli *a priori* measure with $p = q = \frac{1}{2}$.

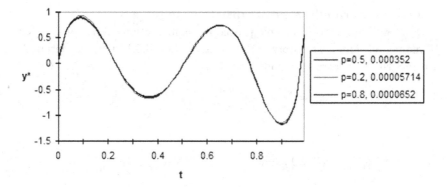

Fig. 7.23 Maxentropic reconstruction for $[a, b] = [-2, 2]$ and $p = 0.5, 0.2, 0.8$ with $q = 1 - p$.

as follows. First we partition $[0,1]$ into an appropriate large number n of subintervals, and denoting by $x_j = x(\frac{j-1}{n})$, $j = 1, ..., n+1$, the (unknown) values of the function there; we interpolate among those values by means of a cubic spline. (See the micro-appendix at the end of the chapter.) By means of this interpolant, we transform the integral equation into the algebraic problem.

One then uses the maxentropic methods to deal with that problem, and once the x_j^* is found, the maxentropic reconstruction $x^*(t)$ is obtained by interpolating among the x_j^* by means of a cubic spline.

Below we exemplify the different issues involved in the process for the data generated by computing the moments of $x_0 = \sin \omega t$ for $\omega = 1, \pi, 2\pi, 4\pi$, with and without the noise terms. In the figures presented below we also list the reconstruction error computed according to

$$\varepsilon = \|Ax^* - y\| = \left(\sum ((Ax^*)_i - y_i)^2\right)^{\frac{1}{2}}.$$

In the following figures we present reconstructions for $\omega = 1$ (figure 7.24) with $m = 2$, and $\omega = \pi$ (figures 7.25 and 7.26) with $m = 2$ and 4 respectively. Note that insufficient data yields a poor reconstruction error! We explicit how much the *a priori* information affects the results. Note that symmetric *a priori* intervals yield better reconstructions.

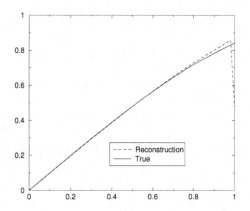

Fig. 7.24 Maxentropic reconstruction with cubic splines from $m = 2$, $w = 1$ with error $\varepsilon = 1.9830 \times 10^{-10}$.

For figures 7.27, 7.28, 7.29 and 7.30 we divided both the data vector and the matrix A by 10 to obtain smaller terms in the exponentials. This trick improves reconstructions most of the time, but not always.

Whereas the data for figures was not contaminated by error, for the next reconstructions (from figure 7.31 to figure 7.36), carried out according to the procedure described in Section 5.6, we added independent, N(0,σ) random variables ε_i, i = number of momenta and $\sigma = 0.01$ or 0.1.

In these figures we present reconstructions for different values of ω, m, σ, tolerance T and *a priori* reconstructions interval. In all cases we list the reconstruction error.

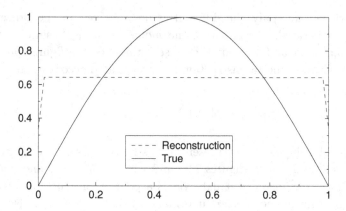

Fig. 7.25 Maxentropic reconstruction with cubic splines from $m = 2$, $w = \pi$ with error $\varepsilon = 1.8340 \times 10^{-8}$.

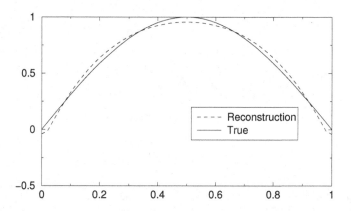

Fig. 7.26 Maxentropic reconstruction with cubic splines from $m = 4$, $w = \pi$ with error $\varepsilon = 4.5339 \times 10^{-8}$.

We have also attempted reconstructions using techniques described in Section 5.7. For that we perturbed the "exact" data with errors distributed according to an $N(\mu, \sigma)$ law and then attempted reconstructions assuming an *a priori* distribution for the noise variable with the exact μ and with some other values of μ.

This time, for some reason the algorithm would not run unless we assumed rather large reconstruction intervals. We present two reconstructions in the next figures to exemplify the issues involved.

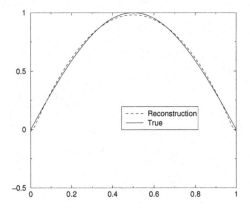

Fig. 7.27 Maxentropic reconstruction with cubic splines from $m = 4$, $w = \pi$ with error $\varepsilon = 7.2104 \times 10^{-9}$.

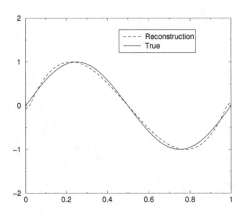

Fig. 7.28 Maxentropic reconstruction with cubic splines from $m = 4$, $w = 2\pi$ with error $\varepsilon = 5.2992 \times 10^{-8}$.

7.2.5 *Fourier versus cubic splines*

This section exemplifies several issues, the most important one being related to what happens when one insists on reconstructing the wrong type of object. We shall begin with the following data

$$\int_0^1 e_i(t)x_0(t)dt + \varepsilon_i = y_i \tag{7.10}$$

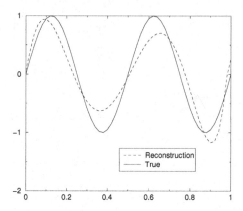

Fig. 7.29 Maxentropic reconstruction with cubic splines from $m = 7$, $w = 4\pi$ with error $\varepsilon = 6.2902 \times 10^{-6}$.

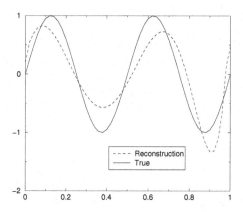

Fig. 7.30 Maxentropic reconstruction with cubic splines from $m = 10$, $w = 4\pi$ with error $\varepsilon = 9.9747 \times 10^{-6}$.

where $x_0(t)$ is the following step function

$$x_0(t) = \begin{cases} 0, & 0 \leq t < 0.25 \\ 0.8, & 0.25 \leq t < 0.5 \\ -0.7, & 0.5 \leq t < 0.75 \\ 0, & 0.75 \leq t \leq 1 \end{cases}$$

and where $e_i(t) = \sqrt{2}\sin(i\pi t)$ for $i = 1, ..., m$. We shall consider both noisy data and non-noisy data. There are (at least) two reconstruction methods readily available to us: First, the time honored Fourier or Hilbert space

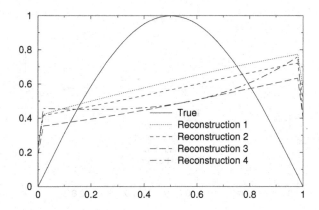

Fig. 7.31 Maxentropic reconstruction with cubic splines from $m = 4$, in $[-5, 5]$, $w = \pi$, $\sigma = 0.1$, $T_1 = 0.5$, $T_2 = 1$, $T_3 = 1.5$ and $T_4 = 2$ with error $\varepsilon_1 = 0.05$, $\varepsilon_2 = 0.10$, $\varepsilon_3 = 0.20$ and $\varepsilon_4 = 0.20$.

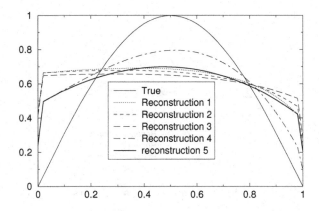

Fig. 7.32 Maxentropic reconstruction with cubic splines from $m = 4$, in $[-5, 5]$, $w = \pi$, $\sigma = 0.01$, $T_1 = 1.5$, $T_2 = 2$, $T_3 = 3$, $T_4 = 4$ and $T_5 = 5$ with the respective error $\varepsilon_1 = 0.020$, $\varepsilon_2 = 0.020$, $\varepsilon_3 = 0.030$, $\varepsilon_4 = 0.040$ and $\varepsilon_5 = 0.050$.

reconstruction method and the method that we developed in the previous section.

According to the first one, the best approximant (in the standard $L_2[0, 1]$-norm) is given by

$$\widetilde{x}(t) = \sum_{i=1}^{m} y_i e_i(t). \qquad (7.11)$$

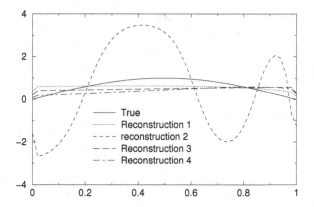

Fig. 7.33 Maxentropic reconstruction with cubic splines from $m = 8$, in $[-5, 5]$, $w = \pi$, $\sigma = 0.1$, $T_1 = 0.5$, $T_2 = 2$, $T_3 = 3$ and $T_4 = 4$ with the respective error $\varepsilon_1 = 0.20$, $\varepsilon_2 = 0.20$, $\varepsilon_3 = 0.30$ and $\varepsilon_4 = 0.40$.

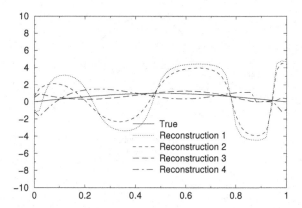

Fig. 7.34 Maxentropic reconstruction with cubic splines from $m = 8$, in $[-5, 5]$, $w = \pi$, $\sigma = 0.01$, $T_1 = 0.5$, $T_2 = 1$, $T_3 = 1.5$ and $T_4 = 2$ with the respective error $\varepsilon_1 = 0.0131$, $\varepsilon_2 = 0.0135$, $\varepsilon_3 = 0.020$ and $\varepsilon_4 = 0.020$.

The second one, would be to proceed as in the previous section to produce an $x_*(t)$ by means of maxentropic cubic spline interpolation. Below we compare the results of both approaches. The L_1-error between signal and reconstruction computed by

$$\delta = \frac{1}{50} \sum_{j=1}^{50} \left| x_0(\frac{j}{50}) - \widehat{x}(\frac{j}{50}) \right|$$

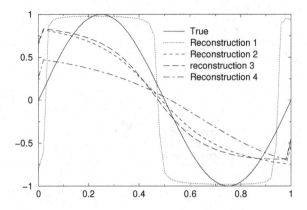

Fig. 7.35 Maxentropic reconstruction with cubic splines from $m = 8$, in $[-1, 1]$, $w = 2\pi$, $\sigma = 0.01$, $T_1 = 0.5$, $T_2 = 1$, $T_3 = 2$ and $T_4 = 3$ with the respective error $\varepsilon_1 = 0.0118$, $\varepsilon_2 = 0.01$, $\varepsilon_3 = 0.02$ and $\varepsilon_4 = 0.03$.

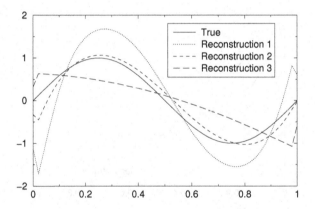

Fig. 7.36 Maxentropic reconstruction with cubic splines from $m = 8$, in $[-2, 2]$, $w = 2\pi$, $\sigma = 0.01$, $T_1 = 0.5$, $T_2 = 1$, $T_3 = 1.5$ and $T_4 = 4$ with the respective error $\varepsilon_1 = 0.005$, $\varepsilon_2 = 0.010$ and $\varepsilon_3 = 0.020$.

where $\widehat{x}(t)$ stands for either $\widetilde{x}(t)$ or $x_*(t)$, is listed. This measure of error is unavailable in practice since $x_0(t)$ is unknown. We display its rate reconstructions.

In the next figures we display the reconstructions for $m = 4$ and 8 respectively. The index 1 labels the maxentropic reconstruction and index 2 labels the L_2-reconstruction. The reconstruction error for the maxentropic method is listed as well.

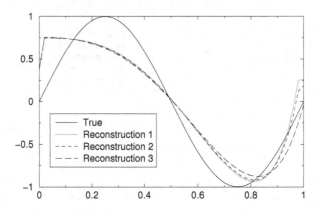

Fig. 7.37 Maxentropic reconstruction with cubic splines and method No. 2, in $[-25, 25]$, from $m = 12$, $w = 2\pi$, $\sigma = 0.01$, $\mu_0 = 0.01$, $\mu_1 = 0$, $\mu_2 = 0.005$ and $\mu_3 = 0.02$ with the respective error $\varepsilon_1 = 0.0274$, $\varepsilon_2 = 0.0326$, and $\varepsilon_3 = 0.745$.

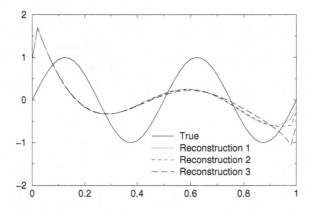

Fig. 7.38 Maxentropic reconstruction with cubic splines and method No. 2, in $[-80, 80]$, from $m = 12$, $w = 4\pi$, $\sigma = 0.01$, $\mu_0 = 0.01$, $\mu_1 = 0$, $\mu_2 = 0.005$ and $\mu_3 = 0.02$ with the respective error $\varepsilon_1 = 0.0246$, $\varepsilon_2 = 0.030$, and $\varepsilon_3 = 0.0734$.

When there is error in the data, and we do not filter it out for the L_2-reconstruction, but we do carry out the maxentropic reconstruction according to Section 5.6, the following is obtained. In both cases the noise was Gaussian and the correct model was assumed known. In the figures below, plots indexed by 1, 2 and 3 correspond to the maxentropic reconstruction with different tolerance and plot number 4 is the Hilbert space reconstruction given by (7.11).

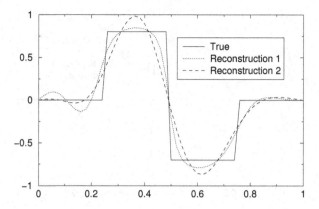

Fig. 7.39 Reconstructions from $m = 8$, the respective L_1 errors are $\sigma_1 = 0.5272$ and $\sigma_2 = 0.6786$ and the reconstruction error is $\varepsilon_1 = 6.1374 \times 10^{-8}$.

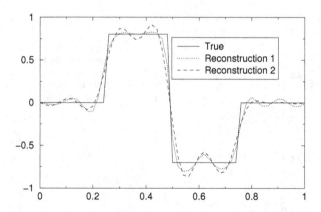

Fig. 7.40 Reconstructions from $m = 14$, the respective L_1 errors are $\sigma_1 = 0.3811$ and $\sigma_2 = 0.5654$ and the reconstruction error is $\varepsilon_1 = 8.2644 \times 10^{-8}$.

7.3 Standard maxentropic reconstruction

We shall now apply the standard maxentropic reconstruction due to Jaynes and briefly recalled in Section 5.7, to the data in the right column of Table 7.1. We cannot apply the method to the data in the first column for we know that such data was produced by a non-positive function.

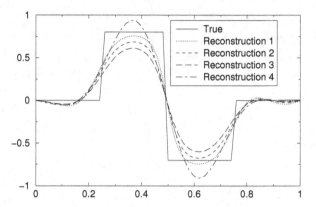

Fig. 7.41 Reconstructions from $m = 6$, $\sigma = 0.1$, $T_1 = 0.5$, $T_2 = 1$ and $T_3 = 1.5$ the respective L_1 errors are $\delta_1 = 0.6088$, $\delta_2 = 0.6269$, $\delta_3 = 0.6502$, and $\delta_4 = 0.6340$, and the reconstruction errors for maximum entropy method are respectively $\varepsilon_1 = 0.05$, $\varepsilon_2 = 0.1$, $\varepsilon_3 = 0.15$.

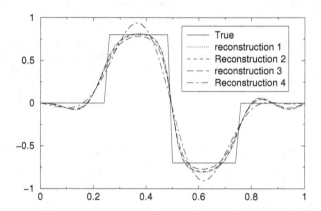

Fig. 7.42 Reconstructions from $m = 6$, $\sigma = 0.01$, $T_1 = 0.25$, $T_2 = 1$ and $T_3 = 3$ the respective L_1 errors are $\delta_1 = 0.5852$, $\delta_2 = 0.5895$, $\delta_3 = 0.5999$, and $\delta_4 = 0.6340$, and the reconstruction errors for maximum entropy method are respectively $\varepsilon_1 = 0.00255$, $\varepsilon_2 = 0.01$, $\varepsilon_3 = 0.03$.

According to the standard ME approach, the function to be reconstructed is

$$x^*(t) = \frac{\exp\left(-\sum_{i=1}^{9} \lambda_i^* t^i\right)}{Z(\lambda^*)} \tag{7.12}$$

where the normalization factor (or partition function) $Z(\lambda)$ is defined by

$$Z(\lambda) = \int_0^1 \exp\left(-\sum_{i=1}^{9} \lambda_i t^i\right) dt$$

for each $\lambda \in \mathbb{R}^9$. To find the right $\lambda^* \in \mathbb{R}^9$, appearing in (7.12) we just have to find the minimum of

$$S(\lambda) = \ln(Z(\lambda)) + \langle \lambda, y \rangle$$

which amounts to finding λ^* such that $x^*(t)$ given by (7.12) satisfies

$$\int_0^1 t^i x^*(t) dt = y_i \qquad (7.13)$$

for $i = 1, ..., 9$. Note that $x^*(t)$ is continuous, positive and a probability density. That is why we only require that moment equations be satisfied for $1 \leq i \leq 9$.

In figure 7.43 we display the reconstructions obtained when we consider 7, 8 or 9 moments as data. The relative reconstruction error is $\frac{\sum |y_i^* - y_i|}{\sum y_i} \simeq 0.92 \times 10^{-4}$ for nine moments.

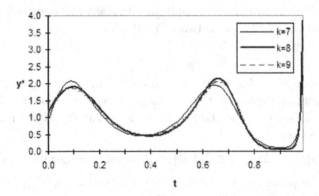

Fig. 7.43 Original function and three reconstructions using standard ME applied to 7, 8 or 9 givens moments.

Regardless of the fact that this scheme is applicable to positive functions only, we see that reconstructions are reasonable and discretization is not needed. The only handicap being that one cannot incorporate *a priori* bounds on the solution into the reconstruction scheme.

7.3.1 Existence and stability

There are issues related to the existence of the λ^* minimizing $S(\lambda)$. The first of them is easy to take care of in the present situation: the domain of $Z(\lambda)$ is the whole $\mathbb{R}^k$ ($k = 7, 8$ or 9). This so happens because of the boundedness of $[0, 1]$. Notice that $S(\lambda)$ is strictly convex on $\mathbb{R}^k$, and to find its minimum one would need to solve

$$-\nabla_\lambda \ln(Z(\lambda)) = y$$

(which is not strictly necessary to find the minimum, it suffices to walk against the gradient). For that it suffices to know whether the data vector is the range of $-\nabla_\lambda \ln(Z(\lambda))$, which in turn leads to finding the range of the mapping $A : \mathcal{C}[0, 1] \to \mathbb{R}^k$ defined by

$$x(t) \to A(x)_i = \int_0^1 t^i x^*(t) dt = y_i.$$

For this problem this is a rather small set (we shall recall some results about it below). For large k it is of size $\sim 2^{-k^2}$. This fact makes the reconstructions difficult when the data contains errors larger than the volume of the region, making it highly likely that it falls out of the region. See Section 7.4 for another way of looking at this issue.

We challenge the reader to apply the standard ME approach to find a distribution like (7.12) that satisfies (7.13) with

$$y_i = \frac{1}{2^i} \quad i = 1, 2, ..., 9.$$

This example is due to Marco Frontini of the Politecnico di Milano. Such moment vector falls out of the range of the map A described above.

We shall now describe the nature of the instability of the reconstructions. Let us set

$$y_i^+ = \sup\{y_i = (Ax)_i : x \in \mathcal{C}[0, 1]\}, \quad y_i^- = \inf\{y_i = (Ax)_i : x \in \mathcal{C}[0, 1]\}.$$

Let us keep the first $k - 1$ moments fixed and vary y_i in (y_k^-, y_k^+). Note that if we differentiate both sides of (7.13) with respect to y_k, we can write

$$\sum_j C^k(i, j) \frac{\partial \lambda_j^*}{\partial y_k} = -\frac{\partial y_i}{\partial y_k} = \delta_{ik}, \quad 1 \leq i \leq k \tag{7.14}$$

where $C^k(i, j) = -\left.\frac{\partial^2 \ln(Z(\lambda))}{\partial \lambda_i \partial \lambda_j}\right|_{\lambda^*}$. When solving for $\frac{\partial \lambda_k^*}{\partial y_k}$ we obtain

$$\frac{\partial \lambda_k^*}{\partial y_k} = -\frac{H_{2(k-1)}}{H_{2k}} \tag{7.15}$$

where for each m, the $\underline{H}_{2m}$ are the determinants

$$\underline{H}_0 = 1, \quad \underline{H}_2 = \begin{vmatrix} 1 & y_1 \\ y_1 & y_2 \end{vmatrix}, ..., \underline{H}_{2m} = \begin{vmatrix} 1 & y_1 & \cdots & y_m \\ y_1 & y_2 & \cdots & y_{m+1} \\ \vdots & & & \vdots \\ y_m & y_{m+1} & \cdots & y_{2m} \end{vmatrix}$$

and $C^k(i,j)$ is the matrix with elements $C^k(i,j) = y_{i+j} - y_i y_j$ for $1 \leq i, j \leq k$. To verify that $\det(C^k(i,j)) = \underline{H}_{2m}$ consider the matrix whose determinant is $\underline{H}_{2m}$ and subtract, from columns 2 to $m+1$, the first column multiplied by $y_1, ..., y_m$ respectively. The submatrix $C^k(i,j)$ should be clearly identifiable. Thus (7.15) becomes clear.

Since

$$\frac{\underline{H}_{2(k-1)}}{\underline{H}_{2k}} = \frac{1}{y_{2k} - y_{2k}^-} > \frac{1}{y_{2k}^+ - y_{2k}^-} \geq \frac{1}{2^{-2(2k-1)}}$$

we obtain $\left|\dfrac{\partial \lambda_k^*}{\partial y_k}\right| > 2^{2k-2}$, which amounts to our assertion about instability in the computation of λ_k^*.

Since $x^*(t)$ given by (7.12) is strictly positive, we can estimate $\delta \ln(x^*(t))$, when y_k changes by δy_k, by means of

$$\delta \ln(x^*(t)) \approx -\left(\sum_{j=i}^{k} \frac{\partial \lambda_j^*}{\partial y_k} t^j\right) \delta y_k.$$

It can be proved that if we write the solutions to (7.14) as $\dfrac{\underline{H}_{2k}^{(j)}}{\underline{H}_{2k}}$, then $\sum_{j=1}^{k} \underline{H}_{2k}^{(j)} t^j = P_k(t)$ where

$$P_k(t) = \begin{pmatrix} 1 & y_1 & \cdots & y_{k-1} & 1 \\ y_1 & y_2 & \cdots & y_k & t \\ \vdots & \vdots & & \vdots & \vdots \\ y_k & y_{k+1} & \cdots & y_{2k-1} & t^k \end{pmatrix}$$

and that $\int_0^1 [P_k(t)]^2 x^*(t) dt = \underline{H}_{2k} \, \underline{H}_{2(k-1)}$.

Therefore

$$\int_0^1 [\delta \ln(x^*(t))]^2 x^*(t) dt \approx \left(\frac{\delta y_k}{\underline{H}_{2k}}\right)^2 \int_0^1 [P_k(t)]^2 x^*(t) dt$$

$$= \frac{(\delta y_k)^2}{y_{2k} - y_{2k}^-} > \frac{(\delta y_k)^2}{y_{2k}^+ - y_{2k}^-} \geq (\delta y_k)^2 2^{2(2k-1)}.$$

Below we provide references to the literature in these topics.

7.3.2 Some convergence issues

In Section 5.5 we mentioned several references where the question of whether increasing the data, improves the approximation to the (unknown) solution of our inverse problem. Let us briefly describe some of that work.

Denote by $x_m^*(t)$ the ME reconstruction of $x(t)$ when the first m - moments are known.

Theorem 7.1. *For any $f \in \mathcal{C}[0,1]$, $x_m^*(t)dt$ converges weakly to $x(t)dt$, i.e.*

$$\int_0^1 x_m^*(t)f(t)dt \to \int_0^1 x(t)f(t)dt, \quad \text{as } m \to \infty.$$

Theorem 7.2. *The corresponding entropies $S_m(x^*)$ converge to the entropy $S(x)$ of the input density $x(t)$, i.e.*

$$-\int_0^1 x_m^*(t)\ln(x_m^*(t))dt \to -\int_0^1 x(t)\ln(x(t))dt.$$

If we invoke the inequality

$$S_m(x^*) - S(x) = \int_0^1 x(t)\ln\left(\frac{x(t)}{x_m^*(t)}\right)dt \geq \frac{1}{4}\left[\int_0^1 |x(t) - x_m(t)|\,dt\right]^2$$

when $x_m(t)$ and $x(t)$ have the same first m-moments, clearly Theorem 7.2 implies the convergence of $x_m(t)$ to $x(t)$ in $L_1([0,1])$.

When $x(t)$ is known to have some regularity, uniform convergence is obtained.

Theorem 7.3. *Assume that the density $x(t)$ is twice continuously differentiable, then $x_m(t)$ converges uniformly to $x(t)$.*

That is, as far as pinning down $x(t)$ this is as good as you can get. A few more related results can be seen in the references mentioned in Section 5.5. When considering the moment problem in a noncompact space, not only quite a bit of non-uniqueness creeps in, but not much is known. More results about these issues appear in [12].

7.4 Some remarks on moment problems

It seems that it was Stieltjes who first set up the problem of finding a positive, finite measure $p(dt)$ on $[0,\infty)$ such that

$$\int_0^\infty t^n p(dt) = y_n, \quad n = 0, 1, \ldots \tag{7.16}$$

where the sequence (of moments, a name apparently borrowed from Mechanics) y_n is given.

The corresponding problem on the whole line, i.e., that of finding a finite, positive measure on $(-\infty, \infty)$ such that

$$\int_{-\infty}^{\infty} t^n p(dt) = y_n, \quad n = 0, 1, \ldots \tag{7.17}$$

was studied by Hamburger, whereas the moment problem on a finite interval, i.e., solve for $p(dt)$ in

$$\int_0^1 t^n p(dt) = y_n, \quad n = 0, 1, \ldots \tag{7.18}$$

is attached to the name of Hausdorff. These problems crop up in many places and are related to a variety of mathematical developments. See references [4]-[7] to name just a few.

Since we will be dealing with positive, finite measures, we can renormalize all of (7.16)-(7.18), assume that $p(dx)$ is in each case a probability distribution and that $y_0 = 1$.

About the **Hausdorff moment problem** (7.18) the following facts are known: Let us denote by $D^k \subset \mathbb{R}_+^k$ the convex hull of the curve $t \to (t, t^2, \ldots, t^k)^*$ obtained as t ranges in $[0,1]$. It is known, see [3] or [7], that

a) If $y^k = (y_1, \ldots, y_k)^*$ is outside D^k, then problem (7.18) cannot admit solution.

b) If $(y_1, \ldots, y_k)^*$ is in the boundary ∂D^k of D^k, then there is a unique convex combination of point masses such that (7.18) holds.

c) If $(y_1, \ldots, y_k)^*$ is in the interior of D^k, there are infinitely many probability distributions admitting $y^{(k)}$ as first k-momenta.

Given a sequence $\{y_k : k \geq 1\}$ finite or not, is to decide whether it can be a sequence of moments of a distribution $p(dt)$. This can be done in a systematic way with the aid of Hankel determinants $\underline{H}_0, \bar{H}_0, \underline{H}_1, \bar{H}_1, \ldots, \underline{H}_k, \bar{H}_k$.

When $k = 2n$ is even

$$\underline{H}_0 = 1, \quad \underline{H}_2 = \begin{vmatrix} 1 & y_1 \\ y_1 & y_2 \end{vmatrix}, \quad \underline{H}_4 = \begin{vmatrix} 1 & y_1 & y_2 \\ y_1 & y_2 & y_3 \\ y_2 & y_3 & y_4 \end{vmatrix}, \ldots,$$

$$\underline{H}_{2n} = \begin{vmatrix} 1 & y_1 & \cdots & y_n \\ y_1 & y_2 & \cdots & y_{n+1} \\ \vdots & \vdots & \ddots & \vdots \\ y_n & y_{n+1} & \cdots & y_{2n} \end{vmatrix}$$

$$\bar{H}_1 = y_1 - y_2, \quad \bar{H}_2 = \begin{vmatrix} y_1 - y_2 & y_2 - y_3 \\ y_2 - y_3 & y_3 - y_4 \end{vmatrix}, \ldots,$$

$$\bar{H}_{2n} = \begin{vmatrix} y_1 - y_2 & \cdots & y_n - y_{n+1} \\ \vdots & \ddots & \vdots \\ y_n - y_{n+1} & \cdots & y_{2n} - y_{2n+1} \end{vmatrix}$$

and when $k = 2n + 1$ is odd

$$\underline{H}_0 = 1, \quad \underline{H}_3 = \begin{vmatrix} y_1 & y_2 \\ y_2 & y_3 \end{vmatrix}, \ldots, \quad \underline{H}_{2n+1} = \begin{vmatrix} y_1 & \cdots & y_{n+1} \\ \vdots & \ddots & \vdots \\ y_{n+1} & \cdots & y_{2n+1} \end{vmatrix},$$

$$\bar{H}_1 = 1 - y_1, \quad \bar{H}_3 = \begin{vmatrix} 1 - y_1 & y_1 - y_2 \\ y_1 - y_2 & y_2 - y_3 \end{vmatrix}, \ldots,$$

$$\bar{H}_{2n+1} = \begin{vmatrix} 1 - y_1 & \cdots & y_n - y_{n+1} \\ \vdots & \ddots & \vdots \\ y_n - y_{n+1} & \cdots & y_{2n} - y_{2n+1} \end{vmatrix}.$$

The following results hold

Theorem 7.4. *The point y^k is in the interior of D^k if and only if all the Hankel determinants $\underline{H}_0, \bar{H}_0, \underline{H}_1, \bar{H}_1, \ldots, \underline{H}_k, \bar{H}_k$, are positive.*

A perhaps simpler criterion to verify is

Theorem 7.5. *The numbers in the sequence $\{1, y_1, \ldots, y_{2n}\}$ are the momenta of a probability distribution on $[0,1]$ if and only if the quadratic forms on $\mathbb{R}^{2n}$*

$$q_1(\xi) = \sum_{i,j=0}^{n} \xi_i \xi_j y_{i+j}, \quad q_2(\xi) = \sum_{i,j=0}^{n-1} \{y_{i+j+1} - y_{i+j+2}\} \xi_i \xi_j$$

are non-negative.

Theorem 7.6. *The sequence $\{1, y_1, \ldots, y_{2n+1}\}$ is the sequence of momenta of a probability distribution on $[0,1]$ if and only if the quadratic forms on $\mathbb{R}$*

$$q_1(\xi) = \sum_{i,j=0}^{n} \xi_i \xi_j y_{i+j+1}, \quad q_2(\xi) = \sum_{i,j=0}^{n} \{y_{i+j} - y_{i+j+1}\} \xi_i \xi_j$$

are non-negative.

The connection between the positivity of the quadratic forms and the corresponding Hankel determinants comes about by noticing that a quadratic form is positive whenever all principal minors of the corresponding (symmetric) matrix are positive.

Actually the Hankel matrices allow us to determine the ranges of the momenta y_k. For example for $k = 2n$ the lower end of the range of possible values of y_k is obtained by letting $\underline{H}_k = 0$, while demanding that smaller determinants be positive, we thus obtain

$$y_k - y_k^- = \frac{\underline{H}_k}{\underline{H}_{k-1}}$$

similarly, for the upper range we obtain

$$y_k^+ - y_k = \frac{\bar{H}_k}{\bar{H}_{k-1}}$$

where for consistency we set $\underline{H}_0 = \bar{H}_0 = \underline{H}_{-1} = \bar{H}_{-1} = 1$.

Defining the *canonical moment sequence* p_k (of $p(dt)$)

$$p_k = \frac{y_k - y_k^-}{y_k^+ - y_k}, \quad q_k = 1 - p_k$$

we see that the mapping $y_k \to p_k$ can be inverted, and the following result can be obtained

$$Vol(D^k) = \int_D \cdots \int_D dy_1 \cdots dy_k = \int_{[0,1]} \cdots \int_{[0,1]} \frac{\partial(y_1....y_k)}{\partial(p_1...p_k)} dp_1 \cdots dp_k$$
$$= \int_{[0,1]} \cdots \int_{[0,1]} \prod_{i=1}^{k-1} (p_i q_i)^{k-i} dp_1 \cdots dp_k = \prod_{k=1}^{n} B(k,k)$$

where $B(\alpha, \beta) = \int_0^1 p^{\alpha-1}(1-p)^{\beta-1} dp = \frac{\Gamma(\alpha)\Gamma(\beta)}{\Gamma(\alpha+\beta)} \quad \alpha, \beta > 0$.

From this the easy estimate for the upper bound of $Vol(D^k)$ follows, that is

$$Vol(D^k) \sim c \times 2^{-k^2}, \quad c = \text{constant}.$$

With this we explain some of the results cited in the previous section.

We conclude this section with some comments on the Stieltjes and Hamburger moment problems.

7.4.1 Some remarks about the Hamburger and Stieltjes moment problems

Certainly, when $\{y_k : k \geq 1\}$ is the moment sequence of a probability distribution $p(dt)$ on $[0,\infty)$ or $(-\infty, \infty)$ we have

Theorem 7.7. *Let $p(dt)$ be a positive measure. Then its associated moment sequence is such that the Hankel determinants $\underline{H}_k$ and $\bar{H}_k$ are non-negative.*

But things can be quite bad as far as uniqueness goes. Consider for example the following result.

Theorem 7.8. *Given any sequence of numbers $\{y_k : k \geq 0\}$, there exist infinitely many functions of bounded variation $F(t)$ on $[0,\infty)$ (or $(-\infty,\infty)$) such that*

$$y_k = \int_0^\infty t^k dF(t) \quad k \geq 0, 1, 2, \ldots.$$

Comment: Recall that bounded variation means that $F(t) = F_1(t) - F_2(t)$ with both $F_1(t)$, $F_2(t)$ non-decreasing and $\int_0^\infty |dF(t)| = \int_0^\infty dF_1(t) + \int_0^\infty dF_2(t) < \infty$.

To put together an example of two increasing functions $F_1(t)$ and $F_2(t)$ having the same moments consider $\varphi(t) = t^{-\frac{1}{4}} \sin(t^{\frac{1}{4}})$ for $t \geq 0$ and 0 for $t < 0$. The change of variables $\tau = t^{\frac{1}{4}}$ allows us to conclude

$$\int_{-\infty}^\infty t^n \varphi(t) dt = 4 \int_0^\infty \exp(-\tau^{\frac{1}{4}}) \tau^{4n+3} \sin(\tau) d\tau = 0.$$

Now set $F_1(t) = \int_0^t \exp(-s^{\frac{1}{4}}) ds$ and $F_2(t) = \int_0^t \exp(-s^{\frac{1}{4}})(1-\sin(s^{\frac{1}{4}})) ds$, and both zero for $t < 0$, the computation above shows that

$$\int_{-\infty}^\infty t^n dF_1(t) = \int_{-\infty}^\infty t^n dF_2(t) = 4(4n+3)!.$$

Another standard example showing the impossibility of reconstructing a density on the real line from a list of moments is the following. Consider the log-normal distribution

$$\rho(x) = \begin{cases} \frac{1}{x\sqrt{2\pi}} \exp\left(-\frac{1}{2}(\ln x)^2\right) & \text{if } x \geq 0, \\ 0 & \text{for } x < 0. \end{cases}$$

Assume that p is in $(-1, 1)$ and consider the family of densities

$$\rho_p(x) = \{1 + p\sin(2\pi \ln x)\} \rho(x).$$

Exercise 7.1. Verify that

i) $\rho_p(x)$ is a probability density.
ii) Verify that $\int x^k \rho_p(x) dx = \int x^k \rho(x) dx$ for all k, (and that these moments are finite!).

More on the indeterminacy issue can be obtained from

Theorem 7.9. *If $\mu_n \geq 1$ and $\mu_n \geq (n\mu_{n-1})^n$ for $n \geq 1$, then the equations*

$$\mu_n = \int_0^\infty t^n dF(t) \tag{7.19}$$

have a non-decreasing solution $F(t)$.

Theorem 7.10. *If $\lambda_0 \geq 1$, $\lambda_2 \geq (2\lambda_1+2)^2$, $\lambda_n \geq (n\lambda_{n-1})^n$ for $n = 2k+1$, $k = 0, 1, \ldots$ and $\mu_n = \lambda_{2n}$, where μ_n is as in Theorem 7.9, then there are at least two essentially distinct non-decreasing functions F_1 and F_2 such that (7.19).*

Proof. By Theorem 7.9 there exists $G_1(t)$ such that

$$\lambda_n = \int_0^\infty t^n dG_1(t) \text{ for } n = 0, 1, 2, \ldots$$

then, since $\mu_n = \lambda_{2n}$ we have

$$\mu_n = \int_0^\infty t^n dF_1(\sqrt{t}) \quad n = 0, 1, \ldots$$

and $F_1(t) = G_1(\sqrt{t})$ is a positive increasing function.

Let now

$$\nu_1 = \lambda_1 + 1 \geq \lambda_0 = \nu_0,$$
$$\nu_2 = \lambda_2 \geq (2\nu_1)^2 = (2\lambda_1 + 2)^2.$$

Thus, again by Theorem 7.9 there is a positive increasing function $G_2(t)$ such that

$$\nu_n = \int_0^\infty t^n dG_2(t) \quad n = 0, 1, 2, \ldots$$

and restricting ourselves to even subscripts we obtain

$$\mu_n = \int_0^\infty t^n dG_2(\sqrt{t}).$$

To see that $F_2(t) = G_2(\sqrt{t})$ is different from $F_1(t) = G_1(\sqrt{t})$ is easy: otherwise

$$\int_0^\infty \sqrt{t} dF_2(t) = \int_0^\infty \sqrt{t} dF_1(t)$$

or $\lambda_1 = \nu_1$ after a change of variables, which is clearly impossible. $\square$

The reader should check with [5-64] or [11-12] for more on these subjects.

7.5 Moment problems in Hilbert spaces

We shall briefly describe some efforts in tackling the Hausdorff, and other generalized moment problems, with the Hilbert space set up described in Chapter 4.

This time the inverse problem consists in solving

$$Ax = y \qquad (7.20)$$

where $A : V \to \mathbb{R}^n$ is given by $(Ax)_i = (\varphi_i, V)$ for $i = 1, 2, ..., n$.

For the time being, the φ_i are to be linearly independent vectors in the Hilbert space V. When $V = L^2([0,1], dx)$ and the scalar product (x, y) is the usual $\int_0^1 \bar{x}(t)y(t)dt$, the φ_i defining the Hausdorff moment problem are $\varphi_i(t) = t^{i-1}$ for $i = 1, ..., n$ (or any collection $\varphi_i(t) = t^{m_i}$ with $m_1 < m_2 < ... < m_n$). In some problems, like in magnetic resonance imaging, the $\varphi_i(t)$ may be trigonometric polynomials.

We shall first mention that even when (φ_i, x) is known for a basis or a Bessel system $\{\varphi_i : i \geq 1\}$ the problem consisting in solving $(\varphi_i, x) = y_i$ $i \geq 1$ in x may not be well defined.

The following example, taken from [22] illustrates the point.

Let $\{\xi_i : i \geq 1\}$ be an orthonormal basis for V. Put $\varphi_i = \frac{\xi_i}{i}$, i.e., the φ_i are Bessel system. Consider $A : V \to \ell_2$ given by $x \to \{(\varphi_i, x) : i \geq 1\}$, and assume that x_0 solves $(\varphi_i, x_0) = \frac{1}{i} := y_i$. Represent x_0 by $x_0 = \sum c_i \varphi_i$. Then $(\varphi_i, x_0) = \frac{c_i}{i^2}$ we obtain $c_i = i^2 y_i = i$. Therefore $x_0 = \sum i\varphi_i = \sum \xi_i$ which cannot be an element in V.

Comment: For a given $y_i \in \ell_2$, the necessary and sufficient condition on $\{\varphi_i : i \geq 1\}$ for $(\varphi_i, x) = y_i$, $i \geq 1$ to have a solution, is that $\{\varphi_i : i \geq 1\}$ be a Riez-Fischer system, i.e., that these exists a bounded linear operator $T : V \to V$ such that $T\varphi_i = \xi_i$ $i \geq 1$, for an orthonormal complete (a basis) system $\{\xi_i : i \geq 1\}$ on V.

But even when it exists (and is unique), the solution to the full Hausdorff moment problem is not continuous on the data. We have already seen a geometric explanation for this. In general, if x_0 satisfies $(Ax_0)_i = y_i$ for $i \geq 1$, and if $\varepsilon = \{\varepsilon_i : i \geq 1\}$ is a vector in ℓ_2 with arbitrary small norm $(\sum \varepsilon_i^2)^{\frac{1}{2}}$, we have to see that $\delta \in V$ such that $A(\delta + x_0) = y + \varepsilon$ can have arbitrary large norm.

Consider $\delta(t) = p^{1/3}[4t^2(1-t^2)]^p$. Since

$$\left(\sum_{n=0}^{\infty} \left|\int_0^1 t^n \delta(t)\right|^2\right)^{1/2} \leq \int_0^1 (1-t^2)^{-1/2} \delta(t) dt = 2^{2p-1}\frac{\Gamma(p+\frac{1}{2})^2}{\Gamma(2p+1)} \sim \frac{\sqrt{\pi}}{2p^{1/6}}$$

and also
$$\int_0^1 [\delta(t)]^2 dt = 2^{4p-1} \frac{\Gamma(2p+\frac{1}{2})\Gamma(2p+1)}{\Gamma(4p+\frac{3}{2})} p^{2/3} \sim \frac{\sqrt{\pi}}{4} p^{1/6}$$

as p becomes large in both cases. Thus the operator $A : L^2[0,1] \to \ell_2$ given by $(Ax)_i = \int_0^1 t^{i-1} x(t) dt$ does not admit a continuous inverse.

Another way of understanding this phenomenon is in terms of the angles between the vectors $\varphi_i(t) = t^{i-1}$ in L^2. The (Hilbert) matrix

$$M_{ij} = (\varphi_i, \varphi_j) = \frac{1}{i+j+1}, \quad i,j \geq 1$$

has eigenvalues decreasing to zero. As a matter of fact, for any given n, the condition number of $M_{ij}^n = \frac{1}{i+j+1}$ for $1 \leq i,j \leq n$ satisfies

$$\text{condition number} = \frac{\text{largest eigenvalue } M^n}{\text{smallest eigenvalue } M^n} \sim e^{2.5(n+1)}.$$

Thus in Hilbert spaces setup we better use a regularization technique to solve the momentum problem. Let us play with notation, to obtain a nice presentation of Tikhonov's regularized solution.

Write $A : V \to \mathbb{R}^n$ as

$$A = \sum_{i=1}^n e_i \otimes \varphi_i^*$$

where e_i stands for the canonical basis in $\mathbb{R}^n$. Apply it to $x \in V$ to verify that $Ax = \sum_{i=1}^n (\varphi_i, x) e_i$. It is easy to verify that

$$K := A^*A = \sum_{i=1}^n \varphi_i \otimes \varphi_j^*, \quad M := AA^* = \sum_{i,j=1}^n M_{ij} e_i \otimes e_j$$

where of course $M_{ij} = (\varphi_i, \varphi_j)$. For $i = 1, ..., n$, denote by v_i the i-th eigenvector of M of eigenvalue μ_i, i.e.,

$$M v_i = \mu_i v_i, \quad v_i = \sum v_{ij} e_j.$$

Set $x^i := \sum_{j=1}^n v_{ij} \varphi_j = A^* v_i$. Notice that

$$(x^i, x^k) = (A^* v_i, A^* v_k) = (AA^* v_i, v_k) = \mu_i (v_i, v_k) = \mu_i \delta_{ik} \quad (7.21)$$

$$Kx^i = (A^*A) A^* v_i = A^* (AA^* v_i) = A^* M v_i = \mu_i A^* v_i = \mu_i x^i. \quad (7.22)$$

Suppose now that you want to solve $Ax = y$ by regularization. Form $A^*Ax = Kx = A^*y$ and perturb K to obtain

$$(K + \lambda I)x = A^*y. \quad (7.23)$$

Using identity (7.22), we can write the spectral representation of $K+\lambda I$ as

$$K + \lambda I = \sum (\mu_i + \lambda) x^i \otimes x^{i*}.$$

Write $x = \sum \alpha_i x^i$, take the scalar product of the identity above by x^i to obtain

$$\alpha_i(\mu_i + \lambda) = (x^i, A^*y) = (Ax^i, y) = (AA^{*v_i, y}) = \mu_i(v_i, y)$$

from which we obtain the representation

$$x = \sum_{i=1}^{n} \frac{\mu_i(v_i, y)}{\mu_i + \lambda} x^i$$

for the solution of the regularized equation (7.23). When $V = L^2[0,1]$, $\varphi_i(t) = t^{i-1}$, $M_{ij} = \frac{1}{i+j+1}$ is the Hilbert matrix of size n, K is an integral operator with kernel $K(s,t) = \sum_{i=1}^{n} s^{i-1} t^{i-1} = \frac{1-(st)^n}{1-st}$ and $Kx^i = \mu_i x^i$ looks like

$$\int_0^1 \frac{1-(st)^n}{1-st} x^i(s) ds = \mu_i x^i(t)$$

where $x^i(t) = \sum_{i=1}^{n} v_{ij} t^{j-1}$.

7.6 Reconstruction of transition probabilities

This constitutes a truly finite dimensional class of problems. Very much dealt with by Markov Chain Monte Carlo (MCMC) simulation nowadays. The problem is stated as follows: One is given a vector $p = (p_1, ..., p_n)$ describing the stationary distribution of a transition probability matrix $\{\Pi_{ij}: i, j = 1, ..., n\}$, that is

$$p_j = \sum_{i=1}^{n} p_i \Pi_{ij} \quad j = 1, ..., n. \tag{7.24}$$

The problem consists of finding the transition probability matrix Π. This matrix is subject to some obvious constraints:

$$0 \leq \Pi_{ij} \leq 1 \quad i,j = 1, ..., n \tag{7.25}$$

$$\sum_{j=1}^{n} \Pi_{ij} = 1 \quad i = 1, ..., n. \tag{7.26}$$

Numerical examples

By appropriate relabeling, this problem can be recast as

$$Ax = y, \quad 0 \leq x_j \leq 1, \quad j = 1,...,n^2. \tag{7.27}$$

Where the vector x is formed by relabeling Π in lexicographic order, and both the $(2n, n^2)$-matrix A and the vector y are determined from (7.25) and (7.26).

We can now form the hypercube $\Omega = [0,1]^{2n}$, $\mathcal{F} = \mathcal{B}([0,1]^{2n})$ and put the usual non-informative a priori uniform measure Q on $(\Omega, \mathcal{F})$ and proceed as usual. As example we present only one reconstruction of transition matrix with stationary probability $p = (0.1022\ 0.0884\ 0.0099\ 0.0857\ 0.12\ 0.037\ 0.0592\ 0.104\ 0.0648\ 0.0323\ 0.0373\ 0.0487\ 0.0226\ 0.066\ 0.1218)$ which was obtained by a trivial simulation. The transition matrix is depicted in Table 7.2. The reconstruction error $\varepsilon = \sum_{i=1}^{n^2} |(Ax^*)_i - y_i|$ is listed in the caption.

Table 7.2 Transition matrix obtained with maxentropic reconstruction, $\varepsilon = 2.3842 \times 10^{-10}$.

0.109	0.092	0.004	0.089	0.131	0.031	0.057	0.111	0.063	0.025	0.031	0.044	0.015	0.065	0.133
0.105	0.09	0.006	0.087	0.123	0.035	0.059	0.106	0.065	0.03	0.035	0.048	0.019	0.067	0.125
0.072	0.07	0.052	0.07	0.073	0.063	0.067	0.072	0.068	0.062	0.064	0.066	0.059	0.068	0.073
0.104	0.09	0.006	0.087	0.121	0.036	0.06	0.105	0.066	0.031	0.036	0.049	0.02	0.067	0.123
0.114	0.094	0.002	0.09	0.142	0.026	0.053	0.117	0.06	0.021	0.026	0.039	0.011	0.062	0.144
0.084	0.079	0.026	0.078	0.091	0.054	0.067	0.085	0.069	0.05	0.054	0.061	0.042	0.07	0.091
0.094	0.085	0.014	0.083	0.105	0.045	0.064	0.095	0.068	0.041	0.046	0.056	0.03	0.069	0.106
0.11	0.092	0.004	0.089	0.132	0.03	0.056	0.112	0.063	0.025	0.031	0.044	0.015	0.064	0.134
0.096	0.086	0.012	0.084	0.108	0.043	0.063	0.097	0.068	0.038	0.043	0.054	0.028	0.069	0.109
0.082	0.078	0.029	0.077	0.088	0.055	0.067	0.083	0.069	0.052	0.056	0.062	0.045	0.07	0.088
0.085	0.079	0.026	0.078	0.091	0.053	0.067	0.085	0.069	0.05	0.054	0.061	0.042	0.07	0.091
0.09	0.082	0.019	0.081	0.098	0.049	0.065	0.09	0.069	0.045	0.049	0.058	0.035	0.07	0.099
0.078	0.075	0.038	0.074	0.081	0.059	0.067	0.078	0.069	0.057	0.059	0.064	0.051	0.069	0.082
0.097	0.086	0.011	0.084	0.109	0.043	0.063	0.098	0.068	0.038	0.043	0.054	0.027	0.069	0.11
0.115	0.094	0.002	0.09	0.143	0.025	0.052	0.117	0.06	0.02	0.026	0.039	0.011	0.062	0.145

We may change the size of the algebraic version of our problem if we take the constraints (7.26) as part of the constraints space C. Relabel Π to form x as above, but now form A as follows

$$A = (P_1 I \ P_2 I \ ... \ P_n I)$$

where I is an $n \times n$-identity. Now Ω is the hyper-triangle T^n where

$$T = \left\{ x \in \mathbb{R}^n : 0 \leq x_i \leq 1, \ i = 1, ..., n \text{ and } \sum_{1}^{n} x_i = 1 \right\}.$$

Again $\mathcal{F} = \mathcal{B}(T^n)$ and $dQ(\xi) = \prod_{i=1}^{n} dQ_i(\xi_i)$ where each $dQ_i(\xi_i)$ is given by

$$dQ_i(\xi_i) = \sum_{j=1}^{n} q_j \varepsilon_{e_j}(d\xi_i^1 d\xi_i^2 ... d\xi_i^n)$$

ε_{e_j} being the Dirac point mass at the e_j-th unit vector in $\mathbb{R}^n$. We choose this instead of a uniform measure on T for obvious computational reasons. From now on, everything is down hill rolling (or the same routine if you will). The resulting reconstruction is displayed in Table 7.3.

Table 7.3 Transition matrix obtained with maxentropic reconstruction, $\varepsilon = 1.1230 \times 10^{-12}$.

0.091	0.11	0.002	0.085	0.127	0.033	0.041	0.108	0.069	0.036	0.021	0.059	0.014	0.065	0.141
0.102	0.084	0.003	0.089	0.127	0.037	0.053	0.109	0.068	0.034	0.029	0.05	0.018	0.069	0.128
0.131	0.012	0.089	0.077	0.081	0.046	0.131	0.077	0.042	0.015	0.118	0.013	0.064	0.058	0.047
0.104	0.079	0.004	0.09	0.126	0.038	0.055	0.109	0.068	0.034	0.031	0.048	0.019	0.069	0.125
0.075	0.151	7E-04	0.077	0.123	0.028	0.03	0.103	0.068	0.039	0.013	0.07	0.009	0.06	0.155
0.133	0.025	0.032	0.089	0.105	0.047	0.106	0.096	0.055	0.022	0.08	0.023	0.046	0.068	0.073
0.124	0.044	0.013	0.093	0.118	0.044	0.081	0.106	0.063	0.028	0.054	0.033	0.032	0.071	0.097
0.089	0.114	0.002	0.084	0.126	0.033	0.04	0.107	0.069	0.037	0.02	0.06	0.013	0.065	0.143
0.12	0.05	0.01	0.093	0.121	0.043	0.075	0.107	0.064	0.029	0.048	0.036	0.029	0.071	0.103
0.134	0.022	0.039	0.088	0.101	0.048	0.111	0.093	0.053	0.021	0.087	0.021	0.049	0.067	0.068
0.133	0.026	0.032	0.089	0.105	0.047	0.105	0.096	0.055	0.022	0.08	0.023	0.046	0.068	0.073
0.129	0.034	0.02	0.092	0.113	0.046	0.093	0.102	0.06	0.025	0.065	0.028	0.038	0.07	0.086
0.134	0.017	0.056	0.084	0.093	0.048	0.12	0.087	0.049	0.018	0.1	0.017	0.056	0.064	0.059
0.119	0.052	0.009	0.093	0.121	0.043	0.074	0.107	0.065	0.029	0.047	0.037	0.029	0.071	0.104
0.074	0.156	7E-04	0.076	0.123	0.027	0.029	0.102	0.067	0.039	0.013	0.071	0.009	0.059	0.157

A host of alternative methods is described in [25].

7.7 Probabilistic approach to Hausdorff's moment problem

A nice application of probabilistic ideas, basically the Law of Large Numbers, provides another approach to the moment problem.

Consider independent, identically distributed random variables $\{X_n : n \geq 1\}$ defined on a probability space $(\Omega, \mathcal{F}, \mathcal{P})$. Let $\theta = E[X]$, $\sigma^2 = E[(X-\theta)^2]$. Then $M_n := \frac{(X_1+...+X_n)}{n}$ such that $E[M_n] = \theta$, $E[(M_n-\theta)^2] = \frac{\sigma^2}{n}$. This makes it plausible to assert that for large values of n, and for a continuous function $u : \mathbb{R} \to \mathbb{R}$, $u(M_n) \sim u(\theta)$.

This result is known as (the weak) law of large number and admits the following generalization.

Let $\{F_{n,\theta} : n \geq 1\}$ be a family distributions having the same expectation

$$E_{n,\theta}[X] = \int_{-\infty}^{\infty} x dF_{n,\theta}(x) = \theta$$

and variance $E_{n,\theta}[(X-\theta)^2] = \sigma_n^2(\theta)$. Put $E_{n,\theta}[u] = \int_{-\infty}^{\infty} u(x)dF_{n,\theta}(x)$.

Theorem 7.11. *Suppose u is bounded and continuous and that $\sigma_n^2(\theta) \to 0$ as $n \to \infty$ (for each θ). Then*

$$E_{n,\theta}[u] \to u(\theta). \tag{7.28}$$

The convergence is uniform in every closed interval in which $\sigma_n^2(\theta) \to 0$ uniformly.

We leave the reader to either look in Feller's classic [23] or work the proof out herself. A nice application of this result is the approximation of continuous functions by polynomials. The result is

Corollary 7.1. *Let $u : [0,1]$ be continuous and let $F_{n,\theta}(x)$ be such that it puts mass $\left(\binom{n}{k}\right)\theta^k(1-\theta)^{n-k}$ at the points $\frac{k}{n}$, $k = 0, ..., n$ of $[0,1]$. Then $\sigma_n^2(\theta) = \frac{\theta(1-\theta)}{n} \to 0$ (uniformly for $0 \leq \theta \leq 1$) and*

$$E_{n,\theta}[v] = \sum_{k=0}^{n} u\left(\frac{k}{n}\right)\binom{n}{k}\theta^k(1-\theta)^{n-k} \to u(\theta)$$

uniformly in $[0,1]$.

Comment: This is the usual Bernstein approximation theorem.

Let us now be given a sequence $y_0 = 1, y_1, y_2, ..., y_n, ...$ of moments of an unknown distribution function $F(dx)$ on $[0,1]$, i.e.,

$$y_k = \int_0^1 x^k dF(x), \quad k \geq 0. \tag{7.29}$$

Clearly the sequence y_k is completely monotone, i.e.,

$$(-\Delta)^n C_k \geq 0, \quad n \geq 0$$

since (verify it)

$$(-\Delta)^n C_k = E[X^k(1-X)^n] = \int x^k(1-x)^n dF. \tag{7.30}$$

In order to recover $dF(x)$ from (7.29), all we have to do is to realize that F is known if its integral of any continuous function u is known, i.e., if $E_F[u] = \int u(x)dF(x)$ is known.

Observe that (7.30) yields for the Bernstein approximate to u

$$E_F[B_{n,u}] := \sum_{k=0}^{n} u\left(\frac{k}{n}\right) \binom{n}{k}(-\Delta)^{n-k} C_k \equiv \sum u\left(\frac{k}{n}\right) p_j^n = E_n[u]. \quad (7.31)$$

Where we introduced the family F_n of discrete probability distributions that puts mass $p_k^n = \left(\binom{n}{k}\right)(-\Delta)^{n-k} C_k$ at $\frac{k}{n}$. Take $u(x) \equiv 1$ and observe that (7.31) implies

$$E_n[1] = \sum_{k=0}^{n} \binom{n}{k}(-\Delta)^{n-k} C_k = E\left[\sum_{k=0}^{n} \binom{n}{k} X^k (1-X)^{n-k}\right] = 1.$$

Now invoke Corollary 7.1 to conclude that

$$\lim E_n[u] \to E[u].$$

Actually, the limit on the left-hand side defines a linear functional on $C([0,1])$ which can be realized as an integral with respect to the desired unknown measure.

This is a nice connection, but of not much help when only a partial list of moments is known, which may be contaminated with error.

7.8 The very basics about cubic splines

Given values $x_j = x(t_j)$ of a continuous function at the end points of the intervals $[t_{j-1}, t_j]$ $j = 1, ..., N$; we can find a cubic polynomial $p(t)$ in each interval such that $p(t_j) = x_j$ in [0,1] such that the resulting function is continuous and has two continuous derivatives.

This can be achieved in terms of the so-called cubic Hermite basis consisting of cubic polynomials $F_0(x)$, $F_1(x)$, $G_0(x)$ and $G_1(x)$ defined on $[0,1]$ and such that:

$$F_0(0) = 1, \quad F_0'(0) = 0, \quad F_0(1) = 0, \quad F_0'(1) = 0,$$
$$G_0(0) = 0, \quad G_0'(0) = 1, \quad G_0(1) = 0, \quad G_0'(1) = 0,$$
$$F_1(0) = 0, \quad F_1'(0) = 0, \quad F_1(1) = 1, \quad F_1'(1) = 0,$$
$$G_1(0) = 0, \quad G_1'(0) = 0, \quad G_1(1) = 0, \quad G_1'(1) = 1.$$

Certainly, in $[t_{j-1}, t_j]$, with $h_j = t_j - t_{j-1}$,

$$p(t) = x_{j-1} F_0\left(\frac{t-t_{j-1}}{h_j}\right) + x_j F_1\left(\frac{t-t_{j-1}}{h_j}\right)$$
$$+ m_{j-1} G_0\left(\frac{t-t_{j-1}}{h_j}\right) + m_j G_1\left(\frac{t-t_{j-1}}{h_j}\right). \quad (7.32)$$

To determine the coefficients m_j, we invoke the continuity of the derivatives across partition points, and impose a second zero derivative at the end points. We thus obtain a linear system for the m_j's in terms of the x_j's. It can be solved to yield $x = Bm$ with

$$B = \begin{pmatrix} \frac{2}{h_1} & \frac{1}{h_1} & 0 & 0 & \cdots & 0 \\ \frac{1}{h_1} & 2\left(\frac{1}{h_1}+\frac{1}{h_2}\right) & \frac{1}{h_2} & 0 & \cdots & 0 \\ \vdots & & & & & \vdots \\ 0 & \cdots & 0 & \frac{1}{h_{n-1}} & 2\left(\frac{1}{h_{n-1}}+\frac{1}{h_n}\right) & \frac{1}{h_n} \\ 0 & \cdots & 0 & 0 & \frac{1}{h_n} & \frac{2}{h_n} \end{pmatrix}$$

which is a tridiagonal matrix with a dominant diagonal. Such systems have unique solution and we are in business, for in this fashion once we write $m = B^{-1}x$ and substitute in (7.32) to obtain an approximant $x_{ap}(t)$ to x, by means of which we transform $\int_0^1 K(s,t)x(t)dt + \varepsilon(s) = y(s)$, $s \in [0,1]$ into a problem like $Ax = y$, $x \in C$.

7.9 Determination of risk measures from market price of risk

7.9.1 Basic aspects of the problem

This section describes a potentially interesting application in actuarial sciences. For a comprehensive review of problems and description of problems and methods, consult the monograph by Kaas *et al.* [29]. The problem may be of interest for a new participant in the market, or by an established participant wishing to price risks consistently.

The basic setup is the following. Risks are usually described by random variables denoting losses. In the following we shall consider positive random variables only. Thus, the collection of risks is a class C which is a homogeneous cone of in the class of square integrable random variables $X : \Omega \to [0, \infty)$, defined on a probability space $(\Omega, \mathcal{F}, P)$.

In [30] Wang *et al.* and in [31], Wirch and Hardy explain why coherent risk measures (measures satisfying "logical" coherence axioms) are interesting, and study their properties. Here we just state

Definition 7.1. A coherent risk pricing measure is a function $\Pi : C \to [0, \infty)$ satisfying the conditions:

- For all $X \in C$, $E[X] \leq \Pi(X) \leq esssup\, X$.

- For all $X \in \mathcal{C}, \lambda \geq 0, a \in \mathbb{R}, \Pi(\lambda X + a) = \lambda \Pi(X) + a$.
- If for all $X, Y \in \mathcal{C}$, it is known that $X \leq Y$, then $\Pi(X) \leq \Pi(Y)$.
- For all $X, Y \in \mathcal{C}, \Pi(X + y) \leq \Pi(X) + \Pi(Y)$.
- $\Pi(1) = 1$.

The first condition assets that n price is less than the expected losses, nor larger than the maximum loss. The second is a homogeneity and translation invariance condition. The third asserts that if risk X produces less losses than risk Y in all market states, then its prices should be lower. The sub-additivity condition four incorporates a diversification requirement, and the last condition asserts that the price of a deterministic loss equals the loss.

It is a simple exercise to verify that for a positive random variable

$$E[X] = \int_0^\infty P(X \geq x) dx.$$

In order to price large losses that occur with small probability, actuaries have proposed a non-linear transformation (distortion) of the probability distribution of a random variable.

Definition 7.2. A distortion function is a measurable function $g : [0, 1] \to [0, 1]$ such that
1) $g(0) = 0$, and $g(1) = 1$.
2) g is a non-decreasing function.

If $F_X(x)$ is the distribution function of a random variable X, we may obtain a new distribution function by setting $F_X^*(x) = 1 - g(P(X > x))$. We have

Definition 7.3. The Choquet integral of X with respect to the distortion function G is defined by

$$H_g(X) = \int_0^\infty g(P(X > x)) dx.$$

A thorough study of this type of non-linear transformations is carried out in Denneberg's [26]. Of interest to us is the following result.

Proposition 7.1. If g is a continuously differentiable, concave distortion function, and we set $\phi(u) = g'(1 - u)$, then

$$H_g(X) = \int_0^\infty g(P(X > x)) dx = \int_0^1 q_X(u) \phi(u) du = \Pi_g(X),$$

is a coherent risk measure, where for $0 < u < 1$, $q_X(u)$ is the left continuous inverse (u-quantile) of F_X defined by $q_X(u) = \inf\{x | F_X(x) \geq u\}$.

Comment: The function ϕ is called a risk aversion function.

Proof. To verify that Π_g satisfies the coherence requirements, see [30] or [31], and to verify the identity it is easier to consider the case where X has a density and use calculus. Begin with the left-hand side and make a change of variable $u = 1 - P(X > x)$ and integrate by parts. $\square$

Comment: The conditions on g translate into the important
$$\int_0^1 \phi(u)du = 1. \tag{7.33}$$

7.9.1.1 Examples of distortion functions

We shall use the following typical examples of risk aversion functions to generate prices for the numerical simulations

1) Dual-power risk aversion function
$$\phi(u) = \nu u^{\nu-1}, \quad \text{with} \quad \nu > 1.$$
2) Proportional hazard risk aversion function
$$\phi(u) = \frac{1}{\gamma}(1-u)^{\frac{1}{\gamma}-1} \quad \text{with} \quad \gamma > 1.$$
3) Wang's risk aversion function
$$\phi_\alpha(u) = e^{-\alpha \Phi^{-1}(u) - \alpha^2/2}.$$

7.9.2 Problem statement

Here we shall consider only the case in which there is no uncertainty in the market price of the risks. It could happen for example that market participants price risks with a like $(1 + \gamma_i)E[X]$ with "randomly" chosen loads γ_i. Or that some use empirical data instead of model distribution functions to compute prices with some of the risk aversion functions listed above, or that the prices contain "randomly" assigned surcharges. This would lead to the case considered in [28]. The interested reader can consider examining that material. The problem we shall consider here is the following.

7.9.2.1 Problem statement

Suppose that we know the prices $\pi_1, \pi_2, ..., \pi_{M-1}$ of a collection $X_1, X_2, ..., X_{M-1}$ of risks. We want to find a continuous, increasing risk aversion function $\phi(u)$ such that
$$\pi_i = \int_0^1 q_{X_i}(u)\phi(u)du \quad \text{for} \quad i = 1,...,M, \tag{7.34}$$

where for $i = M$ we consider $q_{X_M}(u) \equiv 1$ to take care of the integrability condition (7.33).

7.9.2.2 Problem discretization

The second step towards the numerical solution of the problem consists of the discretization procedure. For this we consider a partition of $[0, 1]$ into N parts by points $u_j = j/N$ and put $a_j = (u_j + u_{j-1})/2$ and define $B_{ij} = q_{X_i}(a_j)/N$ and $\phi_j = \phi(a_j)$. The reason for this choice is to avoid possible singularities of q_X at both 0 or 1. The discretized version of our problem is now

$$\boldsymbol{\phi} = \mathbf{B}\boldsymbol{\psi} \qquad \boldsymbol{\phi} \in \mathbf{K}_0 \qquad (7.35)$$

where $\boldsymbol{\phi}$ is the M-vector with components π_i (recall that $\pi_M = 1$), and $\boldsymbol{\phi}$ is the N-vector with components, and $\mathbf{K}_0 = \{\boldsymbol{\phi} \in \mathbb{R}_{++}^N \,|\, \phi_1 < ... < \phi_n\}$ is the constraint set. To obtain a more manageable constraint set we put $\boldsymbol{\phi} = \mathbf{C}\boldsymbol{\psi}$, where $\mathbf{C}$ is the lower diagonal matrix with all entries equal to 1. With this new notation, the problem can be restated as: Find $\boldsymbol{\psi}$ such that

$$\boldsymbol{\phi} = \mathbf{B}\boldsymbol{\psi} \qquad \boldsymbol{\phi} \in \mathbf{K} \qquad (7.36)$$

where $\mathbf{K}$ is a convex subset of $\mathbb{R}_{++}^N$, which we can take to be of the type $[a, b]^N$, and we specify a and b below.

7.9.3 The maxentropic solution

We have already seen that the MEM solution to (7.34) can be obtained as the expected value of the coordinate maps X_j on $[a, b]^N$ with respect to an appropriate distribution, absolutely continuous with respect to dQ given by

$$dQ = \prod_{j=1}^N \left(\frac{1}{2} \{\epsilon_a(dx_j) + \epsilon_b(dx_j)\} \right).$$

We saw above that the solution is of the type

$$\psi^* j = ap_j^* + b(1 - p_j^*),$$

where

$$p_j^* = \frac{e^{-a(\mathbf{A}^*\boldsymbol{\lambda}^*)_j}}{e^{-a(\mathbf{A}^*\boldsymbol{\lambda}^*)_j} + e^{-b(\mathbf{A}^*\boldsymbol{\lambda}^*)_j}}.$$

To obtain $\boldsymbol{\lambda}^*$ we form $Z(\boldsymbol{\lambda}^*) = \prod_j \zeta((\mathbf{A}^*\boldsymbol{\lambda}^*)_j)$, where

$$\zeta(\tau) = \frac{1}{2}(e^{-a\tau} + e^{-b\tau}),$$

and, once again, $\boldsymbol{\lambda}^*$ is found minimizing $\Sigma(\boldsymbol{\lambda}^*) = \ln Z(\boldsymbol{\lambda}^*) + <\boldsymbol{\lambda}^*, \boldsymbol{\pi}>$.

7.9.4 Description of numerical results

We are going to present some of the results obtained in [2]. The interested reader con look in there, and in [3], for more about this.

To generate prices we proceed as follows.
1) We consider risks $X_1 \sim U[0,1]$, $X_2 \sim \text{Pareto}(0,2)$, $X_3 \sim \Gamma(2,4)$ and $X_4 \sim Beta(1,2)$.
2) We shall suppose that the observable prices in the market are computed by three different agents, using a different risk aversion function for each of them. One uses a proportional hazard of parameter $1/2$, the other a dual power function with parameter $1/2$ and the third a Wang distortion function of parameter 0.05.

Using these prices, as inputs, we can reconstruct the distortion function of each agent. In figure 7.44 we plot the reconstructed ϕ^* next to the original spectral function ϕ. The data used for each panel were the prices provided by each of the distortion functions.

In Table 7.4 we display the reconstruction error of each risk when the market prices of 4 liabilities are used to reconstruct one single spectral risk function. This time we considered a $U(0,1)$, a $Pareto(0,2)$, a $Gamma(2,4)$ and a $Beta(1,2)$, and the 4 liabilities were simultaneously priced with a dual power risk aversion function, a proportional hazard risk aversion function and a Wang risk aversion function.

Table 7.4 Error of reconstruction risk price.

Wang		PH		DP	
π	error	π	error	π	error
0.52	0.12×10^{-8}	0.55	0.15×10^{-5}	0.39	0.41×10^{-6}
4.01	0.001×10^{-8}	4.15	0.07×10^{-5}	2.79	0.012×10^{-6}
0.32	0.44×10^{-8}	0.35	0.34×10^{-5}	0.25	0.88×10^{-6}
2.53	0.02×10^{-8}	2.00	0.044×10^{-5}	1.60	0.07×10^{-6}

7.9.4.1 A consistency test

If each market agent would price other risks using the same risk aversion function, a natural question to ask is: How does the price she/he assigns compares with the price obtained with the reconstructed ϕ^*. For that we consider the price of a $U(0,2)$ compute its price as the other agents would do and with the reconstructed ϕ. The prices obtained, and absolute vales of the differences in price are: $\pi_{DP} = 0.7196$ and $|\pi_{DP} - \pi^*| = 0.0002$;

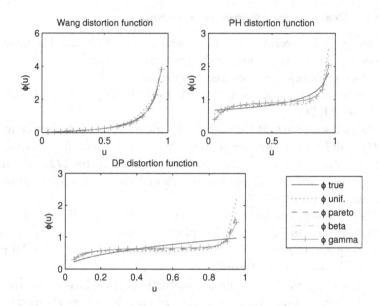

Fig. 7.44 Graph of ϕ^* reconstructed from four different prices.

$\pi_{Wang} = 1.008$ and $|\pi_{Wang} - \pi^*| = 2.11 \times 10^{-5}$ and $\pi_{PH} = 1.02$ with error $|\pi_{PH} - \pi^*| = 2.54 \times 10^{-6}$. Not bad at all.

7.10 Bibliographical comments and references

The material for Section 7.1 follows, as close as you can get, [1] and [2], whereas the contents of Section 7.2 and 7.3 follows [3].

References [4] and [5] relate moment problems to quite a bit of mathematics and in reference [6] you will find an in depth study of the Hausdorff moment problem. Both of [7]-[8] relate moment problems to prediction theory and a different notion of entropy shows there.

We cannot overlook moment systems and polynomials of binomial type and their connection with semigroups explored in [9]-[10].

The classical results we quote at the end of Section 7.4 and further results and references can be found in [11] and [12].

For further guide to numerical aspects and applications consult any of [13]-[17].

To close we mention [18] who explains the choice of self adapting step for the gradient method that we use.

Some references dealing with issues related to the theme of this chapter, i.e., on numerical aspects of linear problems are [19]-[21], and the results in Section 7.6 were lifted from Feller's [23].

The material for Section 7.8 is standard. A very simple exposé can be found in Section 28 of [24].

References [26]-[31] provide background and contain the basic results for the topic of the last section.

References

[1] Gzyl, H. and Velásquez, Y. *"Linear reconstruction problems with convex constraints: influence of the a priori data"*. Applied Mathematics and Computation. Vol. 109 (2000) pp. 189-198.

[2] Gzyl, H. and Velásquez *"Maxentropic reconstruction with uncertainty in the data and varying a priori information"*. Mathem. and Computer Modelling, Vol. 29 (1999) pp. 79-86.

[3] Gzyl, H., Tagliani, A. and Velásquez, Y. *"A comparative study of some reconstruction methods for linear inverse problems"*. Mathematical and Computer Modeling, Vol.30 (1999), pp. 159-167.

[4] Simon, B. *"The classical moment problem as a selfadjoint finite difference operator"*. Advances in Mathematics, Vol. 137 (1998), pp. 82-203.

[5] Dette, H. and Studden, W. *"The theory of canonical moments with applications in statistics, probability and analysis"*. J. Wiley & Sons, New York, 1997.

[6] Karlin, S. and Shapley, L. S. *"Geometry of moment spaces"*. AMS Memories, No. 12, Providence, R.I., 1953.

[7] Gabardo, J. P. *"Extensions of positive-definite distributions and maximum entropy"*. M.A.M.S. Vol. 102, No. 489, 1993.

[8] Landau, H. *"Maximum entropy and the moment problem"*. B.A.M.S. Vol. 16, (1987), pp.47-77.

[9] Feinsilver, P. and Schott, R. *"Algebraic Structures and Operator Calculus: Vol. I Representation Theory"*. Kluwer Acad. Pubs., Dordrecht, 1993.

[10] Di Bucchianco, A. *"Probabilistic and Analytic aspects of the umbral calculus"*. CWI Tract No. 119, Amsterdam, 1997.

[11] Shohat, J. A. and Tamarkin, J. D. *"The problem of moments"*. AMS Survey 1, Providence, R.I. (1953).

[12] Widder, D. V. *"The Laplace Transform"*. Princeton Univ. Press, Princeton, 1946.

[13] Talenti, G. *"Recovering a function from a finite number of moments"*, Inverse Problems, Vol. 3 (1987), pp. 501-517.

[14] Frontini, M. and Tagliani, A. *"Maximum entropy in the Stieltjes and Hamburger moment problem"*. J. Math. Phys., Vol. 35 (1994), pp. 6748-6756.

[15] Frontini, M. and Tagliani, A. *"Stieltjes and Hamburger moment problem"*. Appl. Math. and Comput., Vol. 88 (1997), pp. 39-51.

[16] Anto'lín, J., Zarzo, A. and Angulo, J. C. *"Maximum entropy analysis of electron pair density in many electron systems"*. Phys. Rev. Vol. 50 (1194), pp. 240-246.

[17] Yndurain, F. J. *"The moment problem and applications"*. Lecture Notes, CERN. Ref. T.H. 1521, 1972.

[18] Raydan, M. *"On the Barzilai and Borwein choice of step length for the gradient method"*. J. Num. Analysis. Vol. 13 (1993), pp.321-326.

[19] Varah, J. M. *"Pitfalls in the numerical solutions of linear ill-posed problems"*. SIAM J. Sci. Stat. Comp. Vol. 4 (1983), pp. 164-176.

[20] Varah, J. M. *"A practical examination of some numerical methods for linear discrete ill-posed problems"*. SIAM Rev. Vol. 21 (1979), pp. 100-111.

[21] Essah, W. A. and Delves, L. M. *"The numerical solution of first kind integral equations"*. J. Comp and Math. Vol. 27 (1988), pp. 363-387.

[22] Zwaan, M. *"Moment problems in Hilbert space with applications to magnetic resonance imaging"*. CWI tract No. 89, Amsterdam, 1991.

[23] Feller, W. *"An Introduction to Probability Theory and its Applications"*. Vol. 2, John Wiley, New York, 1971.

[24] Chang, G. and Sederberg, T. *"Over and Over Again"*, Vol. 39 of the New Mathematical Library, M.A.A., 1997.

[25] Guyon, X. *"Random Fields on a Network: Modeling, Statistics and Applications"*, Springer, Berlin (1995).

[26] Denneberg, D. " Non-Additive Measure and Integral". Kluwer Academic Publishers, Dordrecht, (1997).

[27] Gzyl, H. and Mayoral, S. *"Determination of risk pricing measures from marketprices of risk"*. Insurance: Mathematics and Economics, Vol 43 (2008), pp. 437-443.

[28] Gzyl, H. and Mayoral, S. *" A method for determining risk aversion functions from uncertainmarket prices of risk"*, Insurance: Mathematics and Economics, Vol. 47 (2010), pp. 84-89.

[29] Kaas, R., Goovaerts, J.M., Dhaene, J. and Denuit, M. *"Modern Actuarial Science"*, Kluwer Academic Publishers., Dordrecht, (2005)

[30] Wang, S., Young, V., Panjer, H. *"Axiomatic characterization of insurance prices"*. Insurance: Mathematics and Economics Vol.21, (1997) pp. 173-183.

[31] Wirch, J.L. and Hardy, M. *"A synthesis of risk measures for capital adequacy"* Insurance: Mathematics and Economics, Vol. 25 (1999), pp. 337-347.

Chapter 8

Some infinite dimensional problems

In this chapter we shall apply the methodology developed in Chapter 5 to some truly infinite dimensional examples. We shall present several approaches to the problem consisting of solving an integral equation. We shall see that there are different setups within which the problem can be solved. All approaches require familiarity with stochastic calculus. If this is not the case, the reader can skip this chapter.

8.1 A simple integral equation

Consider solving for continuous $x(t)$, $t \in [0,1]$ in

$$\int_0^1 A(s,t)x(t)dt = y(s) \tag{8.1}$$

in which we take $x(t) \in V = \mathcal{C}([0,1])$ without further restrictions, $W = \mathcal{C}(S)$ or $W = L_1(S, \mathcal{B}(S), m)$, where S is some compact, separable metric space say, $\mathcal{B}(S)$ denotes the Borel sets of S and m is some finite Borel measure. The computations in each case are quite similar. We suppose that $A(s,t): S \times [0,1] \to \mathbb{R}$ is continuous. Then (abusing of notation) the operator $A: V \to W$ is continuous.

Since there are no constraints on $x(t)$, the role of Ω and of the constraint set C introduced in Chapter 5 is now played by V. The coordinate maps $X_t: V \to \mathbb{R}$ are the obvious $X_t(w) = w(t)$. We consider $\mathcal{F}$ to be the σ-algebra generated by the cylinder sets

$$\{w \in V : w(t_i) \in B_i, \ i = 1, 2, ..., n\}$$

where $0 \leq t_1 < t_2 < ... < t_n \leq 1$ and $B_i \in \mathcal{B}(\mathbb{R})$, on $(V, \mathcal{F})$ we can define the standard Wiener measure Q.

The two ways of bringing maximum entropy and stochastic into the game are dictated by the two ways thinking of (8.1).

Case 1: The random function approach We may just think of $x(t)$ as a continuous function and search for a measure $P \ll Q$ on $(V, \mathcal{F})$ such that $E_P[X(t)] = x(t)$.

Case 2: The random measure approach Or we may think of $x(t)$ as the density of a measure $x(t)dt$. For this we shall consider the random measure $X(a,b] = X(b) - X(a)$ on $[0,1]$ search for a $P \ll Q$ on $(V, \mathcal{F})$ such that $dE_P[X[0,t]] = x(t)dt$.

We shall develop both points of view, but go with the second as the profitable one, for it allows extensions to more general situations.

8.1.1 *The random function approach*

To recast (8.1) as a problem in entropy maximization, our basic setup consists of $(V, \mathcal{F}, Q)$ as described above, and we want to search for a measure P on $(V, \mathcal{F})$ such that $P \ll Q$ and

$$E_P\left[\int_0^1 A(s,t)X(t)dt\right] = \int_0^1 A(s,t)E_P[X(t)]dt = y(s). \qquad (8.2)$$

Actually, our setup allows for more flexibility in choosing the measure Q. Note that with respect to Q we have $E_Q[X(t)] = 0$. But given $x_0(t) = \int_0^t \xi_0(s)ds$, we can define Q_{x_0} on $(V, \mathcal{F})$ such that $E_{Q_{x_0}}[X(t)] = x_0(t)$. All that is needed is that $\int_0^1 [\xi_0(s)]^2 ds < \infty$. The relation between the integral with respect to Q_{x_0} and the integral with respect to Q is contained in (the Camerom-Martin theorem): for any bounded, $\mathcal{F}$-measurable $H : V \to \mathbb{R}$, we have

$$E_{Q_{x_0}}[H] = E_Q[HM] \qquad (8.3)$$

where $M = \exp\left\{\int_0^1 \xi_0(s)dX(s) - \frac{1}{2}\int_0^1 [\xi_0(s)]^2 ds\right\} = \frac{dQ_{x_0}}{dQ}$.

The $x_0(t)$ plays the role of approximate solution (or best guess at $x(t)$). If $x_0(t)$ happened to satisfy (8.1), then Q_{x_0} is the P satisfying (8.2). But generally we are not so lucky. We shall search for $P \ll Q_{x_0}$ such that (8.2) holds.

To begin our search, we assume that the class

$$\mathcal{P} = \left\{P \ll Q_{x_0} : E_P\left[\int_0^1 A(s,t)X(t)dt\right] = y(s)\right\}$$

is not empty. Then it is a convex set and on it we define the concave entropy functional by

$$S_{Q_{x_0}}(P) = -\int_V \frac{dP}{dQ_{x_0}} \ln\left(\frac{dP}{dQ_{x_0}}\right) dQ_{x_0}$$

whenever $\left|\ln\left(\frac{dP}{dQ_{x_0}}\right)\right|$ is P-integrable or $-\infty$ when not. The next step is to maximize $S_{Q_{x_0}}(P)$ to produce the maxentropic P^* solving our problem. We argued in Chapter 5 that $\frac{dP^*}{dQ_{x_0}}$ should be of the form

$$\frac{dP^*}{dQ_{x_0}} = \frac{1}{Z_{Q_{x_0}}(\lambda)} \exp(-\langle \lambda, AX \rangle). \qquad (8.4)$$

It is here that the specific choice of W enters. When $W = \mathcal{C}(S)$ (with the topology associated with the $\|...\|_\infty$ norm), then $W^* = \mathcal{M}(S) =$ class of measure of total finite variation, and

$$\langle \lambda, AX \rangle = \int_S \lambda(ds) \int_0^1 A(s,t) X(t) dt. \qquad (8.5)$$

When $W = L_1(S, \mathcal{B}(S), m)$, then $W^* = L_\infty(S, \mathcal{B}(S))$ and now

$$\langle \lambda, AX \rangle = \int_S \lambda(ds) \left[\int_0^1 A(s,t) X(t) dt\right] m(ds). \qquad (8.6)$$

Notice that in book cases we can think of $\langle \lambda, AX \rangle$ as

$$\langle \tau, X \rangle = \int_0^1 \tau(t) X(t) dt$$

where $\tau(t) = (A^*\lambda)(t)$ equals $\int_S \lambda(ds) A(s,t)$ or $\int_S \lambda(s) A(s,t) m(ds)$ and is a continuous function (of t) on $[0,1]$. If we define $T(t) = \int_0^t \tau(t') dt'$ we can write

$$\langle \tau, X \rangle = +\int_0^1 X(t) dT(t)$$

$$= -\int_0^1 T(t) dX(t) + (T(1)X(1) - T(0)X(0))$$

$$= \int_0^1 \left(\int_t^1 \tau(t') dt'\right) dX(t)$$

after some rearranging. Along the way made use of the fact that $E_{Q_{x_0}}[X(0)] = 0$. This computation will be used to complete the description of (8.4). For that we need

$$Z_{Q_{x_0}}(\lambda) = E_{Q_{x_0}}\left[e^{-\langle \lambda, AX \rangle}\right]$$

which according to (8.3) is

$$Z_Q\left(e^{-\langle\lambda, AX\rangle}M\right)$$
$$= E_Q\left[\exp-\left\{\int_0^1\left(\xi_0(t)-\int_t^1\tau(t')dt'\right)dX(t)\right\}\right]e^{-\frac{1}{2}\int_0^1[\xi_0(t)]^2dt}$$
$$= \exp\left\{\frac{1}{2}\int_0^1\left(\xi_0(t)-\int_t^1\tau(t')dt'\right)^2dt - \frac{1}{2}\int_0^1[\xi_0(t)]^2dt\right\}$$
$$= \exp\left\{\frac{1}{2}\int_0^1\left(\int_t^1\tau(t')dt'\right)^2dt - \int_0^1\xi_0(s)\left(\int_s^1\tau(t')dt'\right)dt\right\}.$$

These computations are routine for those that have gone through the basics of stochastic calculus. In order to compute the minimum of $\Sigma_{Q_{x_0}}(\lambda) = \ln(Z_{Q_{x_0}}(\lambda)) + \langle\lambda,y\rangle$, we shall reorganize the end point of the above chain.

For not to repeat an almost identical computation, denote both $\lambda(ds)$ and $\lambda(s)m(ds)$ by $\lambda(ds)$. Rewrite

$$\int_t^1\tau(t')dt' = \int_t^1 dt'\int_S\lambda(ds)A(s,t')$$
$$= \int_S\lambda(ds)\int_t^1 A(s,t')dt' := \int_S\lambda(ds)a(s,t)$$

where we set $a(s,t) = \int_t^1 A(s,t')dt'$. All exchanges can be justified by invoking Fubini's Theorem. Put $M(s,s') = \int_0^1 a(s,t)a(s',t)dt$ and verify that

$$Z_{Q_{x_0}}(\lambda) = \exp\left\{\frac{1}{2}\int_S\int_S\lambda(ds)\lambda(ds')M(s,s') - \int_S\int_0^1\lambda(ds)a(s,t)\xi_0(t)dt\right\}$$

and therefore

$$\Sigma_{Q_{x_0}}(\lambda) = \ln\left(Z_{Q_{x_0}}(\lambda)\right) + \langle\lambda,y\rangle$$
$$= \frac{1}{2}\int_S\int_S\lambda(ds)\lambda(ds')M(s,s') + \int_S\lambda(ds)\left\{y(s) - \int_0^1 a(s,t)\xi_0(t)dt\right\}. \quad (8.7)$$

So, if $\lambda^*(ds)$ minimizes (8.7), a standard argument shows that $\lambda^*(ds)$ must satisfy

$$\int_S M(s,s')\lambda^*(ds) = \int_0^1 a(s,t)\xi_0(t)dt - y(s) \quad (8.8)$$

where the symmetry of $M(s,s')$ has been made use of. The kernel $M(s,s')$ in (8.8) defines a map from W^* into W. Of course, the idea is that it is simpler to solve λ^* in (8.8) than to solve (8.1). Assuming $\lambda^*(ds)$ could be explicitly found, what does $x^*(t)$ look like?

For that, let us rewrite the quadratic form in (8.7) slightly. Set $\alpha(t) = \int_S \lambda(ds)a(s,t)$. Then

$$\int_S\int_S\lambda(ds)M(s,s')\lambda(ds) = \int_0^1 dt\left(\int_t^1\alpha(t')dt'\right)^2$$
$$= 2\int_0^1 t\alpha(t)\int_t^1\alpha(t')dt' = 2\int_0^1\alpha(t)\int_0^1 t'\alpha(t')dt'.$$

With these changes (8.8) looks like

$$\int_0^1 dt A(s,t) \left\{ -\int_0^t t' \int_S A(s',t') dt' \lambda^*(ds) \right\} = y(s) - \int_0^1 A(s,t) x_0(t) dt$$

where we rewrote $\int_0^1 a(s,t) \xi_0(t) dt$, (remember that $x_0(t) = \int_0^t \xi(s) ds$) as

$$\int_0^1 A(s,t) \left(\int_0^t \xi_0(t') dt' \right) dt = \int_0^1 A(s,t) x_0(t) dt.$$

Therefore

$$\int_0^1 dt A(s,t) \left\{ x_0(t) - \int_0^t t' \left(\int_S \lambda^*(ds) A(s',t') \right) dt' \right\} = y(s)$$

or equivalently

$$x^*(t) = x_0(t) - \int_0^t t' \left(\int_S \lambda^*(ds) A(s',t') \right) dt' \qquad (8.9)$$

solves (8.1), once $\lambda^*(ds)$ solving (8.8) is available.

8.1.2 The random measure approach: Gaussian measures

Again the basic framework will be as in the preceding section, but this time we shall search for a measure P on $(V, \mathcal{F})$, $P \ll Q_{x_0}$, such that

$$E_P \left[\int_0^1 A(s,t) dX(t) \right] = y(s). \qquad (8.10)$$

The search should be conducted in the class

$$\mathcal{P} = \left\{ P \ll Q_{x_0} : E_P \left[\int_0^1 A(s,t) dX(t) \right] = y(s) \right\}$$

which is convex when not empty, which we assume to be the case. The MEM calls for a $\rho(w) = \frac{dP}{dQ_{x_0}}$ maximizing $S_{Q_{x_0}}(\rho) = -\int_V \rho \ln(\rho) dQ_{x_0}$. As above

$$dP^* = \frac{e^{-\langle \lambda^*, AX \rangle}}{Z_{Q_{x_0}}(\lambda^*)} dQ_{x_0}.$$

To find λ^* we must first compute $Z_{Q_{x_0}}(\lambda) = E_{Q_{x_0}} \left[e^{-\langle \lambda^*, AX \rangle} \right]$. Set

$$\langle \lambda, AX \rangle = \int_0^1 \beta(t) dX(t)$$

with $\beta(t) = \int_S \lambda(ds) A(s,t)$. Under our assumptions on $A(s,t)$ it is simple to verify that $\int_0^1 \beta^2(t) dt < \infty$, so the stochastic integral makes sense. Thus

$$E_{Q_{x_0}} \left[e^{-\langle \lambda^*, AX \rangle} \right] = E_Q \left[e^{-\int_0^1 (\beta(t) - \xi_0(t)) dX(t)} \right] e^{-\frac{1}{2} \int_0^1 \xi_0^2(s) ds}$$

$$= \exp \left\{ \tfrac{1}{2} \int_0^1 \beta^2(t) dt - \int_0^1 \beta(t) \xi_0(t) dt \right\}.$$

After this we must minimize

$$\Sigma_{Q_{x_0}}(\lambda) = \frac{1}{2}\int_S\int_S \lambda(ds)\lambda(ds')N(s,s') + \int_S \lambda(ds)\left\{\int_0^1 A(s,t)\xi_0(t)dt + y(s)\right\}$$

where we introduced the (more or less obvious) notation

$$N(s,s') = \int_0^1 A(s,t)A(s',t')dt$$

which has a simpler inner structure than the corresponding kernel $M(s,s')$ introduced in the previous section. This time the corresponding equation for the $\lambda^*(ds)$ that minimizes $S_{Q_x}(\lambda)$ is

$$\int_S N(s,s')\lambda^*(ds) = \int_0^1 A(s,t)\xi_0(t)dt + y(s).$$

If such equation can be solved, then using the representation for $N(s,s')$ given a few lines above, we obtain, after exchanging some integrals

$$\int_0^1 A(s,t)\left\{\xi_0(t) - \int \lambda^*(ds')A(s',t)\right\}dt = y(t)$$

or equivalently, the maxentropic solution to (8.1) is this time

$$x^*(t) = \xi_0(t) - \int_S \lambda^*(ds)A(s',t). \tag{8.11}$$

Note that the λ^* appearing in (8.11) and (8.9) are quite different objects. The only way of deciding whether the $x^*(t)$ in (8.11) and the $x^*(t)$ in (8.9) coincide is from the possible uniqueness of the solutions to (8.1), and the usual nagging question is: How to rate one solution against the other?

8.1.3 The random measure approach: Compound Poisson measures

A variation of the previous section is provided by choosing $X(t)$ to be a compound Poisson process on [0,1]. This time $\Omega = D([0,1])$ is the class of right continuous functions on [0,1] that have left limits on [0,1], and finitely many jumps in any time interval. The usual σ-algebra generated by the cylinder sets, one can define a measure Q, with respect to which $t \to X(t)$ is a stochastic process that has independent increments, distributed according to compound Poisson law. That is, for each $(a,b] \subset [0,1]$, the random variable $m(a,b) := X(b) - X(a)$ can be written as

$$m((a,b]) = \sum_{N(a)+1}^{N(b)} \xi_k$$

where $t \to N(t)$ is a standard Poisson process on $[0,1]$ with intensity μt, or

$$E_Q\left[e^{-\alpha N(t)}\right] = e^{-\mu t(1-e^{-\alpha})}, \quad \alpha > 0$$

and the $\{\xi_k : k \geq 1\}$ are independent, identically distributed real-valued random variables with common law $q(d\xi) = Q(\xi_k \in d\xi)$. Using this one can prove that

$$E_Q\left[e^{-\alpha N(t)}\right] = \exp\{-\lambda t \int (1 - e^{-\alpha \xi})q(d\xi)$$

and even more generally, for continuous $g : [0,1] \to \mathbb{R}$, define $\int_0^1 g(t)dX(t)$ the obvious way, since it is an integral with respect to the Stieltjes measure $dX(t)$. Then, approximating $\int_0^1 g(s)dX(s)$ by $\sum_{j=1}^{n} g\left(\frac{j-1}{n}\right) m(\left(\frac{j-1}{n}, \frac{j}{n}\right])$ and using the independence of the $m(\left(\frac{j-1}{n}, \frac{j}{n}\right])$ we obtain

$$E_Q\left[e^{-\int_0^1 g(s)dX(s)}\right] = \lim_n \prod_{j=1}^{n} E_Q\left[e^{-g(\frac{j-1}{n})m((\frac{j-1}{n},\frac{j}{n}])}\right]$$

$$= \lim_n \prod_{j=1}^{n} \exp\left\{-\lambda \frac{1}{n} \int \left((1 - e^{-g(\frac{j-1}{n})\xi})q(d\xi)\right)\right\}$$

$$= \exp\left\{-\lambda \int_0^1 ds \int_S \left((1 - e^{-g(s)\xi})q(d\xi)\right)\right\}. \quad (8.12)$$

Now we proceed to the maxentropic solution to (8.10), with $X(t)$ being the compound process just described.

Consider as usual

$$\mathcal{P} = \left\{P \ll Q : E_P\left[\int_0^1 A(s,t)dX(t)\right] = y(s)\right\} \quad (8.13)$$

and the standard reasoning leads to search for a $\lambda^*(ds)$ such that

$$dP^* = \frac{e^{-\langle \lambda^*, AX \rangle}}{Z(\lambda^*)} dQ$$

is in the class $\mathcal{P}$ described in (8.13). Here $\langle \lambda^*, AX \rangle = \int_0^1 g^*(t)dX(t)$ with $g^*(t) = \int_S \lambda^*(ds)A(s,t)$ being a continuous function. In general, $Z(\lambda)$ is given by (8.12) for arbitrary Borel measure $\lambda(ds)$ of finite total variation. The function to be minimized to find λ^* this time is

$$\Sigma_Q(\lambda) = \int_0^1 dt \int_S \left(e^{-g(t)\xi} - 1\right) q(d\xi) + \int_S \lambda(ds)y(s)$$

where $g(t) = \int_S \lambda(ds)A(s,t)$.

The $\lambda^*(ds)$ that minimizes $S_Q(\lambda)$ satisfies for each $s \in S$

$$\int_0^1 dt A(s,t) \left\{ \int_S \xi e^{-g^*(t)\xi} q(ds) \right\} = y(s)$$

from which we can read the maxentropic solution to (8.1). It is

$$x^*(t) = \int_S \xi e^{-g^*(t)\xi} q(ds). \tag{8.14}$$

Exercise 8.1. The reader should work out the prediction implicit in (8.14) for

a) $q(d\xi) = \frac{1}{2} \{\varepsilon_{+h}(d\xi) + \varepsilon_{-h}(d\xi)\}$
b) $q(d\xi) = \mu e^{-\mu \xi} d\xi$, supported on $[0, \infty)$.
c) $q(d\xi) = \frac{1}{\sqrt{2\pi}} \exp\left(-\frac{1}{2}(\xi - \xi_0)^2\right)$.

Comment: We could get "more amusing" results, should we consider process with independent but not necessarily stationary increments.

8.1.4 The random measure approach: Gaussian fields

In this section we repeat the procedure developed in Section 8.1.2 to obtain a maxentropic solution to the following extension of (8.1). Consider the equation

$$\int_T A(s,t)x(t)dt = y(s) \tag{8.15}$$

where, assume that, $A : S \times T \to \mathbb{R}$ is a continuous function. Now T is another compact, metric, separable space, and dt denotes a Borel measure on $(T, \mathcal{B}(T))$.

Let us now denote by $X(A) : \mathcal{B}(T) \to L_2(\Omega, \mathcal{F}, Q)$ a Gaussian field on $(T, \mathcal{B}(T))$, that is for every $A \in \mathcal{B}(T)$, $X(A)$ is a Gaussian random variable, and the map $A \to X(A)$ is such that

i) $E_Q[X(A)X(B)] = c(A \cap B)$ for an appropriate Borel measure $c(dt)$ on $(T, \mathcal{B}(T))$, aptly called the correlation measure of the field. We assume that $c(dt) = \gamma(t)dt$, where dt appears in (8.15).

ii) If $A = \cup_1^\infty A_i$, A_i pairwise disjoint, then $X(A) = \sum_1^\infty X(A_i)$, the sum being convergent in $L_2(\Omega, \mathcal{F}, Q)$.

From these properties we obtain that if $g(t)$ is continuous on T, then

$$E_Q\left[\exp\left(-\int_T g(t)X(dt)\right)\right] = \exp\left(\frac{1}{2} \int_T (g(t))^2 c(dt)\right).$$

From this it should be clear that recasting (8.15) as: Find a measure P on $(\Omega, \mathcal{F})$, $P \ll Q$ such that

$$E_P\left[\int A(s,t)X(dt)\right] = y(s) \qquad (8.16)$$

will lead us to a result similar to (8.11).

After repeating the usual steps, one would be led to a maxentropic P^* on $(\Omega, \mathcal{F})$ given by

$$\frac{dP^*}{dQ} = \frac{\exp(-\langle \lambda^*, AX \rangle)}{Z_Q(\lambda^*)}$$

where $\lambda^*(ds)$ is a measure of finite total variation on $(S, \mathcal{B}(S))$ satisfying

$$\int M(s,s')\lambda^*(ds) = -y(s)$$

where $M(s,s') = \int_T A(s,t)A(s',t)c(dt)$. Once $\lambda^*(ds)$ has been found, the analogue of (8.9) is

$$x^*(t) = -\int_S \lambda^*(ds')A(s',t)\gamma(t).$$

8.1.5 Closing remarks

We rushed a bit too much through the last few sections. This was to avoid being too boring and repetitive. Besides filling in missing details, there are a few loose ends to be tied and possible interesting extensions.

As hinted in Section 8.1.4, the equation (8.1) can be generalized quite a bit: there we may search for a Banach space valued function $t \to x(t)$, such that

$$\int_T A(s,dt)X(t) = y(s)$$

where $A : S \times \mathcal{B}(T)$ is an operator-valued kernel. Certainly the obvious and ubiquitous cases correspond to finite dimensional spaces and/or matrix valued kernels.

Also playing with the *a priori* measures may lead to unexpected connections. Time may tell.

8.2 A simple example: Inversion of a Fourier transform given a few coefficients

Consider the following version of (8.1): Find a continuous, real valued function $x(t)$ on $[0,1]$, such that

$$\int_0^1 e^{2\pi int} x(t) dt = y(n), \quad -N \leq n \leq N. \tag{8.17}$$

Certainly $y(-n) = \overline{y(n)}$. So $T = [0,1]$ and $S = \{-M, ..., M\}$. Now $A(n,t) = \exp(2\pi int)$, $W = \mathbb{C}^{2M+1}$, $W^* = \mathbb{C}^{2M+1}$.

The matrix $M(n, n')$ described in section 8.1.1 is given by

$$M(n, n') = \frac{1}{(2\pi)^2 nm} \{1 + \delta_{m,-n} - \delta_{n,0} - \delta_{m,0}\}, \quad -N \leq m, n \leq N.$$

This is an odd sized squared matrix, with a central row and column of zeros. It is easy to see that it is invertible. Since W^* is finite dimensional, equation (8.8) to determine λ^* becomes

$$\sum_{n=-N}^{N} M(n, n') \lambda_{n'}^* = \int_0^1 a(n,t) \xi_0(t) dt - y(n)$$

where $a(n,t) = \frac{1}{2\pi in}\left(1 - e^{2\pi int}\right)$. From this we can solve for λ^* and substitute in (8.9) to obtain

$$x^*(t) = x_0(t) - \int_0^1 t' (\sum_n \lambda_n^* e^{2\pi int'}) dt'$$

which is the maxentropic solution to (8.17).

The random measure approach described in Section 8.1.2 yields nicer looking results.

The basic setup is above, except that instead of the matrix $M(n, n')$ we have

$$N(n, n') = \int_0^1 A(n,t) A(n, t') dt = \delta_{n+n',0}$$

which is odd square matrix with ones in the antidiagonal, which has its own inverse. The equation determining λ^* is

$$\sum_{-M}^{M} N(n, n') \lambda_{n'}^* = \int_0^1 e^{2\pi int} \xi_0(t) dt - y(n).$$

Therefore
$$\lambda_n^* = \int_0^1 e^{-2\pi i n t}\xi_0(t)dt - y(-n) = \hat{\xi}_0 - y(-n).$$
If we are to substitute in (8.11), i.e., in
$$x^*(t) = \xi_0(t) - \sum_n \lambda_n^* e^{2\pi i n t}$$
we will obtain the following maxentropic solution to (8.17):
$$x^*(t) = \sum_{-M}^{M} y(n)e^{-2\pi i n t} + \xi_0(t) - \sum_{-M}^{M} \hat{\xi}_0(n)e^{-2\pi i n t}$$
and we leave the obvious interpretation to the reader. It is the fact that the probabilistic machinery, combined with the maximum entropy method produces such result, which is fun here.

8.3 Maxentropic regularization for problems in Hilbert spaces

In this section we are going to carry out the analogue of what we did in section 6.5.5. The aim is to obtain, in an infinite dimensional setting, some of the regularization schemes presented in Chapter 4, thus providing more evidence showing that the MEM extends and comprises those approaches.

As in Chapter 4, let V be some separable Hilbert space, and let $A: V \to V$ be a given compact operator on V. Let us take $\Omega = \mathbb{R}^{\mathbb{N}} = \{\xi : \mathbb{N} \to \mathbb{R}\}$, and let $\mathcal{F}$ be the σ-algebra generated by the cylindrical subsets of Ω. Below we shall consider two measures on $(\Omega, \mathcal{F})$. Each of them will yield one of the regularization procedures described in Chapter 4.

8.3.1 Gaussian measures

On each factor $\mathbb{R}$ of Ω we shall place a Gaussian measure as follows: on the n-th factor $(\mathbb{R}, \mathcal{B}(\mathbb{R}))$ we define
$$Q_n(d\xi_n) = \frac{1}{\sqrt{2\pi\sigma_n}} \exp(-\frac{(\xi_n - m_n)^2}{2\sigma_n})d\xi_n$$
(notice that we use σ_n instead of the usual σ_n^2). On $(\Omega, \mathcal{F})$ we put
$$Q(d\xi) = \otimes_{n \geq 1} Q_n(d\xi_n).$$

Since we want to solve $Ax = y$, without convex constraints, we take $C = V$ in the notation of Chapter 5. If we form an orthonormal basis

$\{e_n : n \geq 1\}$ in V, we want to define the coordinate map $X : \Omega \to V$, by the obvious

$$X(\xi) = \sum_{n \geq 1} \xi_n e_n.$$

For $X(\xi)$ to be well defined it suffices that $m = \sum m_n e_n$ be such that $\sum m_n^2 < \infty$ and that

$$E_Q\left[\|X(\xi) - m\|^2\right] = \sum_{n \geq 1} E_Q\left[(\xi_n - m_n)^2\right] = \sum_{n \geq 1} \sigma_n < \infty.$$

Had this not been the case, we would have to identify V with some (obvious) measurable subset of Ω. We could if we so wanted, transport the measure Q onto a measure on $(V, \mathcal{B}(V))$. As a measure on V, it can be characterized by its (Fourier) Laplace transform

$$E_Q\left[e^{-\langle \eta, X \rangle}\right] = \exp\left(-\langle \eta, m \rangle + \frac{1}{2}\langle S\eta, \eta \rangle\right) \qquad (8.18)$$

where $\eta \in V$, and S is a trace class operator, which in the basis $\{e_n : n \geq 1\}$ has trace

$$\sum_{n \geq 1} \langle S e_n, e_n \rangle = \sum_{n \geq 1} \sigma_n < \infty.$$

As usual we have to form the class of measures on $(V, \mathcal{B}(V))$

$$\mathcal{P} = \{P \ll Q : E_P[AX] = A E_P[X] = y\} \qquad (8.19)$$

on which we define the entropy $S_Q(P) = -\int_V \rho(\xi) \ln(\rho(\xi)) dQ(\xi)$, and the usual argument leads us to

$$dP^* = \frac{e^{-\langle \lambda^*, AX \rangle}}{Z_Q(\lambda^*)} dQ(\xi). \qquad (8.20)$$

Replacing η in (8.18) by $A^*\lambda$, to obtain the λ^* in (8.20) we have to minimize

$$S_Q(\lambda) = \ln(Z_Q(\lambda)) + \langle \lambda, y \rangle = \frac{1}{2}\langle \lambda, ASA^*\lambda \rangle + \langle \lambda, (y - y_0)\rangle$$

where we set $y_0 = Am$ for short. To obtain λ^* we need to solve

$$ASA^*\lambda^* = y_0 - y.$$

When ASA^* is invertible, we solve for λ^*. From the same equation we obtain

$$A\left(m - SA^*(ASA^*)^{-1}(y_0 - y)\right) = y$$

which means that
$$x^* = m + SA^*(ASA^*)^{-1}(y - y_0) \qquad (8.21)$$
is the maxentropic solution to $Ax = y$. Comparing with the results of Section 4.4 or 4.6, we see that it provides us with a maxentropic version of the Gaussian regularization/maximum likelihood methods, when there is no error in the data.

To cope with the case where ASA^* is non-invertible, we have to consider a relaxed version of $AX = y$ and look at
$$Ax \in B(y, T). \qquad (8.22)$$

Proceeding as in Section 6.6, we are again led to a representation like (8.20), but λ^* is to be obtained by minimizing
$$S_Q(\lambda) = \frac{1}{2} \langle \lambda, ASA^* \lambda \rangle + \langle \lambda, (y - y_0) \rangle + T \|\lambda\|.$$

The appropriate λ^* has to satisfy
$$ASA^* \lambda^* + T \frac{\lambda^*}{\|\lambda^*\|} = y_0 - y := \delta_y. \qquad (8.23)$$

We shall proceed as in Section 6.5.4, propose a solution $(ASA^* + \alpha I)^{-1} \delta y$, and establish a relationship between the regularization parameter α and the tolerance T.

If $\lambda^* = (ASA^* + \alpha I)^{-1} \delta y$ is to solve (8.23), then α and T have to be related by
$$\alpha \|(ASA^* + \alpha I)^{-1} \delta y\| = T. \qquad (8.24)$$

Lemma 8.1. *The function $\alpha \to F(\alpha) = \alpha^2 \|(ASA^* + \alpha I)^{-1} \delta y\|^2$ defined on $F : [0, \infty)$ is strictly increasing and bounded by $\sum \frac{|\langle \delta y, \eta_n \rangle|^2}{\lambda_n^2}$, where $(\xi_n, \eta_n, \lambda_n)$ is the singular system for A introduced in Section 4.1.*

Proof. Write $F(\alpha)$ as
$$\sum \frac{\alpha^2}{\alpha^2 + \lambda^2} |\langle \delta y, \eta_n \rangle|^2.$$
Since the derivative of $x \to \frac{x}{(x+c)}$ is $\frac{c}{(x+c)^2}$, we can conclude that $F(\alpha)$ is strictly increasing, $F(0) = 0$ and $\lim_{\alpha \to \infty} F(\alpha) = \sum \frac{|\langle \delta y, \eta_n \rangle|^2}{\lambda_n^2} := F(\infty)$. Thus the correspondence $\alpha \to T$ given is bijective for $T < F(\infty)$.

We leave it to the reader to verify that for the appropriate $\alpha(T)$
$$x^* = m + SA^*(ASA^* + \alpha I)^{-1}(y - y_0)$$
solves (8.22). $\square$

8.3.2 Exponential measures

Let us now consider the other theme developed in Section 6.5.4, and define on $(\Omega, \mathcal{F})$ the product measure

$$dQ(\xi) = \otimes_{n\geq 1} \left(\frac{\mu_n}{2}\right) e^{-\mu_n|\xi_n - m_n|} d\xi_n$$

where, as above, $\sum m_n^2 < \infty$ for $\sum m_n e_n$ to be in V. Also $X(\xi) = \sum \xi_n e_n$ lies in V for any $\xi \in \Omega$ if we require $\sum \frac{1}{\mu_n} < \infty$.

Also, for $|\eta_n| < \mu_n$ $\forall n$, we have

$$E_Q\left[e^{-\langle \eta, X \rangle}\right] = \prod_{n\geq 1} \frac{1}{\mu_n^2 - \eta_n^2}$$

whenever the product is convergent. We thus have

$$\mathcal{D}(A, Q) = \left\{\lambda \in V : |(A^*\lambda)_n| < \mu_n, \prod_{n\geq 1} \frac{1}{\mu_n^2 - (A^*\lambda)_n^2} < \infty\right\}.$$

To find a maxentropic solution to $Ax = y$ we have to find $dP^*(\xi)$ as in (8.20) where λ^* is to be found minimizing

$$S_Q(\lambda) = \sum_{n\geq 1} \ln\left(\frac{1}{\mu_n^2 - (A^*\lambda)_n^2}\right) + \langle \lambda, y - Am \rangle$$

over $\lambda \in \mathcal{D}(A, Q)$. As we saw in Chapter 5, if such λ^* were to exist, it would have to satisfy

$$2\sum_{k\geq 1} \frac{A_{nk}(A^*\lambda^*)_k}{\mu_k^2 - (A^*\lambda)_k^2} = (Am - y)_n$$

where $(A^*\lambda)_k = \langle e_k, A^*\lambda \rangle = \langle \lambda, Ae_k \rangle$ and certainly $\langle e_k, Ae_n \rangle = A_{nk}$ in the given basis $\{e_n : n \geq 1\}$.

As at the end of Chapter 6, if A^*A were invertible, we would obtain $x^* = \sum \langle x^*, e_n \rangle e_n$, with

$$\langle x^*, e_n \rangle = \frac{(A^*\lambda^*)_n}{\mu_n^2 - (A^*\lambda^*)_n}$$

$$= \langle e_n, (A^*A)^{-1} A^*(Am - y) \rangle$$

$$= \langle e_n, m \rangle - \langle e_n, (A^*A)^{-1} A^* y \rangle.$$

8.3.3 Degenerate measures in Hilbert spaces and spectral cut off regularization

We saw in Section 4.3 how to obtain approximate solutions to $ax = y$ by neglecting small singular values. Here we shall place the same assumptions on the singular system $(\xi_n, \eta_n, \lambda_n)$ of A, namely the $A(V)$ is dense in V, that A is injective and $\sum \lambda_n^{-2} < M$.

For any $\alpha > 0$, let

$$N(\alpha) = \max\{n : \lambda_n \geq \alpha\}$$

and for each $\alpha > 0$, we shall define a measure Q^α on $(\Omega, \mathcal{F})$ which is non-trivial in the first $N(\alpha)$ factors of $\Omega = \mathbb{R}^\mathbb{N}$. For that set

$$dQ_n(\omega_n) = \begin{cases} \frac{1}{(2\pi\sigma_n)^{1/2}} \exp\left(-\frac{(\omega_n - m_n)^2}{2\sigma_n}\right) d\omega_n, & n \leq N(\alpha) \\ \varepsilon_0(d\omega_n), & n > N(\alpha) \end{cases}.$$

Everything else being pretty much as above. The coordinate maps are now $X(\omega) = \sum_{n=1}^{N(\alpha)} \omega_n \xi_n$ for $\omega = (\omega_1, \omega_2, ...) \in \Omega$, thus clearly

$$E_Q[X] = \sum_{n \leq N(\alpha)} m_n \xi_n, \quad E_Q\left[\|X - m\|^2\right] = \sum_{n \leq N(\alpha)} \sigma_n.$$

Then, as long as we keep $\alpha > 0$ (or $N(\alpha) < \infty$) we do not need to impose further conditions on m and σ.

Note now that if we follow the routine of Section 8.3.1 we would be led to solve

$$ASA^*\lambda^* = y - y_0$$

where again $y_0 = Am$. Having chosen $X(\omega) = \sum \omega_n \xi_n$, since $A = \sum \lambda_n \eta_n \otimes \xi_n^*$ and $A^* = \sum \lambda_n \xi_n \otimes \eta_n^*$, where $\xi_n^*(x) = \langle \xi_n, x \rangle$ and $\eta_n^*(x) = \langle \eta_n^*, x \rangle$ for any $x \in V$, we can solve for λ^*:

$$\lambda^* = \sum_{n \leq N(\alpha)} (\lambda_n^2 \sigma_n)^{-1} \eta_n \langle \eta_n, (y - y_0) \rangle$$

which yields

$$x^*(\alpha) = m + \sum_{n \leq N(\alpha)} \lambda_n^{-1} \xi_n \langle \eta_n, (y - y_0) \rangle$$

which becomes the $R(\alpha)y$ of Section 4.3.

8.3.4 *Conclusions*

Even though we cannot claim that every regularization method to solve $Ax = y$ in a Hilbert space setup has a maxentropic counter part, some of the common ones certainly do.

8.4 Bibliographical comments and references

Material for Section 8.1 is a variation on the theme of results in [1]. The rest of the material is "new" but easily recognizable as "variations on the same theme".

Reference

[1] Gzyl, H. *"Maxentropic reconstructions of Fourier and Laplace transforms"*. Appl. Math. and Computation, Vol. 73 (1995) pp. 181-189.

Chapter 9

Tomography, reconstruction from marginals and transportation problems

9.1 Generalities

Besides being eye catchers, the title describes three classes of problems with quite a bit of mathematics in common.

The easiest to describe is the problem of reconstruction of a finite measure on a product space from its marginals. The simplest version goes as follows. Given $S = S_1 \times S_2$ where S are locally compact, separable, topological spaces and $\mathcal{B}(S) = \mathcal{B}(S_1) \otimes \mathcal{B}(S_2)$ is the σ-algebra generated by the rectangles.

Let $m_i(\cdot) : (S_i, \mathcal{B}(S_i))$ be two finite (or probability) measures.

The problem of reconstruction from marginals consists of finding a finite (or a probability) measure m on $(S, \mathcal{B}(S))$ such that

$$\begin{aligned} m(A_1 \times S_2) &= m_1(A_1) \quad \text{for} \quad A_1 \in \mathcal{B}(S_1) \\ m(S_1 \times A_2) &= m_2(A_2) \quad \text{for} \quad A_2 \in \mathcal{B}(S_2). \end{aligned} \quad (9.1)$$

When there exist measures $\lambda_1(dx_1)$ and $\lambda_2(dx_2)$ such that $m_i(dx_i) = \rho_i(x_i)\lambda(dx_i)$, we may inquire about the existence of a positive function $\rho(x, y)$ measurable with respect to $\mathcal{B}(S)$ such that

$$\begin{aligned} \rho_1(x_1) &= \int_{S_2} \rho(x_1, x_2)\lambda_2(dx_2) \\ \rho_2(x_2) &= \int_{S_1} \rho(x_1, x_2)\lambda_1(dx_1). \end{aligned} \quad (9.2)$$

An interesting variant of problems (9.1) and (9.2) was proposed by Schrödinger during the 1930s. In that version $S_1 = S_2$, $\lambda_1(\cdot) = \lambda_2(\cdot) = \lambda(\cdot)$, and we are given a positive $q(x, y) : S \times S \to [0, \infty)$ and the problem consists of finding positive functions $\phi(x)$ and $\psi(y)$ such that

$$\begin{aligned} \rho_1(x) &= \phi(x) \int q(x, y)\psi(y)\lambda(dy) \\ \rho_2(y) &= \psi(y) \int q(x, y)\phi(x)\lambda(dx). \end{aligned} \quad (9.3)$$

Or in a version that includes the discrete case, we want to search for measures $v_1(dx_1)$ and $v_2(dx_2)$ such that the measure

$$\mu(A_1 \times A_2) = \int_{A_1} \int_{A_2} q(x_1, x_2) v_1(dx_1) v(dx_2) \qquad (9.4)$$

defined on $\mathcal{B}(S)$ satisfies (9.1), i.e.

$$\begin{aligned} \mu(A_1 \times S_2) &= m_1(A_1) \\ \mu(S_1 \times A_2) &= m_2(A_2). \end{aligned} \qquad (9.5)$$

We refer the reader to the papers by Jamison cited at the end for more about these topics.

Problem (9.2) is perhaps a harder case of a problem introduced and solved by Radon in 1917. Consider a bounded convex subset K of the plane. Then any continuous, positive function $f(x_1, x_2)$ with support in K is determined by its integrals along cords cutting K. See [3] for an elementary presentation and connections with Fourier analysis.

In [4] Renyi relates Radon's result to a result by Cramer and Wold that asserts that: Any probability measure on the plane is completely determined when its integral along any line through the origin is known.

Problem (9.1) and usually its discrete, or finite dimensional version, comprise the constraints of another class of problems: transportation and/or other standard problems in linear programming or operations research in general. The first step towards a solution consists of finding feasible solutions, and this is just what problem (9.1) consists of.

The generic formulation that uses most of the set up of (9.1) is the following: Find a probability measure m on $(S, \mathcal{B}(S))$ such that

$$m = \arg\inf \left\{ \int_S c(x_1, x_2) m(dx_1, dx_2) : m \text{ satisfies (9.1)} \right\}. \qquad (9.6)$$

Here $C : S \to \mathbb{R}$ is some (usually, positive) bounded cost function. The quantity $c(x_1, x_2)$ is thought of as the cost of "transporting" goods between origin x_1 where the total output is $m_1(dx_1)$ to destination x_2 where demand is $m_2(dx_2)$. Problem (9.6) consists of finding a transport program that minimizes total cost within the given constraints.

Thus the common thread is again apparent: To produce an optimum m_{op} in (9.6), the class of feasible solutions must be non-empty. To prove it we are again led to problem (9.1) or (9.2).

Actually this situation is common to linear programming problems which in generic abstract setting consist of finding

$$x^* = \arg\inf\{c(x) : x \in K \subseteq V, Ax = y\} \qquad (9.7)$$

where K is some bounded convex set in some Banach space V and $A: V \to W$ is a bounded linear operator between Banach spaces, and $c(x)$ (= $\sum c_j x_j$) is a linear functional.

We see that solving the inverse problem described at the outset of Chapter 5 amounts to prove that the class of feasible solutions to (9.7) is nonempty. Also by appropriately extending A, we will later on provide a maxentropic approach to the optimization problem (9.7).

9.2 Reconstruction from marginals

Let us begin by quoting a related result from Renyi's [4].

Theorem 9.1. *A collection of masses $m_1, ..., m_n$ placed at unknown points $(x_1, y_1), ..., (x_n, y_n)$ in the unit square can be determined from $n+1$ different projections. (Or from their projections on $n+1$ different lines.)*

Note that this time the available data consists of n numbers for each of the $n+1$ different lines. Thus, less data would lead to infinitely many solutions, if a solution can be found.

When the points at which the masses are placed are known, and just the masses are unknown we have the general necessary and existence conditions presented in Strassen's [5].

To state the first result, let $(\Omega_1, \mathcal{F}_1, P_1)$ and $(\Omega_2, \mathcal{F}_2, P_2)$ be two probability spaces and assume a measure $Q(dw_1, dw_2)$ is given on $(\Omega_1 \times \Omega_2, \mathcal{F}_1 \otimes \mathcal{F}_2)$ such that $Q_0(dw_2) = Q(\Omega_1 \times dw_2)$ is σ-finite. We have

Theorem 9.2. *There exists a probability measure $P \leq Q$ (i.e. $P(\Lambda) \leq Q(\Lambda)$ for any $\Lambda \in \mathcal{F}_1 \otimes \mathcal{F}_2$) such that for any $A \in \mathcal{F}_1$ and $B \in \mathcal{F}_2$*

$$P(A \times \Omega_2) = P_1(A), P(\Omega_1 \times B) = P_2(B)$$

if and only if for all $A \in \mathcal{F}_1$, $B \in \mathcal{F}_2$

$$P_1(A) + P_2(B) \leq 1 + P(A \times B).$$

To extend this result a bit assume each Ω_i is a complete separable metric space and let $\Omega = \Omega_1 \times \Omega_2$ as above.

Assume there exist $\hat{\psi}_i : \Omega_i \to [0, \infty)$ continuous and each bounded away from zero. Let $\Pi_i : \Omega \to \Omega_i$ be the obvious projections and consider the Banach spaces

$$V_i := \{f \in \mathcal{C}(\Omega_i) : \|f\|_{\varphi_i} < \infty\}$$

where $\|f\|_{\varphi_i} = \sup\left\{\dfrac{|f(w)|}{\varphi_i(w)} : w \in \Omega_i\right\}$. Similarly define $\varphi : \Omega_1 \times \Omega_2 \to \mathbb{R}$ by $\varphi(x,y) = \varphi_1 \circ \Pi_1 + \varphi_2 \circ \Pi_2$ and consider

$$V := \left\{f \in \mathcal{C}(\Omega_1 \times \Omega_2) : \|f\|_\varphi < \infty\right\}$$

where now $\|f\|_\varphi = \sup\left\{\dfrac{|f(w_1, w_2)|}{\varphi(w_1, w_2)} : (w_1, w_2) \in \Omega_1 \times \Omega_2\right\}$.

To finish let $\mathcal{F}_i = B(\Omega_1)$ and consider the class $\mathcal{P}$ of probability measures on $(\Omega_1 \times \Omega_2, \mathcal{B}(\Omega_1) \otimes \mathcal{B}(\Omega_2))$ on which the topology $\mathcal{T}$ of weak convergence is placed. We are now ready for

Theorem 9.3. *Let $\mathcal{P}_1$ be a closed (in the weak topology) subset of $\mathcal{P}$. Let P_i be probability measures on $(\Omega_i, \mathcal{F}_i)$ such that $\int_{\Omega_i} \varphi_i(w) dP_i(w) < \infty$ for $i = 1, 2$.*

Then there exists $P \in \mathcal{P}_1$ such that $P_i = P \circ \Pi_i^{-1}$ (i.e., having projections P_i) if and only if

$$\int_{\Omega_1} f_1 dP_1 + \int_{\Omega_2} f_2 dP_2 \leq \sup\left\{\int_\Omega (f_1 \circ \Pi_1 + f_2 \circ \Pi_2) dP : P \in \mathcal{P}_1\right\}$$

for any $f_1 \in V_1$ and $f_2 \in V_2$ respectively.

9.3 A curious impossibility result and its counterpart

9.3.1 The bad news

The results we recall below are not impossibility results, but results that allow us to construct non-unique solutions. The result we present below, due to Gutman, Kempemann, Reeds and Shepp, makes one wonder why is it that some reconstruction methods work at all, that is, why is it that we get sensible answers out of the infinitely many possible.

To begin with, recall from Section 9.1, that we can think about marginals as follows: $\Pi_i : S_1 \times S_2 \to S_i$ $\Pi_i(x_1, x_2) = x_i$ and if m_i denotes the i-th marginal of a measure m on $\mathcal{B}(S_1) \otimes \mathcal{B}(S_2)$ then $m_i = \Pi_i(m) = m \circ \Pi_i^{-1}$.

For what comes below we forget about the product structure of S and consider $(S, \mathcal{B}(S))$ as a given measure space and we are given a family (make it finite for concreteness) of measurable maps $\Pi_i : S \to Y_i$, where $(Y_i, \mathcal{F}_i)$ are given measurable spaces for $i = 1, ..., N$. We are given a finite measure $d\lambda$ on $(S, \mathcal{B}(S))$ and we want to find a finite measure $d\mu = gd\lambda$, with $0 \leq g \leq 1$ such that $\pi_i(\mu)$ are preassigned.

Since we are considering measures with bounded densities, it is convenient (to be able to make use of the Krein-Milman Theorem) to bring in $L^\infty(S,\lambda) \subset L^1(S,\lambda)$ and to place upon $L^\infty(S)$ the weak-*-topology. Recall that $L^\infty(S,\lambda)$ and $L^1(S,\lambda)$ are duals.

An important consequence of this choice of set up is that the collection $K = \{g \in L^\infty(S) : 0 \leq g \leq 1\}$ is weak-*-compact, and convex.

Definition 9.1. Let $f \in L^\infty(S,\lambda)$ such that $0 \leq f \leq 1$. Put

$$M(f) = \{g \in K : \Pi_i(g\lambda) = \Pi_i(fd\lambda) : i = 1,...,N\}. \qquad (9.8)$$

That is $M(f)$ is the class of measures, absolutely continuous with respect to λ, having the same "projections" (or generalized marginals) as $fd\lambda$.

We leave it to the reader to convince herself / himself that $M(f)$ is a closed, convex subset of K and therefore by the Krein-Millman Theorem, it is the closed convex hull of its set $E(f)$ of extreme points. Consider now

Definition 9.2.

a) The measure λ on $(S, \mathcal{B}(S))$ is *rich* if for any $f \in K$ there exists $g \in M(f)$ taking only values 0 and 1, that is $g = I_{\{g=1\}}$.
b) The measure λ on $(S, \mathcal{B}(S))$ is said to be *strongly rich* if for any $D \in \mathcal{B}(S)$ with $\lambda(D) > 0$ there exists h in $L^\infty(S,\lambda)$ with $supp(h) \in D$ such that $\Pi_j(hd\lambda) = 0$.

Comments: In both definitions, the richness is relative to the given set of projections.

The following result asserts that elements in $E(f)$ are 0-1 functions,

Theorem 9.4. *For the measure λ to be rich it suffices that it be strongly rich.*

Proof. Take any $g \in E(f)$ and assume that it is not an indicator function, i.e. for some ε, $0 < \varepsilon < 1$, the set $D = \{x \in S : \varepsilon < g(x) < 1-\varepsilon\}$ is such that $\lambda(D) > 0$. By assumption, λ being strongly rich, there exists a measurable h with $|h| \leq C$ such that $\Pi_i(hd\lambda) = 0$. Now take δ such that $0 < \delta < \frac{\varepsilon}{C}$ and consider $g_1 = g + \delta h$ and $g_2 = g - h$. Thus $\Pi_i(g_1 d\lambda) = \Pi_i(g_2 d\lambda) = \Pi_i(gd\lambda)$ since $\Pi_i(hd\lambda) = 0$ and $g = \frac{1}{2}(g_1 + g_2)$ contradicting the extremality of g. $\square$

The result in [9] that can be used to construct non-unique solutions to the standard tomography problem is the following.

Theorem 9.5. *Let $S = [0,1]^n$ be the unit cube in $\mathbb{R}^n$. Let $\Pi_i : S \to Y_i$ be an orthogonal projection into an $(n-1)$-dimensional hyperplane in $\mathbb{R}^n$ for $i = 1, ..., N$. Let $q(x) \in L^1(S, dx)$ and set $d\lambda = q(x)dx$. Then λ is a strongly rich measure.*

This result is Theorem 2 in [9]. We invite the reader to take a look at it there. Some bonuses are to be found: examples of non-rich measures for instance.

9.3.2 The good news

In the previous section we saw that indicator functions were extremals in a convex set, and that in some cases given a g with $0 \leq g \leq 1$ and $\int g d\lambda < \infty$, there exists a set D with finite λ measure such that $\Pi_i(g\lambda) = \Pi_i(I_D \lambda)$ for $i = 1, 2$.

The obvious question arises: what if g is already the indicator of a set?

In this section we shall see that in $\mathbb{R}^2$ (and in $\mathbb{R}^n$ as a matter of fact) some sets are reconstructible from their projections upon the planes $\{x = 0\}$ and $\{y = 0\}$ (or the n-corresponding planes in $\mathbb{R}^n$).

The first result stems from the late '40s see [30], and goes as follows. Let

$$P_1(x_1) = \int I_D(x_1, x_2) dx_2, \quad P_2(x_2) = \int I_D(x_1, x_2) dx_1.$$

Denote by $p_i(u)$ the *nonincreasing rearrangements* of the $P_i(x_i)$, i.e.,

$$p_i(u) = |\{x_i \in \mathbb{R} : P_i(x_i) \geq u\}| \quad \text{for } 0 \leq u < \infty.$$

Again $|A|$ denotes Lebesgue measure of the set A. Then

Theorem 9.6. *A set D in $\mathbb{R}^2$, of finite measure, is a set of uniqueness (or uniquely reconstructed) if and only if p_1 and p_2 are inverses of each other, i.e., $p_1(p_2(u)) = u$ and $p_2(p_1(u)) = u$.*

There is a simple necessary condition for S to be a set of uniqueness. It requires the notion of k-bad configuration which we spell out now: the set of corners of a rectangle, given by $z_1 = (x_1, y_1)$, $z_2 = (x_2, y_2)$, $w_1 = (x_1, y_2)$, $w_2 = (x_2, y_1)$, define a 2-bad configuration for D if $z_1, z_2 \in int(D)$ and $w_1, w_2 \in int(D^c)$. In general, a k-bad configuration for D consists of two collections $z_1, ..., z_k \in int(D)$ and $w_1, ..., w_k \in int(D^c)$, such that any plane $x_j = c$ contains the same number (from 0 to k) of z's and w's, for $j = 1, ..., k$.

Theorem 9.7. *A set $D \subset \mathbb{R}^2$ of uniqueness has no bad configuration.*

Proof. Suppose D has a k-bad configuration consisting of points $z_1,...,z_k \in D$ and $w_1,...,w_k \in D^c$. Pick a small enough δ such that $C_1 = \cup B(z_i, \delta)$ and $C_2 = \cup B(w_i, \delta)$ are contained in $int(D)$ and $int(D^c)$ respectively. Then $E = (D - C_1) \cup C_2$ has the same projections as D. □

Definition 9.3. A set D is said to be additive if there are bounded measurable functions $F_i(x) = f_i(x_i)$ for $i = 1, 2$, such that

$$D = \left\{ x \in \mathbb{R}^2 : \sum_{i=1}^{2} f_i(x_i) \geq 0 \right\}.$$

Example 9.1. The unit ball in $\mathbb{R}^n$ is determined by $f_i(x_i) = \frac{1}{n} - x_i^2$.

Theorem 9.8. *Consider an additive set D of finite (Lebesgue) measure in $\mathbb{R}^2$. Then D is additive if and only if it is a set of uniqueness.*

Proof. (The sufficiency part of the proof can be translated verbatim to any $\mathbb{R}^n$.) Consider any other set G in $\mathbb{R}^2$ such that

$$P_1(x_1) = \int I_D(x_1, x_2) dx_2 = \int I_G(x_1, x_2) dx_2,$$

$$P_2(x_2) = \int I_D(x_1, x_2) dx_1 = \int I_G(x_1, x_2) dx_1.$$

Observe that for any function $g_i(x_i)$, integrable over $D \cup G$, the following clearly holds

$$\int_D g_i(x_i) dx_1 dx_2 = \int g_i(x_i) P(x_i) dx_i = \int_G g_i(x_i) dx_1 dx_2.$$

Since clearly D and G have the same measure (take $g_i(x_i) \equiv 1$ if you want).

Take $g_i(x_i) = f_i(x_i)$, where $D = \{x \in \mathbb{R}^2 : f_1(x_1) + f_2(x_2) \geq 0\}$ and subtract $\int_{D \cap G} f_i(x_i) dx$ from both sides in the identity above to obtain

$$\int_{D - D \cap G} f_i(x_i) dx = \int_{G - D \cap G} f_i(x_i) dx$$

and repeat with $g_i(x) \equiv 1$ to obtain

$$\int_{D - D \cap G} dx = \int_{G - D \cap G} dx.$$

In the next to last identity sum over i to obtain

$$\int_{D - D \cap G} \sum f_i(x_i) dx = \int_{G - D \cap G} \sum f_i(x_i) dx$$

and notice that the right-hand side is strictly negative. Since the right-hand side is positive, this is possible when $|D - D \cap G| = 0$. The former identity yields $|G - D \cap G| = 0$ or: G differs from D by a set of measure zero. □

The necessity part of the theorem is proved as Theorem 2.2 in [31]. There are more nice and interesting results proved in [31]. To round up we present their Theorem 2.3.

Theorem 9.9. *Let D be an open set in $\mathbb{R}^2$ of finite measure. The following are equivalent*

a) *There is no open set $\hat{D}$, different from D having the same projections on the coordinate axes.*
b) *D has no bad configuration for $k \geq 2$.*
   If $|\partial D| = 0$, these conditions are equivalent to
c) *D has no 2-bad configuration.*

We direct the reader to [32] and [33] to see how the notion of additivity has to be modified in dimensions higher than 2.

9.4 The Hilbert space set up for the tomographic problem

We shall follow [7] and [20]-[21] and learn how to find a function $g(x, y)$ whose projections $P_{\theta(j)}g(t)$ coincide with those of the unknown function f.

Recall from Section 2.8 in Chapter 2 that the projection of $f \in L^2(D)$ upon the time

$$L(t, \theta) = \{(x, y) \in R^2 : x\cos\theta + y\sin\theta = t\}$$

is given by

$$P_\theta f(t) = \int_{-\infty}^{\infty} f(t\cos\theta - s\sin\theta, t\sin\theta + s\cos\theta)ds$$
$$= \int_{-\infty}^{\infty} f(t\hat{\nu} + s\hat{\nu}_1)ds \tag{9.9}$$

where $\hat{\nu} = (\cos\theta, \sin\theta)$ and $\hat{\nu}_1 = (-\sin\theta, \cos\theta)$.

Definition 9.4. A ridge function $h(x, y)$ along direction θ is a function

$$h(x, y) = \rho(x\cos\theta + y\sin\theta) \text{ for } \rho \in L^2\left([-1, 1], (1 - t^2)^{1/2}\right).$$

Comment: A ridge function along direction θ is constant along lines perpendicular to the line $L(t, \theta)$.

If $\rho(t)$ determines a ridge function h along θ, then

$$(h, f) = \int_{-1}^{1} \rho(t) P_\theta f(t) dt$$

thus $h \in \mathcal{N}_\theta^\perp$ where $\mathcal{N}_\theta = \operatorname{Ker} P_\theta$. We leave it as an exercise for the reader to verify
$$\int_{-\infty}^{\infty} e^{i\omega t} P_\theta f(t) dt = \hat{f}(\omega \cos \theta, \omega \sin \theta) = \hat{f}(\omega \hat{\nu}). \tag{9.10}$$

Given a collection $\{\theta(i) : i = 1, ..., n\}$ of directions, it determines the space V_r of ridge functions $g(x, y)$ defined by
$$g(x, y) = \sum_{1}^{n} \rho_i(x, y); \quad \rho_i(x, y) \equiv \rho_i(x \cos \theta(i) + y \sin \theta(i)).$$

It is proved in [20] that $V_r = \mathcal{N}^\perp = \sum \mathcal{N}_{\theta(i)}^\perp$ where $\mathcal{N} = \cap \mathcal{N}_{\theta(i)}$ is the intersection of the kernels of the $P_{\theta(i)}$. Thus V_r is a closed subspace of $V = L^2(D)$. The following is from Section 2 of [7]. See Corollary 3.16 in [20] as well.

Theorem 9.10. *Let $g \in V$ be defined by*
$$g = \arg\inf \left\{ \frac{1}{2} \|g\|^2 : P_{\theta(i)} g = P_{\theta(i)} f, \quad i \geq n \right\} \tag{9.11}$$
then $g \in V_r$.

Proof. Denote by g_1 the orthogonal projection of f onto V_r, and set $f_1 = f - g_1$. If ρ_j is a ridge function along $\theta(j)$
$$0 = (f_1, \rho_j) = \int_{-1}^{1} P_{\theta(j)} f_1(t) \rho_j(t) dt$$
and since $\rho_j(t)$ is arbitrary in $L^2([-1, 1])$, we obtain
$$P_{\theta(j)} f(\cdot) = P_{\theta(j)} f_1(\cdot) - P_{\theta(j)} g(\cdot) = 0.$$

Note that if g satisfies (9.10), it will have its norm decreased by projection upon V_r unless it is already in V_r. Thus, since g is unique and a sum of ridge functions, it is the best approximation to f by ridge functions. □

In Sections 3 and 4 of [7] it is shown how to build a ridge function satisfying (9.11) when the angles are equally spaced, and in Section 5 of [20] it is explained how this solution relates to the standard ART (algebraic reconstruction technique). In a nutshell, this technique consists in starting from an arbitrary $g \in V$ and applying repeatedly the projections operator $P = P_1 P_2 ... P_n$, where each P_j is the orthonormal projection on the "hyperplane" $f + \mathcal{N}_{\theta(j)}$. It is shown in [22] that
$$P_j g = g + \frac{\chi(P_{\theta(j)} f - P_{\theta(j)} g)}{P_{\theta(j)} \chi}$$
where $\chi = I_D(x, y)$ is the indicator function of D. Corollary 5.6 in [20] states that for appropriate g_0, for $\varepsilon > 0$
$$\|P^k g_0 - h\|^2 \leq \varepsilon^{2k} \|g_0 - h\|^2$$
where h is the projection of f on V_r.

9.4.1 More on nonuniquenes of reconstructions

Let us now present a short argument, taken from [21], and direct the reader to [23] where the author shows how to construct functions in V whose projections on given directions $\theta(1), ..., \theta(n)$ vanish. Both arguments lead to

Theorem 9.11. *It is impossible to determine uniquely an element $f \in V$ having prescribed projections $P_{\theta(j)}f(t)$, $j = 1, ..., n$.*

To begin, given a finite number of directions $\theta(1), ..., \theta(n)$ construct a polynomial vanishing on lines parallel to the $\theta(j)$, for example

$$q(\xi) = \langle \xi, k_1 \rangle \langle \xi, k_2 \rangle \cdots \langle \xi, k_n \rangle$$

where, as above, $k_j = (-\sin\theta_j, \cos\theta_j)$, and let Q be the differential operator obtained replacing ξ by $-i\left(\frac{\partial}{\partial x}, \frac{\partial}{\partial y}\right)$.

Let $g \in C_0^\infty(D)$ and put $f = Qg$. Again $g \in V$ and from (9.10), for any θ

$$(P_\theta f)\hat{}(\omega) = \hat{f}(\omega n_\theta) = q(\omega n_\theta)\hat{g}(\omega n_\theta) \tag{9.12}$$

where a simple integration by parts leads to (9.12). Therefore $P_{\theta(j)}f(t)$ vanishes.

One can say more. Given $g_0 \in C_0^\infty(D)$ one can solve the linear partial differential equation with constant coefficients $Qg_1 = g_0$. Since g_1 may not necessarily have compact support we proceed as follows: Within the support of g_0 choose a compact set K and a $C_0^\infty(D)$ function φ such that $\varphi \equiv 1$ on K also vanishes off suppg_0.

Since Q is a local operator, take $g = \varphi g_1$ and $f = Qg$, then (i) $P_{\theta(j)}f = 0$ and (ii) $f = g_0$ on K and zero outside suppg_0.

To sum up, given an arbitrary g_0, we can construct an f agreeing with g_0 except on a small ring inside suppg_0, such that $P_{\theta(j)}f = 0$ for all $j = 1, ..., n$.

9.5 The Russian Twist

We shall say only a few words about the problem described in [24]-[26].

Instead of taking X-rays with a CAT scanner, consider the situation in which you have a fixed source-plate configuration and have several radiographs of a given object that has *been* rotated by unknown angles.

That is, if $R_\omega : \mathbb{R}^3 \to \mathbb{R}^3$ describes a rotation, say about the center of mass of the object, and if $\rho_\omega(x) = \rho(R_\omega x)$ describes the rotated object, your data consists of

$$\int_{-\infty}^{\infty} \rho_\omega(x) dx_1 \equiv \int_{-\infty}^{\infty} \rho_\omega(x_1, x_2, x_3) dx_1$$

if we assume that the X-rays are taken along the x-axis.

The problem consists first, of finding the rotation angles (or rotation matrices) and then solve the ordinary reconstruction from projections problems.

In [24]-[25] you will learn how to relate this problem to a moment problem, from which the rotation matrices can be found, to next solve the projection problem. Also a guide to more literature on the subjects is given. In [26]-[27] some allied problems in tomographic reconstruction are examined. In [27] more "realistic" formulation of the tomographic reconstruction problems are examined.

9.6 Why does it work

Given the nonuniqueness results described in Sections 3 and 4, the obvious question is: why do physicians and other scientists seem satisfied with the reconstruction provided by current algorithms? Apart from the jocular: Because they do have other choice and/or they do not know any better, the following result lifted from [10] provides a partial answer.

Theorem 9.12. *Let f_1, f_2 be probability densities supported in the unit square. There exist $N = 2n$ directions $\theta(1), ..., \theta(N)$ such that if $P_{\theta(j)} f_1(t) = P_{\theta(j)} f_2(t)$, for $j = 1, ..., N$, then*

$$\sup_{x \in \mathbb{R}^2} \{f_1 * \varphi_\sigma(x) - f_2 * \varphi_\sigma(x)\} \leq \left[\sqrt{\pi} \sigma^{n+2} 2^{(3n+4)} \Gamma(\frac{n+1}{2})\right]^{-1}. \quad (9.13)$$

The projections are upon lines

$$n_j = \frac{(v_j, 1)}{\sqrt{v_j^2 + 1}} \quad j = 1, ..., n$$

$$n_j = \frac{(1, v_{j-n})}{\sqrt{v_{j-n}^2 + 1}} \quad j = n+1, ..., 2n$$

and $v_j = \cos\left(\frac{\pi(2k-1)}{2n}\right)$ $k = 1, 2, ..., n$. Also $\varphi_\sigma(x) = \dfrac{\exp(-\frac{x^2}{2\pi\sigma})}{2\pi\sigma}$ is the centered Gaussian density of variance σ in $\mathbb{R}^2$.

Comments: The understanding provided by this result is only partial, first due to the way the projection lines come in: we actually need a result holding for any number of arbitrary projection lines. On the other hand, the result described in Section 9.4.1 does not seem to be covered by this result.

Proof. (Sketch. See [10] for more.) The proof hinges on the following auxiliary. □

Lemma 9.1. *Under the assumptions and notation of Theorem 9.8, we have for any* $A > 0$

$$\sup_{\|\eta\|_\infty < A} \left| \hat{f}_1(\eta) - \hat{f}_2(\eta) \right| \le \frac{A^n}{2^{n-2} n!}$$

where $\hat{f}_1(\eta)$, $\hat{f}_2(\eta)$ *are the Fourier transforms of* $f_1(x)$, $f_2(x)$ *and* $\|\eta\|_\infty = \sup\{|\eta_1|, |\eta_2|\}$.

From this lemma, we obtain the following estimate

$$|f_1 * \varphi_\sigma(x) - g * \varphi_\sigma(x)|$$

$$= (2\pi)^{-2} \left| \int_{\mathbb{R}^2} \exp(-i\langle \eta, x \rangle)(\hat{f}_1(\eta) - \hat{f}_2(\eta)) \exp(-\tfrac{\eta^2 \sigma^2}{2}) d\eta \right|$$

$$\le (2\pi)^{-2} \int_0^{2\pi} d\varphi \int_0^\infty \exp(-\sigma^2 r^2) \left| \hat{f}_1(r\cos\varphi, r\sin\varphi) - \hat{f}_2(r\cos\varphi, r\sin\varphi) \right| r dr$$

$$\le (2\pi)^{-2} \int_0^\infty \frac{r^n \exp(-\frac{\sigma^2 r^2}{2})}{2^{n-2} n!} dr = \Gamma(\tfrac{n+2}{2})(2^{\frac{n-2}{2}} \pi \sigma^{n+2} n!).$$

The rest follows from $n! = \Gamma(n+1) = \Gamma\left(\frac{2(n+1)}{2}\right) = \frac{2^{n-1}}{\sqrt{\pi}} \Gamma\left(\frac{n+1}{2}\right) \Gamma\left(\frac{n+2}{2}\right).$

Theorem 9.8 can be improved in a different direction. Consider the metric defined on the class of probability distribution that are absolutely continuous with respect to Lebesgue measure on $\mathbb{R}^2$, i.e., on $\mathcal{C} = \{\rho \in L_1(R^2) : \rho_1 \ge 0, \|\rho\|_1 = 1\}$.

For any two such densities, define

$$\lambda(P_1, P_2) = \min_{T > 0} \max \left\{ \sup_{\|\eta\| \le T} |\hat{\rho}_1(\eta) - \hat{\rho}_2(\eta)|, \frac{1}{T} \right\} \quad (9.14)$$

where $\hat{\rho}_i(\eta) = \int \exp(i\langle \eta, x \rangle) \rho_i(x) dx$ is the Fourier transform of ρ.

Theorem 9.13. *Let* $\theta(1), ..., \theta(n)$ *be angles in* $\mathbb{R}^2$ *such that no two* $n_i = (\cos\theta(i), \sin\theta(i))$ *are collinear. Let* $\nu_i = (-\sin\theta(i), \cos\theta(i))$ *and assume that* $\sup \rho_1 \subset D = \{x \in R^2 : \|x\| < 1\}$. *Assume that* ρ_2 *has the same marginals as* ρ_1 *along* $\theta(1), ..., \theta(n)$.

Put $s = 2[\frac{1}{2}(n-1)]$, then

$$\lambda(P_1, P_2) = \left(\frac{2}{s!}\right)^{\frac{i}{s+1}}. \qquad (9.15)$$

Proof. Since Lebesgue measure on $\mathbb{R}^2$ and $\lambda(P_1, P_2)$ given by (9.13) are invariant under rotations, we may assume that:

i) The directions n_i are not parallel to the coordinate axes.
ii) For $a \neq 0$, $b \neq 0$, there exist a pair of directions, say n_{j_1} and n_{j_2} such that $n_{j_1} = (a, b)$ and $n_{j_2} = (a, -b)$ are symmetric with respect to the horizontal axis. Since

$$\int t^k P_{\theta_j} \rho(t) = \int_{\mathbb{R}^2} \langle x, n_\theta \rangle^k \rho(x) dx$$

we have for any k and any θ_j

$$\int_{\mathbb{R}^2} \langle x, n_j \rangle^k \rho_1(x) dx = \int_{\mathbb{R}^2} \langle x, n_j \rangle^k \rho_2(x) dx. \qquad (9.16)$$

Since ρ_1 has compact support, it has moments of all orders and (9.15) can be used (in conjunction with i) and ii)) to prove that ρ_2 also has moments of all orders. Also all moments of ρ_1 and ρ_2 of order less than or equal to n agree. To see this set $\eta_j = (\cos\theta_j, \sin\theta_j)$ in (9.15) to obtain

$$\sum_{l=0}^{k} (\cos\theta_j)^l (\sin\theta_j)^{k-l} [\mu_{l,k-l}(\rho_1) - \mu_{l,k-l}(\rho_2)] = 0$$

where we are using the obvious notation

$$\mu_{l,k}(\rho) = \int_{\mathbb{R}^2} x_1^l x_2^k \rho(x_1, x_2) dx_1 dx_2.$$

Since no two directions $\theta(1), ..., \theta(n)$ are collinear, the numbers $z_j = \frac{\cos\theta_j}{\sin\theta_j}$ are distinct and

$$\sum_l z_j^l [\mu_{l,k-l}(\rho_1) - \mu_{l,k-l}(\rho_2)] = 0.$$

In other words, the polynomial of degree k

$$p(z) := \sum_l^k z^l [\mu_{l,k-l}(\rho_1) - \mu_{l,k-l}(\rho_2)]$$

has n distinct roots $z_1, z_2, ..., z_n$. If $n \geq k+1$ this will be possible only if the $k+1$ coefficients are zero, i.e., if

$$\mu_{l,k-l}(\rho_1) = \mu_{l,k-l}(\rho_2) \quad l = 0, ..., k, \quad k = 0, ..., n-1.$$

Which implies that for any unit vector $\eta \in \mathbb{R}^2$

$$\int_{\mathbb{R}^2} \langle \eta, x \rangle^k \rho_1(x) dx = \int_{\mathbb{R}^2} \langle \eta, x \rangle^k \rho_2(x) dx \quad k = 0, ..., n-1. \qquad (9.17)$$

This can be rewritten in terms of q derivatives of characteristic functions as

$$\left. \frac{\partial^k}{\partial t^k} \rho_1(t\eta) \right|_{t=0} = \left. \frac{\partial^k}{\partial t^k} \rho_2(t\eta) \right|_{t=0}, \quad k = 0, ..., n-1.$$

Therefore, making use of the Taylor expansion theorem, up to order s

$$\rho_1(t\eta) - \rho_2(t\eta) = \sum_{l=0}^{s-1} \frac{t^l}{l!} \left(\rho_1^l(t\eta)|_{t=0} = \rho_2^l(t\eta)|_{t=0} \right) + \frac{1}{s!} \rho_1(\xi\eta) - \rho_2(\xi\eta) t^s$$

where $\xi \in (0,t)$. Note furthermore that, since s is even

$$|\rho_i^s(\xi\eta)| = \left| \int_{-\infty}^{\infty} t^s P_\eta \rho_1(t) \exp(i\xi t) dt \right| \leq \int t^s (P_\eta \rho_1)(t) dt$$

$$= \int_{-1}^{1} t^s (P_\eta \rho_1)(t) dt \leq 1.$$

From this, it follows that for any $t \in \mathbb{R}$

$$|\rho_1(t\eta) - \rho_2(t\eta)| \leq 2 \frac{t^s}{s!}.$$

Choose $T > 0$ satisfying $2 \frac{t^s}{s!} = \frac{1}{T}$, i.e., $T = \left(\frac{s!}{2} \right)^{\frac{1}{s+1}}$, and therefore

$$\sup_{\|\eta\| < T} |\rho_1(t\eta) - \rho_2(t\eta)| \leq \left(\frac{2}{s!} \right)^{\frac{1}{s+1}}.$$

$\square$

We recommend the reader to take a glance at [12] and [14] for more results along this line.

9.7 Reconstructions using (classical) entropic, penalized methods in Hilbert space

In [29], besides more references to efforts to solve tomographic problems, you will find the full details about the following approach.

Again consider D to be the unit circle, centered at the origin and $V = L^2(D)$. For a finite collection of directions, divide the plane in strips A_{jm} : $m = 1, ..., M(j)$, where j labels the directions. Might as well assume the strips to have equal width.

Let $f(x)$ represent the density of X-ray absorbers (called attenuation function in [29]). This time the data consists of the numbers

$$D_{jm} = I_{jm}(f) := \int_{A_{jm}} f(x)dx. \tag{9.18}$$

To find f the authors define the following functional on V

$$G(f) = -\Sigma(f) + \gamma \sum_{jm}(G_{jm} - I_{jm}(f))^2 \tag{9.19}$$

where $I_{jm}(f)$ is defined in (9.17) and

$$\Sigma(f) = -\int_D |f|\ln(|f|A)dx \tag{9.20}$$

where A is the area of D.

They prove that G is strictly convex, continuous functional on V from which the following result is obtained.

Theorem 9.14. *The functional G satisfies*

i) *It is lower semicontinuous on $L^2(D)$.*
ii) *If Σ is a convex and weakly (sequentially) compact subset of V (for example, $\Sigma = \{f \in V : 0 \leq f \leq B\}$ for some fixed B), then G has a unique minimizer in Σ for any (measured) data G_{jm}.*
iii) *The solutions are weakly continuous in the data.*

These are all proved in detail in [29]. Here we shall present their characterization of the minimizing solution.

Definition 9.5. By a cell we mean the interior of a maximal region which is not crossed by the boundaries of any of the strips $A_{jm} : j = 1, ..., n; m = 1, ..., M(j)$.

Theorem 9.15. *For the situation described in ii) above, the optimizing solution is (almost everywhere) constant on each cell.*

Proof. Let Δ be an arbitrary cell, let f^* be the minimizer found in ii) and set

$$f_\Delta^* = \frac{1}{A(\Delta)} \int_\Delta f^*(x)dx$$

where $A(\Delta)$ is the area of the cell Δ. Let

$$f_\Delta(x) = \begin{cases} f^*(x) & x \in D\backslash\Delta \\ f_\Delta^* & x \in \Delta \end{cases}$$

and observe that for any $j = 1, ..., n$; $m = 1, ..., M(j)$ such that $\Delta \subset A_{jm}$

$$\begin{aligned} I_{jm}(f_\Delta) &= \int_{A_{jm}} f_\Delta(x)dx \\ &= \int_{A_{jm}\setminus\Delta} f^*(x)dx + \int_\Delta f^*_\Delta dx \\ &= \int_{A_{jm}\setminus\Delta} f^*(x)dx + f^*_\Delta A(\Delta) \\ &= \int_{A_{jm}\setminus\Delta} f^*(x)dx + \int_\Delta f^*(x)dx \\ &= \int_{A_{jm}} f^*(x)dx = I_{jm}(f^*). \end{aligned}$$

To verify that f_Δ would lower the value of G, notice that since $f_\Delta(x) = f^*_\Delta$ in Δ,

$$\begin{aligned} \int_\Delta (f^* \ln f^* - f_\Delta \ln f_\Delta) dx &= \int_\Delta f^*(\ln f^* - \ln f_\Delta) dx \\ &= \int_\Delta f^* \{\ln f - \ln f_\Delta - 1 + \tfrac{f_\Delta}{f^*}\} dx \\ &= \int_\Delta f^* \{-\ln \tfrac{f_\Delta}{f^*} - 1 + \tfrac{f_\Delta}{f^*}\} dx \geq 0 \end{aligned}$$

where the last inequality follows from the fact that $(x - 1 - \ln x) \geq 0$ for $x \geq 0$, therefore

$$G(f^*) \geq G(f_\Delta).$$

But from ii) in Theorem 9.14 we known that the minimizer is unique, thus $f^* = f_\Delta$, i.e., in D. This being valid for any odd cell, we conclude that $f^* = f^*_\Delta$ in each Δ. □

Comment: If we consider the partition

$$\Pi = \{\Delta^k_{jm} : k = 1, ..., N(j,m),\ m = 1, ..., N(j),\ j = 1, ..., N\}$$

of D into cells, determined by the strips $A_{jm} : j = 1, ..., n;\ m = 1, ..., M(j)$ it is clear that

$$I_{jm}(f) = \int_{A_{jm}} f(x)dx = \sum_{k=1}^{N(j,m)} \int_{\Delta^k_{jm}} f(x)dx = \int_{A_{jm}} f^\Pi(x)dx$$

where $f^\Pi(x)$ is the function defined by $f^\Pi(x) = f^*_{\Delta^k_{jm}}$ in Δ^k_{jm}.

In other words, a function and its average over cells provide the same tomographic data. The nice aspect of (9.12) is the observation that piecewise constant functions have higher entropy.

To finish, note that we are dismissing the issue of what happens at the cell boundaries since we are integrating with respect to Lebesgue measure in $\mathbb{R}^2$, and cell boundaries have zero area.

9.8 Some maxentropic computations

Consider the following simplified setup, corresponding to either reconstruction from two marginals or two X-ray images on perpendicular lines. We shall consider

$$V = L^2([0,1]^2) \quad W = \left(L^2([0,1]^2)\right)^2 \approx R^2 \otimes L^2([0,1]^2),$$

i.e., W consists of two-dimensional vectors whose components are in square integrable functions in [0,1].

Consider $A: V \to W$ defined by

$$Af(x_1, x_2) = e_1 \int_0^1 f(x_1, x_2) dx_2 + e_2 \int_0^1 f(x_1, x_2) dx_1$$

$$= e_1 P_1 f(x_1) + e_2 P_2 f(x_2)$$

where the e_i denote the canonical basis in $\mathbb{R}^2$. Since $W^* = W$, we have for

$$g(x_1, x_2) = e_1 g_1(x_1) + e_2 g_2(x_2) = \begin{pmatrix} g_1(x_1) \\ g_2(x_2) \end{pmatrix},$$

and an easy computation shows that

$$(A^*g)(x_1, x_2) = g_1(x_1) I_{[0,1]}(x_2) + g_2(x_2) I_{[0,1]}(x_1)$$

which can obviously be identified with $g_1(x_1) + g_2(x_2)$.

Consider now a probability space $(\Omega, \mathcal{F}, Q)$ such that a Gaussian field $X(S)$ is defined on $[0,1]^2$. Assume X to be centered and $E[X(S_1)X(S_2)] = \int_{S_1 \cap S_2} dx$ for any $S_1, S_2 \in \mathcal{B}([0,1]^2)$.

We want to find a measure $P \ll Q$ on $(\Omega, \mathcal{F})$ such that

$$E_P[\langle g, AX \rangle] = \langle g, h \rangle \tag{9.21}$$

where $h(x) = \begin{pmatrix} h_1(x_1) \\ h_2(x_2) \end{pmatrix}$ is the X-ray image of an unknown $f_u \in V$. From this we shall obtain that

$$E_P[X(S)] = \int_S f^*(x) dx$$

where $f^*(x)$ is the maxentropic reconstruction that we are after.

As we saw in Chapter 8, the desired P has the form $dP = Z(g)^{-1} \exp(-\langle g, AX \rangle) dQ$, where

$$Z(g) = E_Q[\exp(-\langle g, AX \rangle)]$$

$$= \exp\left(\tfrac{1}{2} \sum_{i,j=1}^2 \int g_i(x_i) g_j(x_j) dx_1 dx_2\right).$$

The P that realizes

$$\max\{S_Q(P) : E_P[\langle g, AX \rangle] = \langle g, h \rangle,\ g \in W^*\}$$

is obtained by finding

$$g^* = \arg\inf \{\ln Z(g) + \langle g, h \rangle : g \in W^*\}. \tag{9.22}$$

We leave it as an exercise for the reader to verify that such g^* has to satisfy

$$\begin{aligned} g_1^*(x_1) + \int_0^1 g_2^*(x_2)dx_2 + h_1(x_1) = 0 \\ \int_0^1 g_1^*(x_1)dx_1 + g_2^*(x_2) + h_2(x_2) = 0. \end{aligned} \tag{9.23}$$

Since $\int_0^1 h_1(x_1)dx_1 = \int_0^1 h_2(x_2)dx_2 = \int_0^1 \int_0^1 f_u(x_1, x_2) dx_1 dx_2$, we leave it to the reader to verify that

$$g_1^*(x_1) = h_1(x_1) - \tfrac{1}{2} \int_0^1 h_1(x_1)dx_1$$

$$g_2^*(x_2) = h_2(x_2) - \tfrac{1}{2} \int_0^1 h_2(x_2)dx_2$$

and that (9.21) or (9.22) yield

$$f^*(x_1, x_2) = h_1(x) + h_2(x) - \frac{1}{2}\left[\int_0^1 h_1(x_1)dx_1 + \int_0^1 h_2(x_2)dx_2\right] \tag{9.24}$$

written in a lengthier but more symmetric notation.

Comments: Notice that f^* gives a zero value to $\|Af - h\|$ when f ranges over V and $h = Af_u$. It realizes $\inf\left\{\tfrac{1}{2}\|Af - h\|^2 : f \in V\right\}$. We saw in Chapter 8 why this is not surprising.

Also, $f^*(x_1, x_2)$ is what is called the *"back propagated solution"* to the tomographic reconstruction problem. It would be a ridge function except for the constant term.

Exercise 9.1.

a) Modify Q such that $X(S)$ satisfies

$$E_Q[X(S)] = \int_S f_0(x)dx.$$

b) Carry out the procedure outlined above.
c) Repeat for the case where $X(S)$ is a Poisson random measure.

Conjecture: If we could prove that for any collection given non-collinear unit vectors $n_1, n_2, ..., n_k$ the following is true. For any i,

$$\sum_{j \neq i} \frac{1}{|\langle n_j, n_i^\perp \rangle|} = c$$

then, if $h_i(t) = P_{n(i)} f_u(t) = \int f_u(tn(i) + sn(i)^\perp) ds$ are the projections of unknown $f_u(x)$ (supported in $[0,1]^2$), then

$$f^*(x) = \sum h_j(\langle n_j, x \rangle) - \frac{c}{n} \left(\int h_j(t_j) dt_j \right). \tag{9.25}$$

Proof. Compute

$$P_{n(i)} f^*(t) = h_i(t) + \sum_{j \neq i} \int h_j \left(t \langle n_j, n_i \rangle + s \langle n_j, n_i^\perp \rangle \right) ds - c \int h_i(t)$$

$$= h_i(t) + \sum_{j \neq i} \frac{1}{|\langle n_j, n_i^\perp \rangle|} \int h_j(t) dt - c \int h_i(t)$$

where we used the fact that $\int h_i(t) dt = \int f_u(x) dx$ for all i and $h_i(t)$ vanishes off $(-1, 1)$ for any i. $\square$

Open question related to the conjecture: Notice that (9.25) is a sum of ridge functions. How does it relate to the representation that Logan and Shepp find in [7]? Notice also that (9.25) is a "back-propagated solution".

9.9 Maxentropic approach to reconstruction from marginals in the discrete case

We will now explore different aspects of the maxentropic solution of the problem consisting of finding a function x_{ij}, $i = 1, ..., m$; $j = 1, ..., n$ such that

$$a_{ij} \leq x_{ij} \leq b_{ij} \qquad 1 \leq i \leq m, 1 \leq j \leq n$$

$$\sum_{j=1}^{n} x_{ij} = p_i, \sum_{i=1}^{m} x_{ij} = q_i \qquad 1 \leq i \leq m, 1 \leq j \leq n. \tag{9.26}$$

Again, as in Chapter 7, we shall see that the reconstruction obtained in this inverse problem depends on the *a priori* information brought in.

The first issue to address is to bring problem (9.26) into standard format. By appropriate labeling, lexicographical for example, (9.26) can be recast into the usual form

$$Ax = y, \quad x \in \mathbb{R}^N, \quad y \in \mathbb{R}^M$$

where A is an $M \times N$ matrix, with $N = nm$ and $M = n + m$. If we may choose $y_1 = q_1, ..., y_n = q_n, y_{n+1} = p_1, ..., y_{n+m} = p_m$, we would end with an A looking like

$$\begin{pmatrix} I & I & \cdots & I \\ \underline{u} & \underline{0} & \cdots & \underline{0} \\ \underline{0} & \underline{u} & \cdots & \underline{0} \\ \vdots & \vdots & \ddots & \vdots \\ \underline{0} & \underline{0} & \cdots & \underline{u} \end{pmatrix}$$

where the I are $m \times n$ identity matrices and there are n of them and $\underline{u} = (1, 1, ..., 1)$ and $\underline{0} = (0, ..., 0)$ are n-dimensional row vectors. There are m such rows below the blocks of unit matrices.

In the next two sections we shall consider two maxentropic approaches to solve the discrete version of (9.3). We aim at comparing the effect of the *a priori* information in the two procedures.

9.9.1 Reconstruction from marginals by maximum entropy on the mean

Consider the square brain depicted in Table 9.1. The softer cortex occupies the three outer layers and has density varying uniformly in [0.1,0.2] and a denser block with of larger density varying randomly with uniform distribution in [0.2,0.9].

Table 9.1 Data matrix.

0.1	0.2	0.12	0.19	0.15	0.121	0.1	0.2	0.12	0.19	0.15	0.121	0.132	1.894
0.15	0.15	0.14	0.2	0.16	0.2	0.15	0.15	0.14	0.2	0.16	0.2	0.156	2.156
0.2	0.17	0.13	0.1	0.14	0.1	0.2	0.17	0.13	0.1	0.14	0.1	0.192	1.872
0.19	0.18	0.15	0.13	0.145	0.11	0.19	0.18	0.15	0.13	0.145	0.11	0.156	1.966
0.11	0.1	0.16	0.11	0.123	0.1564	0.11	0.1	0.16	0.11	0.123	0.1564	0.11	1.6288
0.1	0.2	0.12	0.23	0.25	0.55	0.69	0.56	0.22	0.64	0.15	0.121	0.19	4.021
0.15	0.15	0.14	0.58	0.22	0.64	0.28	0.28	0.36	0.78	0.16	0.2	0.11	4.01
0.2	0.17	0.13	0.28	0.36	0.56	0.22	0.64	0.22	0.64	0.14	0.1	0.16	3.82
0.19	0.18	0.15	0.26	0.59	0.28	0.36	0.26	0.59	0.28	0.2	0.15	0.14	3.63
0.11	0.1	0.16	0.6	0.68	0.26	0.59	0.6	0.68	0.26	0.1	0.2	0.2	4.54
0.1	0.2	0.12	0.8	0.89	0.78	0.56	0.8	0.89	0.78	0.11	0.19	0.1	6.32
0.15	0.15	0.14	0.9	0.88	0.85	0.66	0.56	0.22	0.64	0.1564	0.11	0.11	5.5264
0.2	0.17	0.13	0.1	0.14	0.1	0.11	0.17	0.13	0.1	0.14	0.1	0.1658	1.7558
0.2	0.15	0.2	0.12	0.19	0.15	0.121	0.132	0.18	0.15	0.13	0.145	0.11	1.978
0.15	0.15	0.14	0.2	0.16	0.145	0.11	0.19	0.18	0.15	0.13	0.145	0.2	2.05
0.19	0.18	0.15	0.13	0.145	0.11	0.19	0.18	0.15	0.13	0.145	0.11	0.14	1.95
0.11	0.1	0.16	0.11	0.123	0.1564	0.11	0.1	0.16	0.11	0.123	0.1564	0.145	1.6638
2.6	2.7	2.44	5.02	5.346	5.2688	4.731	5.272	4.68	5.39	2.4024	2.4148	2.5168	

If we take two radiographies of that head along the obvious axes, we obtain two images corresponding to row sums and column sums. A look at the data suggests the two zone structures. From this data one can infer approximate mean values for the soft cortex and the hardened core as well as range of variation of the data. Do it!

In Table 9.2 we present a reconstruction using the techniques presented in Chapter 6, applied to the reconstruction problem reformulated as described a few lines above. The results obtained using a uniform *a priori* measure on the constraint space, whereas in Table 9.3 we present a reconstruction using a Bernoulli prior with a bias toward larger values.

Table 9.2 Reconstruction using uniform *a priori* measure. $\varepsilon = 2.1544 \times 10^{-9}$.

0.1517	0.1581	0.1417	0.145	0.1481	0.1459	0.1438	0.1459	0.1436	0.1463	0.1394	0.1401	0.1464
0.1707	0.1749	0.1625	0.1655	0.1665	0.1662	0.1645	0.1662	0.1643	0.1666	0.1603	0.1611	0.1667
0.15	0.1564	0.14	0.1432	0.1444	0.1441	0.1421	0.1442	0.1419	0.1446	0.1378	0.1385	0.1447
0.1573	0.1633	0.1472	0.1506	0.1518	0.1515	0.1494	0.1515	0.1492	0.1519	0.1447	0.1455	0.1521
0.129	0.1339	0.1228	0.1247	0.1254	0.1252	0.124	0.1252	0.1239	0.1255	0.1216	0.122	0.1256
0.1542	0.1605	0.1441	0.4327	0.4845	0.4718	0.3914	0.4723	0.3846	0.4918	0.1417	0.1425	0.1489
0.1542	0.1604	0.144	0.4312	0.4827	0.4701	0.3902	0.4706	0.3835	0.49	0.1417	0.1424	0.1489
0.1535	0.1598	0.1434	0.4057	0.4522	0.4406	0.3698	0.4411	0.364	0.4589	0.141	0.1418	0.1482
0.1527	0.1591	0.1426	0.3807	0.4211	0.4109	0.3503	0.4113	0.3454	0.4271	0.1403	0.141	0.1474
0.1558	0.162	0.1457	0.5053	0.564	0.5503	0.4528	0.5509	0.4438	0.5717	0.1433	0.1441	0.1506
0.1623	0.1678	0.1526	0.7658	0.7861	0.7819	0.7401	0.782	0.7345	0.7884	0.1501	0.151	0.1574
0.1588	0.1647	0.1487	0.6502	0.6963	0.6864	0.5981	0.6869	0.5877	0.7016	0.1463	0.1471	0.1536
0.1403	0.1466	0.1316	0.1343	0.1353	0.1351	0.1333	0.1351	0.1332	0.1354	0.1298	0.1304	0.1356
0.1582	0.1641	0.1481	0.1515	0.1527	0.1524	0.1504	0.1524	0.1501	0.1529	0.1457	0.1465	0.153
0.1634	0.1687	0.1538	0.1572	0.1583	0.158	0.156	0.1581	0.1558	0.1585	0.1514	0.1522	0.1586
0.156	0.1622	0.1459	0.1493	0.1505	0.1503	0.1482	0.1503	0.1479	0.1507	0.1435	0.1443	0.1508
0.1321	0.1376	0.1252	0.1273	0.1281	0.1279	0.1266	0.1279	0.1264	0.1282	0.1238	0.1243	0.1283

Table 9.3 Reconstruction using Bernoulli *a priori* measure, with $p = 0.85$, $q = 0.15$. $\varepsilon = 1.6444 \times 10^{-9}$.

0.1629	0.1687	0.1529	0.1355	0.1385	0.1363	0.1346	0.1363	0.1344	0.1366	0.1504	0.1512	0.1578
0.1802	0.184	0.1729	0.1569	0.1579	0.1577	0.1559	0.1577	0.1557	0.1581	0.1709	0.1715	0.1766
0.1612	0.1671	0.151	0.1339	0.1348	0.1346	0.133	0.1346	0.1328	0.135	0.1486	0.1494	0.156
0.1682	0.1735	0.1586	0.141	0.1421	0.1418	0.1401	0.1418	0.1399	0.1422	0.1562	0.157	0.1634
0.1393	0.1456	0.13	0.1174	0.118	0.1179	0.1168	0.1179	0.1167	0.1181	0.128	0.1286	0.1344
0.1389	0.1453	0.1297	0.4465	0.4963	0.4843	0.4046	0.4848	0.3975	0.5032	0.1277	0.1283	0.134
0.1389	0.1452	0.1296	0.4449	0.4947	0.4827	0.4032	0.4832	0.3961	0.5016	0.1276	0.1283	0.134
0.1383	0.1446	0.1291	0.4186	0.4661	0.4545	0.3796	0.455	0.373	0.4728	0.1271	0.1278	0.1334
0.1377	0.144	0.1286	0.3925	0.4371	0.4261	0.3566	0.4266	0.3506	0.4435	0.1266	0.1273	0.1328
0.1404	0.1468	0.131	0.5196	0.5721	0.5597	0.4727	0.5602	0.4645	0.5791	0.1289	0.1296	0.1354
0.1465	0.1529	0.1365	0.7769	0.8045	0.7986	0.7466	0.7988	0.7405	0.8078	0.1342	0.135	0.1412
0.1433	0.1498	0.1336	0.6616	0.7064	0.6963	0.6167	0.6967	0.6083	0.712	0.1314	0.1321	0.1382
0.1514	0.1579	0.1412	0.1256	0.1264	0.1262	0.1249	0.1263	0.1247	0.1266	0.1388	0.1396	0.1461
0.169	0.1743	0.1596	0.142	0.143	0.1428	0.141	0.1428	0.1409	0.1432	0.1572	0.158	0.1643
0.1738	0.1785	0.1651	0.1478	0.1489	0.1486	0.1468	0.1487	0.1467	0.149	0.1628	0.1636	0.1695
0.167	0.1725	0.1574	0.1398	0.1408	0.1406	0.1389	0.1406	0.1387	0.141	0.1549	0.1557	0.1622
0.1428	0.1493	0.1331	0.1196	0.1203	0.1201	0.119	0.1201	0.1189	0.1204	0.131	0.1317	0.1377

In both cases the reconstruction error is measurable by $\sum |p_i^* - p_i| + \sum |q_j^* - q_j|$, where the p_i and q_j are respectively the given row and column sums and p_i^* and q_j^* are the corresponding quantities computed from the reconstructed matrix. Even though the reconstruction errors are small, the matrix producing the data and the reconstructed matrix may differ as shown in figures 9.1 and 9.2, where $x_i - x_i^*$ (the matrix difference in lexicographically order) is plotted.

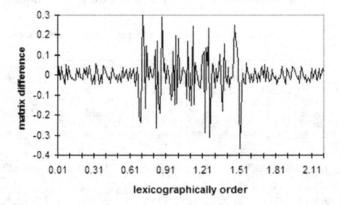

Fig. 9.1 Difference between data and reconstruction, using uniform a priori measure.

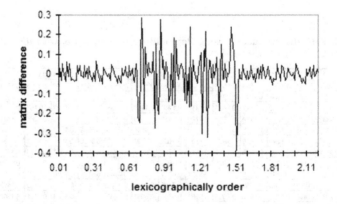

Fig. 9.2 Difference between data and reconstruction, using Bernoulli a priori measure, with $p = 0.85$, $q = 0.15$.

9.9.2 Reconstruction from marginals using the standard maximum entropy method

Since the entries of the matrix to be reconstructed are positive, after normalization by division by $\sum p_i$ (or $\sum q_j$) we can think of the problem as consisting of reconstructing a joint probability given the marginals, a class of problems in which the standard maximum entropy technique proposed by Jaynes has proved successful.

To restate, we must find $p_{ij} \geq 0$ such that

$$\sum_{ij} p_{ij} = 1$$

$$\sum_{j=i}^{n} p_{ij} = p_i, \quad \sum_{i=1}^{m} p_{ij} = q_j. \tag{9.27}$$

The standard way consists of picking an *a priori* distribution q_{ij} reflecting some preliminary information about the solution, and then minimizing

$$S(p,q) = -\sum_{ij} p_{ij} \ln\left(\frac{p_{ij}}{q_{ij}}\right)$$

which, after setting $p_{ij} = x_{ij} q_{ij}$ becomes

$$S(p,q) = -\sum_{i,j} x_{ij} \ln x_{ij} q_{ij} \tag{9.28}$$

where the x_{ij} have to satisfy

$$\sum_{i,j} x_{ij} q_{ij} = 1$$

$$\sum_{j=1}^{n} x_{ij} q_{ij} = p_i \tag{9.29}$$

$$\sum_{i=1}^{m} x_{ij} q_{ij} = q_j.$$

Again, the usual argument leads to a representation

$$P_{ij} = q_{ij} \frac{\exp(-\lambda_i^* + \mu_j^*)}{Z(\lambda^*, \mu^*)} \tag{9.30}$$

where

$$Z(\lambda, \mu) = \sum_{i,j} \exp(-\lambda_i^* + \mu_j^*) q_{ij}. \tag{9.31}$$

We choose q_{ij} to be equal to 0.20 for i,j corresponding to the outer layers and equal to 0.85 for i,j corresponding to the inner block (this choice

Table 9.4 Reconstruction using standard maximum entropy.

0.1656	0.1497	0.1363	0.1451	0.143	0.1284	0.1431	0.1271	0.1463	0.1474	0.1481	0.1544
0.1885	0.1704	0.1551	0.1652	0.1628	0.1462	0.1629	0.1446	0.1666	0.1677	0.1686	0.1757
0.1637	0.1479	0.1347	0.1435	0.1414	0.1269	0.1415	0.1256	0.1446	0.1456	0.1464	0.1526
0.1719	0.1553	0.1415	0.1507	0.1485	0.1333	0.1486	0.1319	0.1519	0.153	0.1537	0.1602
0.1424	0.1287	0.1172	0.1248	0.123	0.1105	0.1231	0.1093	0.1258	0.1267	0.1274	0.1328
0.132	0.1193	0.4617	0.4917	0.4846	0.4351	0.4849	0.4304	0.4957	0.1175	0.1181	0.123
0.1316	0.119	0.4604	0.4903	0.4832	0.4339	0.4835	0.4292	0.4944	0.1171	0.1177	0.1227
0.1254	0.1133	0.4386	0.4671	0.4603	0.4134	0.4606	0.4089	0.4709	0.1116	0.1122	0.1169
0.1192	0.1077	0.4168	0.4439	0.4374	0.3928	0.4377	0.3886	0.4475	0.106	0.1066	0.1111
0.149	0.1347	0.5213	0.5551	0.5471	0.4913	0.5474	0.486	0.5597	0.1326	0.1333	0.1389
0.2075	0.1875	0.7257	0.7728	0.7616	0.6839	0.7621	0.6765	0.7791	0.1846	0.1856	0.1934
0.1814	0.164	0.6345	0.6757	0.666	0.598	0.6664	0.5916	0.6813	0.1614	0.1623	0.1691
0.1535	0.1387	0.1263	0.1345	0.1326	0.1191	0.1327	0.1178	0.1357	0.1366	0.1373	0.1431
0.173	0.1563	0.1423	0.1516	0.1494	0.1341	0.1495	0.1327	0.1528	0.1539	0.1547	0.1612
0.1792	0.162	0.1475	0.1571	0.1548	0.139	0.1549	0.1375	0.1584	0.1595	0.1603	0.1671
0.1705	0.1541	0.1403	0.1494	0.1473	0.1322	0.1474	0.1308	0.1507	0.1517	0.1525	0.1589
0.1455	0.1315	0.1197	0.1275	0.1257	0.1128	0.1257	0.1116	0.1285	0.1294	0.1301	0.1356

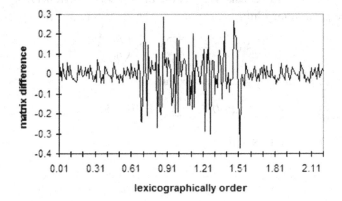

Fig. 9.3 Difference between data and reconstruction.

is consistent with the information given by the projections), and then proceeded to minimize $\ln Z(\lambda, \mu) + \langle \lambda, p \rangle + \langle \mu, q \rangle$. Once λ^*, μ^* realizing the minimum are found, they are substituted into (9.31) and Table 9.4 is obtained.

This time the reconstruction error is 5.3652×10^{-9}, and in figure 9.3 we display the difference between data and reconstruction.

In general one should try different q_{ij}'s or back up one's choice by some non-mathematical criterion.

The techniques in this section seem to have been first tried in [34].

9.10 Transportation and linear programming problems

Consider the variational problem: Find x^* such that

$$x^* = \arg\inf\{(c,x) : x \in \mathcal{C} \subset V, Ax = y\} \qquad (9.32)$$

where $A : V \to W$ is our old linear bounded operator, the constraint set $\mathcal{C}$ is a convex subset of V, and $c : V \to \mathbb{R}$ is a given element of V^*. Take a look at [18] and [28] for the connection of problem (9.32) with the problem described at the outset or in Section 2.7 of Chapter 2, and in any book on linear programming.

To sift problem (9.32) through the maximum entropy crib we proceed as follows: extend A to $\tilde{A} : V \to \tilde{W} = W \times \mathbb{R}$ by setting

$$\tilde{A}x = \begin{pmatrix} A \\ c \end{pmatrix} x = \begin{pmatrix} Ax \\ (c,x) \end{pmatrix} \qquad (9.33)$$

and to be on the safe side, we must assume that (9.32) is well set, which amounts to

$$\gamma_0 = \inf\{(c,x) : x \in \mathcal{C}\} > -\infty. \qquad (9.34)$$

The strategy consists of finding, for every $\gamma > \gamma_0$, a maxentropic solution $x^*(\gamma)$ to the inverse problem

$$\tilde{A}x = \begin{pmatrix} y \\ \gamma \end{pmatrix}, \; x \in \mathcal{C} \qquad (9.35)$$

and then let $\gamma \downarrow \gamma_0$.

Recall that $x^*(\lambda)$ was obtained as $E_{P^*}[X]$, where $X : \Omega \to \mathcal{C}$ was appropriately constructed random variable on $(\Omega, \mathcal{F}, Q)$ and $P^* \ll Q$ has a density $\frac{dP^*}{dQ} = \frac{\exp(\langle \lambda^*, AX \rangle)}{Z(\lambda^*)}$, where λ^* minimizes

$$S(\lambda) = \ln Z(\tilde{\lambda}) + \langle \lambda, y \rangle + \langle \mu, \gamma \rangle \qquad (9.36)$$

over $\tilde{W}^* = W^* \times \mathbb{R}$ for every $y = \begin{pmatrix} y \\ \gamma \end{pmatrix}$ in $\tilde{W}$.

Note that when the γ in (9.35) is smaller than γ_0, we cannot expect (9.36) to have a minimum. When $\gamma > \gamma_0$ and $x \in \text{int}(\mathcal{C})$, the $x^*(\gamma)$ can be found as in Chapter 5.

In [35] such ideas were formalized in a somewhat restricted setting and were put to test in a simpler model in [36]. From the numerical point of view, such an approach is tricky, for when $\gamma \to \gamma_0$, $x^*(\gamma) \to \partial \mathcal{C}$ and $\lambda^*(\gamma)$ minimizing (9.36) tends to ∞, so once you get close enough to γ_0, numerical routines become unwieldy, and one has to conform oneself with suboptimal solutions. These solutions may be a starting point for other algorithms. See [37] for an unrelated approach that also makes use of maxentropic arguments.

Here we shall apply the maxentropic approach to solve the linear programming problem that appeared in Chapter 7. The problem consists of finding an $x^* \in \mathbb{R}^n$ ($n = 100$) satisfying

$$x^* = \arg\inf\{\sum x_i : Ax = y, \ x \in [0,1]^n\}$$

where $y \in \mathbb{R}^2$ and the matrix A is specified there. The vector c mentioned above has all of its components equal to 1. The transposed of the extended matrix $\tilde{A}$ is $\tilde{A}^* = [A^* \ c]$ where the added part is just the column vector c. Now $\lambda \in \mathbb{R}^3$ and $\tilde{A}^*\lambda = A^*\hat{\lambda} + \lambda_3 c$ where $\hat{\lambda} \in \mathbb{R}^2$ is the vector formed with the first two components of $\lambda \in \mathbb{R}^3$.

This time the auxiliary sample space and the constraint space coincide: $\Omega = [0,1]^n$ and we consider an *a priori* probability Q on Ω made up of a product of Bernoulli measures $dQ = p\varepsilon_0(d\xi) + q\varepsilon_1(d\xi)$. We shall favor $p > q$ since we want to minimize the ℓ, norm of x.

We vary the third component γ of the extended data vector

$$y = \begin{pmatrix} y_1 \\ y_2 \\ \gamma \end{pmatrix}$$

until the maxentropic routine breaks down. In figure 9.4 we present the results.

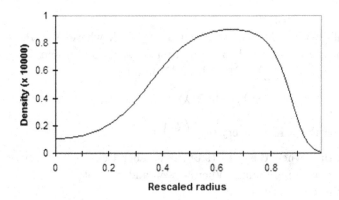

Fig. 9.4 Maxentropic approach to a linear programming problem, with $p = 0.75$, $q = 0.25$, $\gamma = 50.2199$. The reconstruction error $\varepsilon = 1.7831 \times 10^{-11}$.

Compare this with the various reconstructions obtained in Section 7.1.

9.11 Bibliographical comments and references

Some interesting connections between the problem of reconstruction from marginals as posed by Schrödinger and a class of processes defined on a finite time interval is explored in [1]-[2]. Different aspects and applications of the problem of reconstruction from marginals are presented in [4]-[33]. A description of the results in many of these was already given along the course of the chapter. That of the others can be partially guessed from their title.

A presentation and generalization of the results of Section 9.2 using large deviation theory is to be found in [39].

References

[1] Jamison, B. *"Reciprocal Processes"*. Zeit. Wahrsch. v Gebiete, Vol. 30, (1974), pp.65-86.

[2] Jamison, B. "The Markov Processes of Schroedinger". Zeit. Wahrsch. v. Gebiete, Vol. 32, (1975) pp. 323-331.

[3] Dym, H. and McKean, H. *"Fourier Series and integrals"*. Academic Press, New York, 1972.

[4] Renyi, A. *"On projections of probability distributions"*. Acta Math. Acad. Sci. Hung., Vol. 3, (1952) pp. 131-142.

[5] Strassen, V. *"The existence of probability measures with given marginals"*.

[6] Vasershtein, L. N. *"Markov processes over denumerable product of spaces, describing large systems of automata"*. Problems in Info Transm. Vol. 5, (1969) pp.64-72.

[7] Logan, B. F. and Shepp, L. A. *"Optional reconstruction of a function from its projections"*. Duke Math. Journal. Vol. 42, (1975) pp. 645-652.

[8] Logan, B. F. *"The uncertainty principle in reconstructing functions from projections"*. Duke Math. Journal. Vol. 42, (1975) pp. 661-706.

[9] Gutman, S., Kempermann, J., Reeds, J. and Shepp, L. *"Existence of probability measures with given marginals"*. The Annals of Prob. Vol. 19, (1991) pp. 1781-1797.

[10] Khalfin, L. and Klebanov, L. *"A solution of the computer tomography paradox and estimating the difference between densities of measures with the same marginals"*. The Annals of Prob. Vol. 22, (1994) pp. 2235-2241.

[11] Dall'Aglio, G. et al. (eds.). *"Advances in probability with given marginals"*. Kluwer Acad. Pubs., Dordrecht (1991).
[12] Klebanov, L and Rachev, S. *"The method of moments in computer tomography"*. Math. Scientist. Vol. 20, (1995) pp. 1-14.
[13] Applegate, D, Reeds, J., Steinberg, S. Shepp, L. and Shor, P. *"Some problems in probabilistic tomography"*. Theory of Prob. Appl. Vol. 41, (1996) pp. 199-209.
[14] Klebanov, L. and Ratchev, S. *"Computer Tomography and Quantum Mechanics"*. Tech. Rep. 296. U.C. St. Barbara, Dept. Stat, 1995.
[15] Hansel, G and Trallic, J. *"Mesures marginales et Theoreme de Ford-Fulkerson"*. Zeit. Warsch. v Geb., Vol. 43, (1978) pp. 245-251.
[16] Keller, H.G. *"Duality theorems and probability metrics"* in Ionifescu, M (ed) Proceed. VII-Romanian Prob. Conf. Brasov, Romanian (1984) pp. 211-220.
[17] Haneveld, W. *"Distributions with known marginals and duality of mathematical programming with applications to PERT"* in Ionifescu, M (ed) Proceed. VII-Romanian Prob. Conf. Brasov, Romanian (1984) pp. 221-236.
[18] Levin, V. *"General Monge-Kantorovich problem and its applications in measure theory and mathematical economics"* in Leifman, L. (ed.) "Functional Analyisi, Optimization and Mathem. Economics", Oxford. U.P., Oxford, 1990.
[19] Ramachadran, D and Rüschendorf, L *"A general duality for marginal problems"*, Theory Prob. & Related Fields, Vol 101, (1995) pp. 311-319.
[20] Hamaker, C and Solomon, D *"The angles between the null space of X rays"*. J. Math Anal. Appl. Vol. 62, (1978) pp. 1-23.
[21] Smith, K. Solomon, D. and Wagner, S. *"Practical and mathematical aspects of the problem reconstructing objects from radiographs"*. Vol. 83, (1977) pp. 1227-1270.
[22] Solomon, D. *"TheX-ray transform"*. J. Math. Anal. Appl. Vol. 56, (1976) pp. 61-83.
[23] Louis, A. K. *"Ghosts in tomography – The null space of the Radon Transform"*. Math. Meth. Appl. Sci. Vol. 3, (1981) pp. 1-10.
[24] Goncharov, A.B. *"Three dimensional reconstruction of arbitrary arranged identical particles given their projections"*. Transl. of Math. Monographs. Vol. 81, (1990) pp. 67-95.
[25] Gelfand, M. S. and Goncharov, A. *"Spatial rotational aligment of identical particles given their projections: Theory and practice"*. Transl.

of Math. Monographs. Vol. 81, (1990) pp. 97-122.
[26] Palanodou, V. *"Some singular problems in tomography"*, Transl. Math Monographs, Vol. 81, (1990) pp. 123-140.
[27] Popov, D, Sokova, E. and Susko, D. *"Mathematical models in two dimensional Radon Tomography"*. Am. Math. Soc. Transl. (2) Vol. 162, Providence 1994.
[28] Cuesta-Albertos, J., Matran, C. Ratchev, S and Rüschendorf, L. *"Mass transportation problems in probability theory"*. Math. Scient., Vol.21, (1996) pp. 34-72.
[29] Klaus, M. and Smith, R. *A Hilbert space approach to maximum entropy reconstruction"*. Math. Methods in Appl. Sci. Vol. 10, (1988) pp. 397-406.
[30] Lorentz, G.G. *"A problem of plane measure"*. Amer. J. Math., Vol. 71, (1949) pp. 417-426.
[31] Fishburn, P., Lagarias, J., Reeds, J. and Shepp, L. *"Sets uniquely determined by projection on axes I. Continuous case"*. SIAM J. Appl. Math., Vol. 50, (1990) pp. 288-306.
[32] Kemperman, J. *"Sets of uniqueness and systems of inequalities having a unique solution"*. Pacific J. of Mathematics, Vol. 148 (1991), pp. 275-301.
[33] Kemperman, J. *"On sets that are uniquely determined by a restricted set of integrals"*. TAMS, Vol. 332 (1990), pp. 417-458.
[34] Gzyl, H. and Herrero, B. *"A maxentropic procedure for reconstructing a distribution from its marginals"*. Math. and Comp. Modelling, Vol. 24. No. 2 (1996) pp. 83-89.
[35] Kuba and Volcic, A. 1*"Characteristics of measurable plane sets which are reconstructible from their two projections"*. Inverse problems, Vol. 4, (1988) pp. 513-527.
[36] Gamboa, F. and Gzyl, H. *"Linear programming with maximum entropy"*. Math. and Comp. Modelling, Vol. 13, (1990) pp. 49-52.
[37] Gzyl, H. and Moreno, F. *"Maxentropic analysis of linear programming problems"*. Vol. 26, (1997) pp. 107-116.
[38] Huan-wne, T. and Li-wei, Z. *"A maximum entropy method for linear programming"*. Chin. J. Num. Math. & Appl. Vol. 17, (1995), pp. 54-65.
[39] Cattiaoux, P. and Gamboa, F. *"Large deviations and variational theorems for marginal problems"*. Bernoulli, Vol. 5, No. 1 (1999), pp. 81-108.

Chapter 10

Numerical inversion of Laplace transforms

10.1 Motivation

Consider a measurable function $x(t)$, with $t \in [0, \infty)$ such that $|x(t)| \leq A \exp(\alpha_0 t)$ for some $\alpha_0 > 0$. Then for any complex $z = \alpha + i\beta$, with $\alpha > 0$ it is easy to see that

$$\tilde{x}(z) = \int_0^\infty \exp(-zt) x(t) dt \tag{10.1}$$

is a well-defined analytic function of z in the semiplane $\Re z > \alpha_0$.

Even better, when $x(t)$ is in $L_1[0, \infty) \cap L_2[0, \infty)$, $\tilde{x}(z)$ is analytic in $\Re z > \alpha_0$ and continuous up to $\Re z \geq \alpha_0$. Below we shall go into the problem of inverting (10.1), and in the rest of this section we shall examine two instances where we know $\tilde{x}(z)$ at several points in the half plane $\Re z > \alpha_0$ and we want either to find $\tilde{x}(z)$ there, or to find $x(t)$ for all t.

The first example is taken from [1]. When considering linear systems, described by (in the simplest case)

$$\frac{d^2 x}{dt^2} + a\frac{dx}{dt} + bx = f(t) \tag{10.2}$$

where the forcing term is $f(t) = \exp(i\omega t)$, the stationary state response of the systems is of the type $x_s(t) = Y(i\omega) \exp(i\omega t)$, where

$$Y(s) = (s^2 + as + b)$$

is called the transfer function or characteristic function of the system described by (10.2). Engineers (see [1] or references on related topics) have a way of measuring $|Y(i\omega)|$ and the phase $N(i\omega)$ in $Y(i\omega) = |Y(i\omega)| \exp(N(i\omega))$ for various values of ω. They represent the results in plots called Bode plots, and then infer the values of a and b from these.

In a variety of physico-chemical systems, one is interested in the energy distribution function among micro states, and the only thing accessible to observation is

$$\zeta(T) = \int_0^\infty \exp(-\frac{\beta}{T}\varepsilon)\rho(\varepsilon)d\varepsilon.$$

Here $\zeta(T)$ is related to measurable quantity (the free energy) for values of T. The problem then is to find $\rho(\varepsilon)$, the energy distribution function.

To finish the preamble, consider the scalar wave equation in $\mathbb{R}^3$

$$\frac{\partial^2 u}{\partial t^2} = \sum \sigma_y(x)\frac{\partial^2 u}{\partial x_i \partial x_j} + \sum b_i(x)\frac{\partial u}{\partial x_j} - c(x)u + s(t,x). \tag{10.3}$$

Assume $u(0,x)$ and $\frac{\partial u}{\partial t}(0,x)$ vanish (i.e., the problem has homogeneous initial condition), is usually interested in the solution of (10.3) for all $t > 0$. To find one, procedure consists of taking the Laplace transform of (10.3) to bring at to the form

$$(\alpha^2 - \mathcal{L})\tilde{u}(\alpha,x) = \tilde{s}(\alpha,x) \tag{10.4}$$

where $\tilde{u}(\alpha,x)$ and $\tilde{s}(\alpha,x)$ denote the Laplace transforms of $u(t,x)$ and $s(t,x)$ respectively, and

$$\mathcal{L} = \sum \sigma_{ij}(x)\frac{\partial^2}{\partial x_i \partial x_j} + \sum b_i(x)\frac{\partial}{\partial x_i} - c(x).$$

It so happens that the Green function for (10.4) can be explicitly computed in terms of path integrals with respect to a diffusion associated with the pair (σ, b) in (10.3) and an auxiliary Brownian motion in $\mathbb{R}^3$. This connection between elliptic problems and diffusions is widely known, see [2] and a possibility of using it for studying wave problems was attempted in [3]. Such representations are quite amenable to Monte Carlo Simulation, actually they are trivially parallelizable. The problem that is left open consists then of obtaining $u(t,x)$ from the values of $\tilde{u}(\alpha,x)$ computed (with error) for some values of α.

10.2 Basics about Laplace transforms

Even though Laplace transforms can be defined for rather fancy objects, see [4] for example, here we shall be concerned mainly with Laplace transform of measures defined on the line, or even better, on the half line $[0,\infty)$.

Consider the discrete measure on $[0, \infty)$ with commutative mass function $F(x) = \sum_{n \leq x} a_n$ with $a_0 = 0, a_1, a_2, \ldots$ given. One can write $dF(t) = \sum_{n=1}^{\infty} a_n \varepsilon_n(dt)$. Then

$$\tilde{F}(s) := \int_0^{\infty} \exp(-st) dF(t) = \sum a_n \exp(-ns). \tag{10.5}$$

If we set $\exp(-s) = z$ and put $H(z) = \sum a_n z^n$, then $H(z)$ is analytic function of z in the unit circle if $\tilde{F}(s)$ is an analytic function in $\Re s > 0$. Set $s_0 = 0$, $s_n = \sum_{i=1}^{n} a_i$ and notice that

$$\frac{H(z)}{1-z} = \sum_{n \geq 1} s_n z^n$$

from which, Cauchy's integral formula yields

$$s_n = \frac{1}{2\pi i} \int_C \frac{H(z)}{(1-z) z^n} dz \tag{10.6}$$

where C denotes the unit circle run counterclockwise. What (10.6) asserts is that given $\tilde{F}(s)$ (or $H(z)$) one can recover $F(t)$ by another integral transform. This is taken from Widder's classic [7-12]. Let us now go to [5] for the statement and proof of

Theorem 10.1. *Let $\tilde{\phi}(z)$ be analytic in the strip $\alpha < \Re z < \beta$ of the complex z plane. Assume that in the narrower strip $\alpha + \delta \leq \Re z \leq \beta - \delta$, $|\tilde{\phi}(z)| \leq k(|z|)$, where $k \in L_1([0, \infty))_+$ is assumed to exist. Then for real x and fixed y*

$$\phi(x) = \frac{1}{2\pi i} \int_{y-i\infty}^{y+i\infty} \tilde{\phi}(s) \exp(xs) ds \tag{10.7}$$

exists and in the strip $\alpha < \Re x < \beta$ the equation

$$\tilde{\phi}(z) = \int_{-\infty}^{\infty} \phi(x) \exp(-xz) dx \tag{10.8}$$

holds.

Corollary 10.1. *If $\beta = \infty$ and $\tilde{\phi}(z)$ is analytic in the whole $\Re z > \alpha$ half plane, and the rest is as in the statement of the theorem, then $\phi(x) = 0$ for $x < 0$ and (10.8) is to be replaced by*

$$\tilde{\phi}(z) = \int_0^{\infty} \exp(-xz) \phi(x) dx \tag{10.9}$$

for $\Re z > \alpha$. In other words (10.9) can be inverted by (10.7).

10.3 The inverse Laplace transform is not continuous

The following comment and example show that although the Laplace transform is continuous, the inverse transform is not. This is a real source of frustration when inverting from approximate data.

Let $x_1(t), x_2(t)$ both satisfy $|x_i(t)| \leq M \exp(at)$ for some $M, a > 0$, or let both be in $L_1([0,\infty)) \cap L_2([0,\infty))$. Then for $\Re z > a$

$$\begin{aligned} |\tilde{x}_1(z) - \tilde{x}_2(z)| &= \left|\int_0^\infty (x_1(t) - x_2(t)) \exp(-zt) dt\right| \\ &\leq \int_0^\infty |x_1(t) - x_2(t)| \exp(-at) \exp(-(z-a)t) dt \\ &\leq \sup_{t>0} |x_1(t) - x_2(t)| \exp(-at) \int_0^\infty \exp(-(\Re z - a)t) dt \\ &< \frac{\varepsilon}{\Re z - a}. \end{aligned}$$

Thus if $\sup_{t>0} |x_1(t) - x_2(t)| \exp(-at) \leq \varepsilon$, the difference in Laplace transforms stays small in $\Re z - a > \sigma$ for every $\sigma > 0$.

A similar computation shows that when $\|x_1 - x_2\|_p < \varepsilon$ then $|\tilde{x}_1(t) - \tilde{x}_2(t)| < M\varepsilon$ uniformly in compact regions of the complex right half plane. That is, the Laplace transform is continuous in many setups.

Consider now the two following examples: Consider first $x_\omega(t) = A \sin(at)$. Its Laplace transform is $\tilde{x}_\omega(z) = \frac{\omega A}{z^2 + \omega^2}$ for $\Re z > 0$.

Even though $x_\omega(t)$ has constant amplitude A, which can be small or large, its oscillations get higher and higher as ω increases, but $|\tilde{x}_\omega(z)| \leq \frac{A}{\omega}$ (for $\Re z > 0$). Thus small changes in $\tilde{x}_\omega(z)$ can produce widely different reconstructions.

Consider now $x_a(t) = A \frac{at^{-\frac{3}{2}}}{2\sqrt{\pi}} \exp(-\frac{a^2}{4t})$ together with $\tilde{x}_a(\alpha) = A \exp(-a\sqrt{\alpha})$ with $\alpha > 0$. Notice that for small a, $\tilde{x}_a(\alpha)$ stays bounded by A but $x_a(t)$ has a maximum at $t = \frac{a^2}{6}$, of size $\frac{cA}{a^2}$, with $c > 0$. Thus even when A is small, the hump in $x_a(t)$ can be large.

10.4 A method of inversion

Below we shall consider some methods of inversion of Laplace transforms which consist of expanding the unknown function on some basis of appropriate functions. To begin, let us sketch the following result from [7].

Theorem 10.2. *If the Laplace transform $\tilde{x}(z)$ of $x(t)$ vanishes on an infinite sequence of points located equidistantly on a line parallel to the $\Re z$-axis, i.e., if $\tilde{x}(z_0 + n\sigma) = 0$, $\sigma > 0$, $n = 1, 2, \ldots$ with z in the convergence region, then $x(t) = 0$.*

Inversion of Laplace transforms

Proof. Define $f(t) = \int_0^t \exp(-z_0 s)x(s)ds$, integrate by parts to obtain

$$\tilde{x}(z_0 + n\sigma) = n\sigma \int_0^\infty \exp(-n\sigma t)f(t)dt$$

and now let $\tau = \exp(-\sigma t)$, $t = -\frac{1}{\sigma}\ln\tau$, $\varphi(\tau) = f(-\frac{1}{\sigma}\ln\tau)$. Thus

$$\tilde{x}(z_0 + n\sigma) = n\sigma \int_0^1 \tau^{n-1}\varphi(\tau)d\tau.$$

If $\tilde{x}(z_0 + n\sigma) = 0$, then the uniqueness of the Hausdorff moment problem, which follows from the Hahn-Banach Theorem (see Theorem C.31) implies that $\psi \equiv 0$.

Using this Papoulis proposed in [7] three ways to tackle the problem of inverting

$$\tilde{x}(z_0 + n\sigma) = \int_0^\infty \exp(-(z_0 + n\sigma)t)x(t)dt.$$

If we replace $x(t)$ by $\exp(-z_0 t)x(t)$ if need be, we want to invert

$$\tilde{x}(\sigma n) = \int_0^\infty \exp(-n\sigma t)x(t)dt \qquad (10.10)$$

with $\sigma > 0$, $n = 1, 2, \ldots$. □

10.4.1 Expansion in sine functions

Consider the change of variables $\exp(-\sigma t) = \cos\theta$ which maps $[0, \infty)$ onto $[0, \frac{\pi}{2})$. Equation (10.10) becomes

$$\sigma\tilde{x}(\sigma n) = \int_0^{\frac{\pi}{2}} (\cos\theta)^{n-1} \sin\theta \, \xi(\theta) d\theta \qquad (10.11)$$

where $\xi(\theta) = x(-\frac{1}{\sigma}\ln\cos\theta)$. Consider now $n = 2k + 1$, $k = 0, 1, 2, \ldots$, replace $\xi(\theta)$ by $\xi(\theta) - \xi(0)$, modify $\tilde{x}(\sigma n)$ appropriately (how?) if need be, and expand $\xi(\theta)$ (in $(-\pi, \pi)$) but restrict yourself to $(0, \frac{\pi}{2})$ as

$$\xi(\theta) = \sum_{k=0}^{\infty} c_k \sin(2k+1)\theta. \qquad (10.12)$$

Using a binomial expansion verify that

$$(\cos\theta)^{2n}\sin\theta = \left(\frac{\exp(i\theta)+\exp(-i\theta)}{2}\right)^{2n}\left(\frac{\exp(i\theta)-\exp(-i\theta)}{2i}\right)$$
$$= \tfrac{1}{2^{2n}}\left\{\sin 2(n+1)\theta + \cdots + \left\{\left(\binom{2n}{k}\right) - \left(\binom{2n}{k-1}\right)\right\}\sin[(2n-k)+1]\theta + \cdots\right.$$
$$\left. \cdots + \left\{\left(\binom{2n}{n}\right) - \left(\binom{2n}{n-1}\right)\right\}\sin\theta\right\}\tfrac{1}{2k}.$$

Since the odd sines are orthogonal in $[0, \frac{\pi}{2}]$ and $\int_0^{\frac{\pi}{2}} (\sin(2n+1)\theta)^2 \, d\theta = \frac{\pi}{4}$, inserting (10.12) and the trigonometric identity we obtain the following system for the C_k in terms of the $\tilde{x}(\sigma n)$:

$$\sigma \tilde{x}[(2n+1)\sigma] = \frac{1}{2^{2n}} \left(\frac{\pi}{4}\right) \left\{ \left[\binom{2n}{n}\right) - \left(\binom{2n}{n-1}\right)\right] C_0 + \cdots \\ \cdots + \left[\binom{2n}{k}\right) - \left(\binom{2n}{k-1}\right)\right] C_{n-k} + \cdots + C_n \right\}. \quad (10.13)$$

Once N coefficients C_k are found, one can compare

$$\xi_N(\theta) = \sum_{k=0}^{N} C_k \sin(2k+1)\theta$$

with $\xi(\theta)$. The identity

$$\xi_N(\theta) = \frac{4}{\pi} \int_0^{\frac{\pi}{2}} \xi(y) \frac{\sin\{\frac{1}{2}(4N+3)(\theta-y)\}}{\sin\frac{1}{2}(\theta-y)} dy$$

comes in handy. From information about the class of functions in which $\xi(\theta)$ (or $x(t)$) is expected to be, we can use the identity to compare $\xi_N(\theta)$ to $\xi(\theta)$.

10.4.2 *Expansion in Legendre polynomials*

Introducing the change of variables $\tau = \exp(-\sigma t)$ which maps $[0, \infty)$ onto $(0, 1]$ and again set $\xi(\tau) = x(-\frac{1}{\sigma} \ln \tau)$, everything else as above. For $n = (2k+1)$ we obtain

$$\sigma \tilde{x}[(2k+1)\sigma] = \int_0^1 \tau^{2k} \xi(\tau) d\tau.$$

To use the orthogonality properties of the Legendre polynomials on $[-1, 1]$ we extend $\xi(\tau)$ in an even way to $[-1, 1]$ by setting $\xi(-\tau) = \xi(\tau)$ for $\tau > 0$. This function can be expanded in terms of even Legendre polynomials as

$$\xi(\tau) = \sum_{k=0}^{\infty} C_k P_{2k}(\tau). \quad (10.14)$$

If we undo the change of variables and computed the Laplace transform $\Phi_{2k}(z)$ of $P_{2k}(\exp(-\sigma t))$, we would obtain

$$\Phi_{2k}(z) = \frac{Q_{2k}(z)}{z(z+2\sigma)\ldots(z+2k\sigma)}$$

where $Q_{2k}(z)$ is a polynomial in z of degree less than $2k$. Since

$$\int_0^1 x^{2n} P_{2k}(x) dx = 0 \quad \text{for } n < k$$

undoing the change of variables we would get

$$\Phi_{2k}[(2n+1)\sigma] = 0 \quad n = 0, 1, ..., k-1.$$

In other words, the roots of $\Phi_{2k}(z)$ are $(2n+1)\sigma$, $n = 0, ..., k-1$. Therefore we can write

$$\Phi_{2k}(z) = \frac{(z-\sigma)(z-3\sigma)...(z-(2k-1)\sigma)A}{z(z+2\sigma)...(z+2k\sigma)}. \tag{10.15}$$

An elementary computation (a change of variables plus the dominated convergence theorem if you will) shows that

$$\lim_{z \to \infty} z\Phi_{2k}(z) = A = P_{2k}(1)$$

when $z \to \infty$ along (a line parallel to) the real axis. Thus $A = 1$ and

$$\Phi_{2k}(z) = \frac{(z-\sigma)(z-3\sigma)...(z-(2k-1)\sigma)}{z(z+2\sigma)...(z+2k\sigma)}.$$

If we now transform both sides of the expansion (10.14) and bring in (10.15) we obtain

$$\tilde{x}(z) = \frac{C_0}{z} + \sum_{k=1}^{\infty} \frac{(z-\sigma)(z-3\sigma)...(z-(2k-1)\sigma)}{z(z+2\sigma)...(z+2k\sigma)} C_k$$

and if now we replace z by $\sigma, 3\sigma, ..., (2k-1)\sigma, ...$ we obtain the system

$$\sigma\tilde{x}[(2k+1)\sigma] = \frac{C_0}{2k+1} + \frac{2kC_1}{(2k+1)(2k+3)} + \cdots + \frac{2k(2k-2)\cdot 2 C_k}{(2k+1)(2k+3)...(4k+1)}. \tag{10.16}$$

Papoulis offers the following criterion for choosing σ: If we want to concentrate on the interval $(0,T)$ to determine $x(t)$, then we choose σ such that $\exp(-\sigma T) = \frac{1}{2}$.

Another way of thinking about (10.16) is as an alternative to the moment problems dealt with in Chapter 7.

10.4.3 *Expansion in Laguerre polynomials*

If we do not want to change domains, we may expand $x(t)$ in terms of $\varphi_k(t) = \exp(-\frac{t}{2})L_k(t)$, (where $L_k(t)$ are Laguerre polynomials) which are orthogonal in $L_2([0,\infty))$. We write, for $\Re z > 0$, $\tilde{\varphi}_k(z) = \frac{z^k}{(z+1)^{k+1}}$ for the Laplace transform of $\varphi_k(t)$. Since the derivatives of $\tilde{\varphi}_k(z)$ of order less than k vanish at $z=0$, we have $\int_0^\infty t^n \varphi_k(t)dt = 0$ for $n \leq k-1$. Writing

$$x(t) = \sum_{k=0}^{\infty} C_k \varphi_k(t)$$

for $x(t)$ in $L_2([0,\infty))$ we obtain

$$\tilde{x}(z) = \sum_{k=0}^{\infty} C_k \tilde{\varphi}_k(z) = \sum_{k=0}^{\infty} A_k z^k \qquad (10.17)$$

if we expand $\frac{z^k}{(z+1)^{k+1}}$ as $z^k \sum \left(\binom{n+k}{k}\right)(-z)^n$ and rearrange. The connection between A's and C's being

$$A_k = \sum_{j=0}^{k}(-1)\left(\binom{k}{j}\right)C_{k-j}$$

from which

$$C_k = \sum_{j=0}^{k}\left(\binom{k}{j}\right)A_{k-j}.$$

Exercise 10.1. Use induction to prove it.

Thus, if we know $x(n\sigma)$ for a collection $1 \leq n \leq N$, we can solve for A_k for $1 \leq k \leq N$ and then for the corresponding C_k's.

10.5 From Laplace transforms to moment problems

We saw above, in Section 4.2 to be precise, how the problem of inverting the Laplace transform of a function $x(t)$, given the values of its Laplace transform $\tilde{x}(s)$ at points $s_n = s_0 + rn$ leads to the problem of reconstructing a function from its moments. To recall assume z_0 is real and $r > 0$. Set $t = -\frac{1}{r}\ln s$ and $\xi(s) = s^{(\frac{z_0}{r}-1)}x(-\frac{1}{r}\ln s)$ to again obtain

$$\int_0^1 s^n \xi(s)ds = y_n = \tilde{x}(s_0 + rn) \quad n = 0, ..., N. \qquad (10.18)$$

Thus the problem now consists in solving (10.18) for $\xi(s)$. The basic differences in the methods proposed in [8]-[11] are two: First, the orthonormal basis chosen for expanding $\xi(s)$. These references consider the shifted Legendre polynomials as basis for $L_2[0,1]$. On the other hand, they consider minimization of appropriately weighted and penalized norms instead of the standard L_2-norm.

The standard Legendre polynomials P_n are defined in $[-1,1]$, which can be mapped onto $[0,1]$ by means of $\sigma = 2s - 1$, which maps $P_n(\sigma)$ onto $\hat{P}_n(s) = P_n(2s - 1)$.

In [9] the authors expand $\xi(s) = \sum_{j=0}^{N} \alpha_j \hat{P}_j(s)$ and then proceed to minimize the quadratic form

$$q(\alpha) = \lambda(\alpha, \Delta\alpha) + \frac{1}{m}\|H\alpha - y\|^2$$

which comes from inserting the expansion $\xi(s) = \sum_{j=0}^{N} \alpha_j \hat{P}_j(s)$ into

$$Q(\xi) = \lambda \left\|\xi^{(2)}\right\|^2 + \frac{1}{m}\|A\xi - y\|^2$$

where the operator A is defined by the left-hand side of (10.18) and Δ is readily derived by taking second derivatives and square norm of $\xi(s) = \sum_{j=0}^{N} \alpha_j \hat{P}_j(s)$.

The approach in [11] consists in solving the following variational problem: Find $\xi^* \in L_2[0,1]$ such that

$$\xi^* = \arg\inf\left\{\frac{1}{2}\int_0^1 |\xi(s)|^2 : \int_0^1 s^n \xi(s)ds = y_n, \quad n = 0, ..., N\right\}.$$

Again, after expansion in terms of the shifted Legendre polynomials $\hat{P}_n(s)$, canceling the expansion at order N, minimization problem in $\mathbb{R}^{n+1}$ is obtained and solved.

10.6 Standard maxentropic approach to the Laplace inversion problem

There is a particular case of our problem that can be approached by the standard maximum entropy method. This problem appears in many applications of probabilistic modeling in which one has a differential equation describing the evolution of the transition probabilities of the system. In some cases such equations cannot be directly integrated, but when they are Laplace transformed, some sort of an approximation (or even) exact solution can be found for a collection of values of the transformed variable. Thus are confronted with the problem of finding a continuous function $x(t)$ defined for $t \geq 0$, given that

$$\int_0^\infty \exp(-\alpha_n t)x(t)dt = \tilde{x}(\alpha_n) \quad n = 1, ..., M+1 \qquad (10.19)$$

and $x(t) \geq 0$, continuous and

$$\int_0^\infty x(t)dt = 1, \quad \int_0^\infty tx(t)dt = m < \infty. \qquad (10.20)$$

The first of the conditions in (10.20) just embodies the fact that $x(t)$ is a probability density. In our previous sections it corresponded to the fact that $\int_0^\infty \exp(-\alpha_0 t)|x(t)|\,dt < \infty$, that is, $\tilde{x}(\cdot)$ is continuous up to the abscissa of (absolute) convergence. When $x(t) \in L_1([0,\infty))$, $s_0 = 0$.

The real constraint in (10.20) is the knowledge of the mean m of the distribution $x(t)$. This may either be assumed given or be inferred from the behavior of $\tilde{x}'(\alpha)$ at $\alpha = 0$. We shall include the first of the constraints in (10.20) where it really belongs, namely in (10.19), where we assume that $\alpha_1 = 0$.

As above, setting $\alpha_n = nr$ for $n = 0, 1, ..., M$ and $\xi(s) = \dfrac{1}{rs} x(-\tfrac{1}{r}\ln(s))$ we obtain instead of (10.19) and (10.20)

$$\int_0^1 s^n \xi(s)\,ds = y_n := \tilde{x}(\alpha_n) \tag{10.21}$$

$$\int_0^1 \ln(s)\xi(s)\,ds = -mr \tag{10.22}$$

as the system to solve for $\xi(s)$.

As we saw in Chapter 5, the standard maximum entropy method leads to

$$\xi^*(s) = \frac{1}{\zeta(\lambda^*)} \exp\left(-\langle \lambda^*, \Phi(s)\rangle\right) \tag{10.23}$$

where $\Phi : [0,1] \to \mathbb{R}^{M+2}$ is the vector-valued function with components $\phi_i(S) = s^i$ when $0 \leq i \leq M$ and $\phi_{M+1}(s) = \ln(s)$.

As always, for $\lambda \in \mathbb{R}^{M+2}$

$$\zeta(\lambda) = \int_0^1 \exp\left(-\langle \lambda, \Phi(s)\rangle\right) ds$$

which is defined whenever $\lambda \in \mathcal{D}(\Phi)$, with

$$\mathcal{D}(\Phi) = \{\lambda \in \mathbb{R}^{M+2} : \zeta(\lambda) < \infty\}$$

which now happens to be $\mathbb{R}^{M+1} \times (-\infty, 1)$. The λ^* appearing in (10.23) is obtained by minimizing

$$\Sigma(\lambda) = \ln \zeta(\lambda) + \langle \lambda, d\rangle \tag{10.24}$$

where $d \in \mathbb{R}^{M+2}$ has components $y_n = \tilde{x}(\alpha_n) = \tilde{x}(rn)$ for $0 \leq n \leq M$ and $y_{M+2} = -rm$. The minimization is to be carried over the set $\mathcal{D}(\Phi)$. Once λ^* and then $\xi^*(s)$ are obtained, the corresponding $x^*(t)$ is

$$x^*(t) = r \exp(-rt) \xi^*(\exp(-rt)). \tag{10.25}$$

10.7 Maxentropic approach in function space: The Gaussian case

In this section we develop a variant of the method used in Section 8.2 of Chapter 8 to deal with Fourier transforms. The variation on the theme relative to the material presented there, consists of a choice of the *a priori* measure. This section requires knowledge of stochastic calculus.

Let us consider $(\Omega, \mathcal{F})$ to be respectively $\mathcal{C}([0,1]) = \{\omega : [0,1] \to \mathbb{R}\}$ the continuous real-valued functions defined on [0,1] and $\mathcal{F}$ is the smallest σ-algebra making the coordinate maps $X_t : \Omega \to \mathbb{R}$ measurable. $X_t : \Omega \to \mathbb{R}$ is defined by $X_t(\omega) = \omega(t)$. Let us denote by Q the Wiener measure on $(\Omega, \mathcal{F})$. Such measure is defined on the cylinder sets by

$$Q(X_{t_1} \in A_1, ..., X_{t_n} \in A_n)$$
$$= \int_{A_1} ... \int_{A_n} \rho_{t_1}(x_1) \rho_{t_2-t_1}(x_2 - x_1)...\rho_{t_n-t_{n-1}}(x_n - x_{n-1}) dx_1...dx_n$$

where $\rho_t(x) = \frac{\exp(-\frac{x^2}{2ta})}{\sqrt{2\pi ta}}$, $0 \leq t_1 < t_2 < ... < t_n \leq 1$ and $A_1, ..., A_n$ are Borel sets in $\mathbb{R}$. It is easy to see that $Q(X_0 = 0) = 1$, thus Q concentrates its mass on the trajectories that start from zero.

This is nice since we want to solve (10.21) for $\xi(s)$ and we know that $x_1(t) \to 0$ as $t \to \infty$, and that $\int_0^\infty x(t)dt = 1$, thus $\int_0^1 \xi(s)ds = 1$ and $\xi(s) \to 0$ as $s \to 0$.

To motivate the procedures below note that if $k \in L_2([0,1])$ and if we define the translation operator $T_k : \Omega \to \Omega$ by $T_k(\omega)(t) = \omega(t) + \int_0^t k(s)ds$, it induces a measure $T_k^{-1}(Q) := Q_k$ on $(\Omega, \mathcal{F})$, absolutely continuous with respect to Q, with Radon-Nikodym derivative given by the famous Cameron-Martin formula $\frac{dQ_k}{dQ} = \exp\left(\int_0^1 k(s)dX(s) - \frac{a}{2}\int_0^1 k^2(s)ds\right)$. It is important for us since

$$E_{Q_k}[X(t)] = a \int_0^t k(s)ds \qquad (10.26)$$

or, equivalently, if $b \in L_2([0,1])$

$$E_{Q_k}\left[\int_0^1 b(s)dX(s)\right] = a \int_0^1 b(s)k(s)ds. \qquad (10.27)$$

Exercise 10.2. Go from (10.26) to (10.27) and back!

Let us now put $\psi : [0,1] \to \mathbb{R}^{M+1}$ to be the vector-valued function with components $\psi_n(s) = s^{n-1}$ for $n = 1, 2, ..., M+1$. As said above, we do not worry about the last constraint for we are considering measures on $(\Omega, \mathcal{F})$ which assign probability 1 to $\{X_0 = 0\}$.

The problem now consists of searching for a measure P^* on $(\Omega, \mathcal{F})$, $P^* \ll Q_k$ such that

$$E_{P^*}\left[\int_0^1 \psi(s)dX(s)\right] = \mu \qquad (10.28)$$

where μ is the vector in $\mathbb{R}^{M+1}$ with components $\mu_n = y_{n-1}$ defined in (10.21). If such a P can be found, according to (10.26) and (10.27) we would have that

$$\xi^*(s) = \frac{1}{a}\frac{d}{ds}E_{P^*}[X(s)] \qquad (10.29)$$

satisfies (10.21).

To carry on the maxentropic procedure we set

$$\mathcal{P}(\mu) = \left\{P \ll Q_k : E_P\left[\int_0^1 \psi(s)dX(s)\right] = \mu\right\}$$

which we better assume to be nonempty. On $\mathcal{P}(\mu)$ we define the entropy

$$S(P, Q_k) = -\int \frac{dP}{dQ_k}\ln\left(\frac{dP}{dQ_k}\right)dQ_k$$

and seek a P^* that maximizes $S(P, Q_k)$. The usual argument leads to a P^* with Radon-Nikodym derivative

$$\frac{dP^*}{dQ_k} = \frac{\exp\left(-\left\langle\lambda^*, \int_0^1 \psi(s)dX(s)\right\rangle\right)}{Z(\lambda^*)} \qquad (10.30)$$

where $\lambda^* \in \mathbb{R}^{M+1}$. The exponent can be rewritten as $\int_0^1 \langle\lambda, \psi(s)\rangle\, dX(s)$ where $\langle\lambda, \psi(s)\rangle = \sum_0^M \lambda_{n+1}s^n$. The vector λ^* appearing in (10.30) is to be found maximizing

$$Z(\lambda) = \int_\Omega \exp\left(-\int_0^1 \langle\lambda, \psi(s)\rangle\, dX(s)\right)dQ_k$$

over $\mathbb{R}^{M+1}$. To compute $Z(\lambda)$ we just use the definitions:

$$Z(\lambda) = E_{Q_k}\left[\exp\left(-\int_0^1 \langle\lambda, \psi(s)\rangle\, dX(s)\right)\right]$$

$$= E_{Q_k}\left[\exp\left(\int_0^1 \{k(s) - \langle\lambda, \psi(s)\rangle\}\, dX(s)\right)\right]\exp\left(-\frac{a}{2}\int_0^1 k^2(s)ds\right)$$

$$= \exp\left(\frac{a}{2}\int_0^1 \{k(s) - \langle\lambda, \psi(s)\rangle\}^2\, dX(s) - \frac{a}{2}\int_0^1 k^2(s)ds\right)$$

$$= \exp\left(\frac{a}{2}\int_0^1 \langle\lambda, \psi(s)\rangle^2\, ds - a\int_0^1 \langle\lambda, \psi(s)\rangle k(s)ds\right).$$

All that lies behind these computations is the independence of the increments $X(b) - X(a)$ over disjoint intervals, plus the standard result asserting that $\int_{-\infty}^{\infty} \exp(bx) \frac{\exp(-\frac{x^2}{2})}{\sqrt{2\pi}} dx = \exp(\frac{b^2}{2})$. We let the reader to work out the details. To find the right λ we minimize

$$\Sigma(\lambda) = \ln Z(\lambda) + \langle \lambda, \mu \rangle$$

over $\mathbb{R}^{N+1}$. A simple computation shows that

$$\lambda^* = C^{-1}\left(\langle \psi, k \rangle - \frac{1}{a}\mu\right) \tag{10.31}$$

where $\langle \psi(s), k(s) \rangle = \int_0^1 \psi(s) k(s) ds$ is in $\mathbb{R}^{M+1}$ and $C := \int_0^1 \psi(s)(\psi(s))^t ds$ or if you prefer, with components $C_{ij} = [r(i+j-1)]^{-1}$ for $i, j = 1, 2, ..., N+1$. This is a standard lousy matrix to invert: its smallest eigenvalue is too small when M is about 13, as we saw in Chapter 7.

To complete, the variance a in our Gaussian measure is arbitrary, and we can set it equal to 1 for the time being.

Also, the $k(s)$ that we have been dragging along is arbitrary. We can set $k(s) = \xi_0(s)$ where $\xi_0(s)$ is the output of any of the previous procedures to solve the same momentum problem (10.21). With these choices (10.29) looks like

$$\xi^*(s) = \xi_0(s) - \sum_{i=1}^{n} \psi_i(s) \lambda_i^*. \tag{10.32}$$

Exercise 10.3. Interpret (10.32).

When we undo all changes of variables, we obtain something like

$$x^*(t) = x_0(t) - \sum \lambda_n^* \exp(-r(n+1)t). \tag{10.33}$$

10.8 Maxentropic linear splines

The following approach can be considered a refinement of the discretization techniques presented in Chapter 7. Let us carry it in the specific context of finding a continuous $\xi(s)$ satisfying (10.21) and the constraint $\xi(0) = 0$.

If we split $[0,1]$ into N (say $N = 100$) intervals $(j-1, j]$, let us denote the values of the unknown function $\xi(s)$ at $s_j = \frac{j-1}{N}$ by $\xi\left(\frac{j-1}{N}\right) = \xi_j$ and let $j = 1, 2, ..., N+1$. Let $\xi_\ell(s)$ be the polygonal function that takes values $\xi_\ell\left(\frac{j-1}{N}\right) = \xi_j$ at the points s_j. For example, when $\frac{j-1}{N} \leq s \leq \frac{j}{N}$, then $\xi_\ell(s) = \xi_{j-1} N \left(\frac{j}{N} - s\right) + \xi_j N \left(s - \frac{j-1}{N}\right)$. When such an approximant is inserted into (10.21), we obtain the linear system for the vector $\xi \in \mathbb{R}^{N+1}$, with components $\xi_1 = 0, \xi_2, ..., \xi_{N+1}$

$$A\xi = y \qquad (10.34)$$

where recall $y_i = \tilde{x}(\alpha_i)$ for $i = 1, 2, ..., M+1$.

Exercise 10.4. Substitute $\xi_\ell(s)$ in (10.21) and obtain the elements of the matrix A.

To continue we should make up our minds as to where the vector ξ is to be searched for, that is, we have to decide on the convex set $C \subset \mathbb{R}^{N+1}$ of convex constraints imposed on ξ.

To relate the present results to the situation described in Section 10.6, let us consider $C = [0, \infty)^{N+1} = \mathbb{R}_+^{N+1}$, that is let us search for positive solutions to (10.34). We saw in Chapters 7 and 8 how to deal with this problem. One possibility consists of taking $\Omega = C$, set $\mathcal{F} = \mathcal{B}(C)$ the Borel sets of C and take as an *a priori* measure a product of gamma measures, i.e., $dQ = \otimes dQ_j$

$$dQ_j(u_j) = \frac{\mu^\omega u_j^{\omega-1}}{\Gamma(\omega)} \exp(-\mu u_j) du_j, \quad j = 1, ..., N+1$$

where $\mu > 0$ and $\omega > 1$ are parameters. To simplify the notations assume $\omega = 1$. To exploit the product nature of the setup, notice that the Laplace transform of dQ_j is

$$z_j(\tau_j) = \int_0^\infty \mu \exp\left(-(\tau_j + \mu)u\right) du = \frac{\mu}{(\tau_j + \mu)}$$

from which $Z(\lambda) = \prod_{j=1}^{N+1} z_j\left((A^*\lambda)_j\right)$ for $\lambda^* \in \mathbb{R}^{M+1}$. The function $\Sigma(\lambda)$ to be minimized now is

$$\Sigma(\lambda) = \sum_{j=1}^n \ln\left(\frac{\mu}{((A^*\lambda)_j + \mu)}\right) + \sum_{i=1}^{M+1} \lambda_i y_i$$

and the domain in which $Z(\lambda)$ is defined is $\bigcap_{j=1}^{N+1} \{(A^*\lambda)_j + \mu > 0\}$.

Once the right minimizer is found, each ξ_j is found as above: form

$$dP^* = \bigotimes_{j=1}^{N+1} \frac{\mu \exp\left(-(A^*\lambda)_j + \mu\right) u_j}{z_j\left((A^*\lambda)_j\right)} du_j$$

and compute
$$\xi_j^* = E_{P*}[X_j]$$
where, recall $X_j(u) = u_j$ whenever has components $u_1, ..., u_{N+1}$.

Exercise 10.5. Complete the missing details.

Exercise 10.6. Carry out the maxentropic procedure when each component ξ_j is required to lie in a preassigned range.

10.9 Connection with the complex interpolation problem

Let us now add a few words about the connection between inverting Laplace transforms from finite data and the Pick-Nevanlinna interpolation problem, mentioned in Chapter 2.

Let $x(t)$ be our unknown function and let us assume that we know
$$\tilde{x}(\alpha_i) := y_i = \int_0^\infty \exp(-\alpha_i t) x(t) dt \quad i = 1, ..., N \qquad (10.35)$$
for a finite number of complex numbers in the right half complex α-plane. We want to find $x(t)$.

Consider the maps
$$T : U \to \mathbb{C}_+ \quad w \to T(w) = \tfrac{1+w}{1-w}$$
$$T^{-1} : \mathbb{C}_+ \to U \quad \alpha \to T^{-1}(\alpha) = \tfrac{\alpha-1}{\alpha+1}$$
mapping bijectively the unit circle $U = \{w : |w| \leq 1\}$ onto the right-half complex plane $\{\alpha : \Re\alpha > 0\}$. To pull back $\tilde{x}(\alpha)$ to U we compose $\tilde{x} \circ T$ to obtain
$$\varphi(w) = \int_0^\infty \exp(-T(w)t) x(t) dt. \qquad (10.36)$$

Let us write $T(w) = 1 + \tfrac{2w}{1-w}$ and recall that
$$\sum L_n(t) w^n = \exp\left(-\tfrac{tw}{1-w}\right)/(1-w)$$
where $L_n(t) = \exp(t) \tfrac{d^n}{dt^n}(x^n \exp(-t))$ are the Laguerre polynomials, to obtain from (10.35)
$$\frac{\varphi(w)}{1-w} = \sum_{n \geq 0} \frac{w^n}{n!} \int_0^\infty \exp(-t) L_n(2t) x(t) dt$$

or if you prefer

$$\varphi(w) = \tfrac{1}{2}(1-w)\sum_{n\geq 0}\frac{w^n}{n!}\int_0^\infty \exp(-\tfrac{t}{2})L_n(t)x(\tfrac{t}{2})dt$$

$$= \tfrac{1}{2}(1-w)\sum_{n\geq 0}\frac{w^n}{n!}x_n$$

where the x_n are the coefficients of the expansion of $x(\tfrac{t}{2})$ in terms of the system $\{\exp(-\tfrac{t}{2})L_n(t):n\geq 0\}$, orthonormal and complete in $L_2([0,\infty),dx)$.

We can rewrite the last identity as

$$\varphi(w) = \frac{1}{2}x_0 + \frac{1}{2}\sum_{n\geq 1}\frac{w^n}{n!}(x_n - x_{n-1}) \tag{10.37}$$

therefore, if we have any means of reconstructing an analytic function $\varphi(w)$ from the knowledge of its values $\tilde{x}(\alpha_i)$ at points $T^{-1}(\alpha_i) = w_i$ inside the unit disk, by means of (10.37) we relate the series expansion of $\varphi(w)$ about $w=0$ to the coefficients of the expansion of $x(\tfrac{t}{2})$ in the basis $\{\exp(-\tfrac{t}{2})L_n(t):n\geq 0\}$.

The famous *Pick-Nevanlinna* theorem mentioned in Chapter 2 asserts that there exists c and $c_1,...,c_L$ with $|c|=1$ and $|c_i|<1$ such that

$$\varphi(w) = c\prod_{i=1}^{L}\frac{w-c_i}{1-\bar{c}_i w} \tag{10.38}$$

satisfies $\varphi(w_i) = y_i$ for $i=1,2,...,N$. A nice proof of this result can be seen in Rudin's [5-2] and it involves the basics summarized in the appendices, but it is a non-constructive or non-algorithmic proof, for it does not tell us how to compute the coefficients appearing in (10.38).

But the results we have piled up so far give us a neat way to say something about the complex interpolation problem, namely, suppose that we wanted to find $\varphi(w)$ defined on U such that $\varphi(w_i) = y_i$, where y_i are given complex numbers in U.

If we could prove that $\varphi_o(T^{-1}(\alpha))$, which is now a holomorphic function on $\mathbb{C}_+$ and continuous in $\overline{\mathbb{C}}_+$, is a Laplace transform, we could invoke the techniques developed above to invert it, then by means of (10.37) we could find $\varphi(w)$.

10.10 Numerical examples

We shall illustrate for the purpose of comparison, how some of the methods of inverting Laplace transforms work in a specific example. The test function we want to invert is $x(t) = \sin(\omega t)$ for various values of ω. The chosen function is continuous and bounded, so its Laplace transform $\tilde{x}(\alpha) = \frac{\omega}{\alpha^2+\omega^2}$ is defined (and analytic) in the $\Re\alpha > 0$ part of the complex α-plane, but it is not continuous up to the imaginary axis for $x(t)$ is not integrable. To make it integrable we multiply it by $\exp(-t)$ say and obtain $\tilde{x}(\alpha) = \frac{\omega}{(\alpha+1)^2+\omega^2}$ as the new transform.

The function $\sin(\omega t)$ has an interesting feature: It has an infinite number of oscillations. When $[0, \infty)$ is mapped onto a finite interval, these oscillations pile up at some end point, and it is impossible to capture them by attempting to reconstruct the function from its values at finitely many points.

To use the standard maximum entropy we need to shift and normalize the function. In this case we consider $x(t) = N \exp(-t)[1 + \sin(\omega t)]$ which has Laplace transform

$$\tilde{x}(\alpha) = N \left\{ (\alpha+1)^{-1} + \omega[\omega^2 + (\alpha+1)^2]^{-1} \right\}.$$

The first figures are divided in to groups of two, display the reconstructions of $\exp(-t)\sin(\omega t)$ for $\omega = 2\pi, 8\pi$ and $\sin(\omega t)$ for $\omega = 2\pi, 8\pi$, using expansions in terms of sine functions provide good agreement for "small" times, as time becomes larger the discrepancy increases. This is well "dramatized" in figures 10.3 and 10.4.

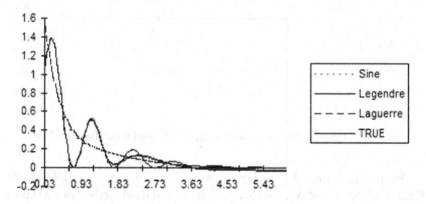

Fig. 10.1 Reconstruction of $\exp(-t)\sin(2\pi t)$ using sine, Legendre and Laguerre bases.

The errors for figures 10.1 and 10.3 are $0.1360 \times 10^{-2}, 0.6551 \times 10^{-4}$ and 0.2625×10^{-1} for the sine, Legendre and Laguerre bases respectively, and the errors for figures 10.2 and 10.4 are $0.1360 \times 10^{-2}, 0.1307 \times 10^{-2}$ and 0.1765×10^{-1} for the sine, Legendre and Laguerre bases respectively.

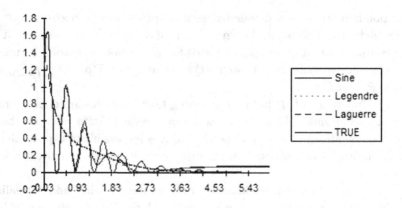

Fig. 10.2 Reconstruction of $\exp(-t)\sin(8\pi t)$ using sine, Legendre and Laguerre bases.

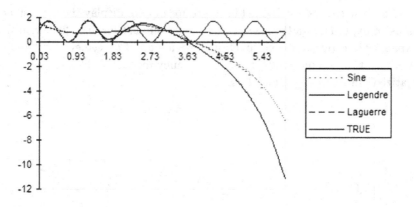

Fig. 10.3 Reconstruction of $\sin(2\pi t)$ using sine, Legendre and Laguerre bases.

Figures 10.5 and 10.6 display reconstructions of $N\exp(-t)[1+\sin(\omega t)]$ for $\omega = 2\pi, 8\pi$ and of $N[1+\sin(\omega t)]$ using the standard method of maximum entropy. Again, even though for long period of time the "misconstruction" is large, it is not as bad as that provided by the expansion with sine and Legendre bases.

The errors for figures 10.5 and 10.6 are 0.1602×10^{-1} and 0.2178×10^{-1}.

The comments about the reconstruction error should be mollified. In practice what we should really compare is the real data versus the reconstructed data. That is if $x^*(t)$ denotes the reconstruction of $x(t)$ from data

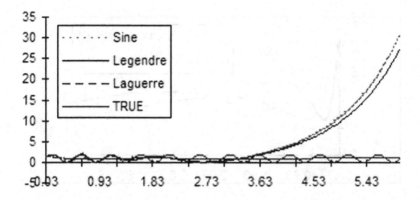

Fig. 10.4 Reconstruction of $\sin(8\pi t)$ using sine, Legendre and Laguerre bases.

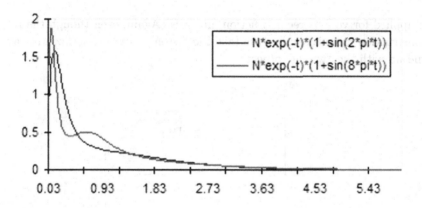

Fig. 10.5 Reconstruction of $N[1 + \sin(2\pi t)]\exp(-t)$ and $N[1+\sin(8\pi t)]\exp(-t)$ using standard maximum entropy.

$y_i = \widetilde{x}(\alpha_i)$, what we should compare is

$$\varepsilon = \left(\sum_{i=1}^{n} (\widetilde{x}^*(\alpha_i) - x(\alpha_i))^2\right)^{1/2}.$$

In figures 10.7 and 10.8 we present a reconstruction by means of first order spline interpolation and maximum entropy in the mean of the Laplace transforms of $x(t) = \exp(-\beta t)H(t - t_0)$ and $x(t) = \exp(-\beta t)[H(t) - H(t - t_0)]$ which are given respectively by $\widetilde{x}(\alpha) = \frac{\exp(-t_0(\alpha+\beta))}{(\alpha+\beta)}$ and $\widetilde{x}(\alpha) = \frac{1-\exp(-t_0(\alpha+\beta))}{(\alpha+\beta)}$. Here $\beta, t_0 > 0$ and the method

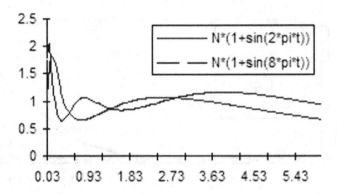

Fig. 10.6 Reconstruction of $N\left[1+\sin(2\pi t)\right]$ and $N\left[1+\sin(8\pi t)\right]$ using standard maximum entropy.

is applied for various reconstruction intervals. Again, even though the reconstruction error is small, if we take the "wrong" a priori setup, we may end up with the "wrong" object.

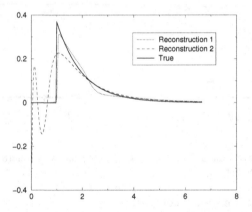

Fig. 10.7 Maxentropic reconstruction with first order splines from $m = 5$ (0,2,4,6,8) $\beta = 1$, $t_0 = 1, \sigma_1 = \sigma_2 = 0.6931$, for different intervals of reconstruction $I_1 = [0,1]$, $I_2 = [-1,1]$ with errors $\varepsilon_1 = 1.4438 \times 10^{-6}$, $\varepsilon_2 = 1.7652 \times 10^{-6}$.

In figures 10.9 and 10.10 we present reconstructions of $x(t) = \exp(-\beta t)\sin(t)$ and $x(t) = \sin(t)$. These were produced by first transforming the problem into a moment problem in [0,1], and then undoing the change of variables back to $[0,\infty)$.

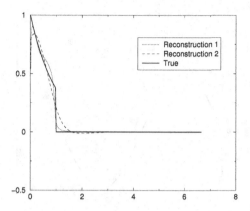

Fig. 10.8 Maxentropic reconstruction with first order splines from $m = 5$ (0,2,4,6,8) $\beta = 1$, $t_0 = 1, \sigma_1 = \sigma_2 = 0.6931$, for different intervals of reconstruction $I_1 = [0, 1.5]$, $I_2 = [-2, 2]$ with errors $\varepsilon_1 = 1.2994 \times 10^{-4}$, $\varepsilon_2 = 1.4573 \times 10^{-6}$.

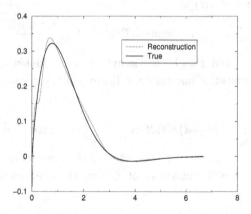

Fig. 10.9 Maxentropic reconstruction with first order splines from $m = 10$ (0,1,2,3,4,5,6,7,8,9), $\omega = 1$, $\sigma = 0.6931$ with error $\varepsilon = 1.7567 \times 10^{-5}$.

When $\sin(t)$ is replaced by $\sin(\omega t)$, things are similar, but worse: when reconstructing the Laplace transform of $\exp(-\beta t)\sin(t)$, the decay damps the oscillations and discrepancies with the original. When the $\exp(-\beta t)$ factor is removed, the discrepancies become apparent. But reconstructions errors are small nevertheless!

One obvious thought is that the difficulty lies in the nondecaying oscillatory nature of the datum $x(t) = \sin(\omega t)$. To test it we tried $x(t) =$

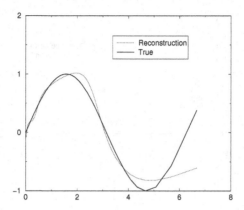

Fig. 10.10 Multiplying by $\exp(\beta t)$ the functions showed in figure 10.2.

$\exp(-\beta t)\sin(t)$ for $0 \leq t \leq T$ and $x(t) = 0$ for $t > T$. In this case the Laplace transform of $x(t)$ is

$$\tilde{x}(\alpha) = \frac{\omega}{(\alpha+\beta)^2 + \omega^2}[1 + \exp(-\alpha T)\cos(\omega T)] - \frac{\alpha \sin(\omega T)\exp(-\alpha T)}{(\alpha+\beta)^2 + \omega^2}.$$

If we set $\omega = 2n\pi$ and $T = 1$, that simplifies a bit. Below, in figures 10.11 and 10.12 we present a maxentropic linear splines reconstruction of the solution $x(t)$ to

$$\int_0^1 \exp(-\alpha t) x(t) dt = \frac{\omega}{\alpha^2 + \omega^2}[1 + \exp(-\alpha)]$$

where we take $\alpha_n = n\sigma$, $n = 0, 2, ..., 28$, i.e., 15 values of α. We present the reconstructions for different values of σ. Note the changes as σ varies.

10.11 Bibliographical comments and references

A quite recent reference about applications of Laplace transform to systems analysis is [13], where inputs to a linear systems are to be determined from measurements of the Laplace transform of the output contaminated by noise.

A work in which basic complex analysis is applied to moment problems, Laplace transforms and the material of section 9 is [14]. Reference [15] is a distilled version of [7-12], quite nice indeed.

Reference [16] has quite a suggestive title. They try to explain why regularization is a must for such problems.

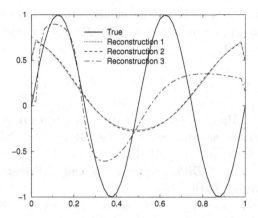

Fig. 10.11 Maxentropic reconstructions with first order splines from $m = 15$ $(0, 2, 4, ..., 26, 28)$, $\omega = 2\pi$, with different σ's values. Respectively values of σ's and errors are listed: $\sigma_1 = 0.009$, $\varepsilon_1 = 3.28 \times 10^{-6}$; $\sigma_2 = 0.09$, $\varepsilon_2 = 0.0014$; $\sigma_3 = 0.9$, $\varepsilon_3 = 0.0025$.

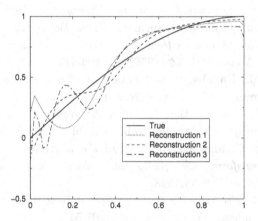

Fig. 10.12 Maxentropic reconstructions with first order splines from $m = 15$ $(0,2,4,...,26,28)$, $\omega = 8\pi$, with different σ's values. Respectively values of σ's and errors are listed: $\sigma_1 = 0.009$, $\varepsilon_1 = 1.22 \times 10^{-7}$; $\sigma_2 = 0.09$, $\varepsilon_2 = 4.68 \times 10^{-5}$; $\sigma_3 = 0.9$, $\varepsilon_3 = 6.25 \times 10^{-5}$.

A few items devoted to the problem of numerical computation of the complex inversion formula, specially when there is noise in the data are [17]-[22], and in [23] a direct application of the standard maximum entropy routine is invoked.

More or less recent work on the numerical aspects of the Nevanlina-Pick interpolation problem can be seen in [31] and to finish consider (the suggestive title of) [16].

References

[1] Powers, W. *"Modeling linear systems by frequency response methods"*, in *"Differential equation models"*, Ed. Raun, M., Coleman, C. and Drew, D., Springer Verlag, Berlin, 1983.

[2] Varadhan, S.R. *" Diffusion processes and Differential Equations"*. Springer-Verlag, Berlin, 1980.

[3] Gzyl, H. and Jiménez, R. *"Diffusions and Waves"*. J. Math. Anal. & Apply., Vol. 174 (1993) pp. 498-517.

[4] Vladimirov, V. *"Generalized Functions in Mathematical Physics"*. Mir Pubs, Moscow, 1979.

[5] Courant, R. and Hilbert, D. *"Methods of Mathematical Physics"*. Vol. 2. J. Wiley-Interscience, New York, 1962.

[6] Doetsch, G. *"Introduction to the Theory and Applications of the Laplace Transformation"*. Springer-Verlag, Berlin, 1974.

[7] Papoulis, A. *"A new method of inversion of the Laplace transform"*. Q. Apply. Math., Vol. 14 (1956) pp. 405-414.

[8] Bellman, R., Kallaba, R. and Locket, J. *"Numerical inversion of the Laplace Transform"*. Elsevier, New York, 1966.

[9] Brianzi, P. and Frontini, M. *"On the regularized inversion of the Laplace transform"*. Inverse Problems, Vol. 7 (1991) 355-368.

[10] Rodriguez, G. and Seatzu, S. *"On the numerical inversion of the Laplace transform with equally spaced data"*. Atti. Sem. Fis. Univ. Modena, XL (1992) 539-548.

[11] Talenti, G. *"Recovering a function from a finite number of moments"*. Inverse Problems, Vol. 3 (1987) pp. 501-517.

[12] Gzyl, H., Tagliani, A. and Velásquez, Y. *"Numerical reconstructions of Laplace transforms by various maxentropic procedures"*.

[13] Dey, A., Martin, C. and Ruymgaart. *"Input recovery from noise output data, using regularized inversion of the Laplace transform"*. IEEE Trans. Inf. Th. Vol. 44 (1998) pp. 1125-1130.

[14] Henrici, P. *"Applied computational complex analysis, Vol. 2"*. J. Wiley& Sons, New York, 1977.

[15] Widder, D. *"An Introduction to Transform Theory"*. Acad. Press, New York, 1971.

[16] Craig, I. and Thompson, W. *"Why Laplace transform are difficult to invert"*. Computers in Physics, Vol. 8 (1994) pp. 648-659.

[17] Longman, I. *"Application of best rational function approximation for Laplace transform inversion"*. J. Comp. and Apply. Math., Vol. 1 (1975) pp. 17-23.

[18] Levin, D. *"Numerical inversion of the Laplace transform by accelerating the convergence of Bromwick's integral"*. Vol. 1 (1975) pp. 247-250.

[19] Lyness *"A modification of Weeks method for numerical inversion of the Laplace transform"*. Math. Comp., Vol. 47 (1988) pp. 313-321.

[20] Ang, D., Lund, J. and Stenger, F. *"Complex variable and regularization methods of inversion of the Laplace transform"*. Math. Comp. Vol. 53 (1989) pp. 589-608.

[21] Essah, W. and Delves, L. *"On the numerical inversion of the Laplace transform"*. Inv. Prob., Vol. 4 (1988) pp. 705-724.

[22] Al-Suaibi, A. *"On the inversion of the Laplace transform by the use of a regularized displacement operator"*. Inv. Prob., Vol. 13 (1997) pp. 1153-1160.

[23] Wagner, U. and Geyer, A. *"A maximum entropy method for inverting Laplace transforms of probability density functions"*. Biometrika, Vol. 82 (1995) pp. 888-892.

[24] Gaver, D. *"Observing stochastic process and approximate transform inversion"*. Operat. Res. Vol. 14 (1966) pp. 444-456.

[25] Davies, B. and Martin, B. *"Numerical inversion of the Laplace transform: a survey and comparison of methods"*. J. Comp. Phys. Vol. 33 (1979) pp. 1-32.

[26] Piessens, R. *"A bibliography on numerical inversion of the Laplace transform and applications"*. J. Comp. Appl. Math. Vol. 2 (1975) pp. 115-126.

[27] Piessens, R. and Dang, N. *"A bibliography on numerical inversion of the Laplace transform and applications: a supplement"*. J. Comp. Appl. Math. Vol. 2 (1976) pp. 225-227.

[28] Schonberg, I. *"Remarks concerning a numerical inversion of the Laplace transform due to Bellman, Kallaba and Locket"*. J.M.A.A. Vol. 43 (1973) pp. 823-828.

[29] d'Motoni, P. *"Stabilization and error bounds for the inverse Laplace transform"*. Num. Funct. Anal. and Optim. Vol. 3 (1981) pp. 265-283.

[30] Antolín, J. *"Maximum entropy formalism and analytic extrapolation"*. J. Math. Phys. Vol. 31 (1990) pp. 791-797.

[31] Koç, Ç. and Chen, G. *"A fast algorithm for scalar Nevanlina-Pick interpolation"*. Numer. Mathem. Vol. 64 (1993) pp. 115-126.

Chapter 11

Maxentropic characterization of probability distributions

11.1 Preliminaries

So far we have examined the curious reasonable and satisfactory effectivity of the method of maximum entropy in dealing, with a reasonable degree of satisfaction, with ill-posed linear inverse problems with non-linear constraints.

But so far we have devoted no space to one of the original roles of the method: That of providing a unified variational framework, within which probability distributions of different nature were obtained and explained.

The appeal of such approach is related to Jaynes' original formulation of the maximum entropy method, to obtain equilibrium distributions in classical and quantum statistical mechanics.

In the classical literature on the subject, see [5-24]-[5-35] there you will find many instances and a long guide to thousands of further references to the problem: Find a probability distribution P on some measure space $(S, \mathcal{B})$, which is absolutely continuous with respect to some (usually σ-finite) measure on Q on $(S, \mathcal{B})$ and such that

$$E_P[\phi_i] = \int_S \phi_i dP = m_i \qquad i = 1, ..., N. \tag{11.1}$$

In statistical mechanics the functions ϕ_i defined on the state space S are called "observables", whereas in statistics the functions ϕ_i are called "statistics" and are regarded simply as random variables defined on the sample space S.

In the standard (Jaynes') formulation, the method of maximum entropy allowed us to choose a member of an exponential family

$$dP(\lambda) = \frac{\exp(-\langle \lambda, \phi \rangle)}{\zeta(\lambda)} dQ \tag{11.2}$$

where $\zeta(\lambda)$ is the usual normalization factor (called partition function in the physical literature)

$$\zeta(\lambda) = \int_S \exp(-\langle \lambda, \phi \rangle) dQ. \tag{11.3}$$

The parameter λ^* that makes (11.2) satisfy (11.1) is a minimizer of

$$\Sigma(\lambda) = \ln(\zeta(\lambda)) + \langle \lambda, m \rangle. \tag{11.4}$$

That is the routine followed by users of the standard maxentropic method. Below we shall apply MEM to a few particular cases. To establish, once more the basic notation let us begin with an S that is either finite or countable, and let us rewrite (11.1) as

$$\Sigma \phi_i(j) P_j = \Sigma A_{ij} P_j = m_i \ i = 1, ..., N \tag{11.5}$$

and assume as well that the first row of the matrix A is made up of ones, i.e., $A_{1j} = 1$ for all j.

Let us denote by I the range of the index j and consider

$$C := \{p : I \to [0, \infty), \sum p_j < \infty\} = \ell_{1+}$$

the positive integrable functions on I. Define $A : C \to \mathbb{R}^n$ by the obvious $Ap = m$ and extend A to all of $\ell_1(I)$. Thus instead of problem (11.1) we now have the problem: Find $p \in \ell_1(I)$ such that

$$Ap = m, \quad p \in C. \tag{11.6}$$

We shall employ this twist on the classical maximum entropy method to deal with a problem proposed by Kapur and Kesawan in [5-29] consisting of characterizing maxentropically all the discrete probability distributions in the following list

$$\begin{aligned} p_j &= (\mu_1 + \mu_2 \xi_j)^{-1} \\ p_j &= (1 + \exp(\mu_1 + \mu_2 \xi_j))^{-1} \\ p_j &= (\alpha + \exp(\mu_1 + \mu_2 \xi_j))^{-1} \\ p_j &= (\exp(\mu_1 + \mu_2 \xi_j) - 1)^{-1} \end{aligned} \tag{11.7}$$

where μ_1 and μ_2 are parameters such that

$$\sum p_j = 1 \quad \sum \xi_j p_j = m \tag{11.8}$$

where $\{\xi_j : j \in I\}$ and m are given. Here $I = \{1, ..., N\}$ or $I = \mathbb{N}$. Clearly (11.8) corresponds (11.5) with $A_{1j} \equiv 1$ and $A_{2j} = \xi_j$ for every $j \in I$.

We shall again see how the choice of setup determines the answer to the problem.

11.2 Example 1

Consider, in the notation of Chapter 5, $\Omega = [0,a]^I$ with the obvious product σ-algebra $\mathcal{F} = \otimes \mathcal{B}([0,a])$ and a priori measure Q on $(\Omega, \mathcal{F})$ given by $dQ = \otimes dQ_j(x_j)$

$$dQ_j(x_j) = s_j \varepsilon_0(dx_j) + t_j \varepsilon_a(dx_j)$$

and we do not assume that $s_j + t_j = 1$ but it is better that they are positive. Also for $X : \Omega \to C$, defined by the obvious $\xi = (\xi_1, ..., \xi_j, ...)$ as element of $\Omega \to \xi = (\xi_1, ..., \xi_j, ...)$ as element of C, to be well defined we need

$$E_Q[\|X\|_1] = a \sum t_j < 1$$

which we assume to be the case.

We have seen repeatedly that the method of maximum entropy in the mean leads to

$$dP = \frac{\exp(-\langle \lambda, AX \rangle)}{Z(\lambda)} dQ \qquad (11.9)$$

where $\lambda^* \in \mathbb{R}^2$ is to be found minimizing

$$\Sigma(\lambda) = \ln(Z(\lambda)) + \langle \lambda, m \rangle \qquad (11.10)$$

and $Z(\lambda)$ is given by

$$Z(\lambda) = \int_\Omega \exp(-\langle \lambda, AX \rangle) dQ = \prod_{j \in I} \zeta_j((A^*\lambda)_j) \qquad (11.11)$$

with

$$\zeta_j(\tau_j) = \int_0^a \exp(-\tau_j x_j) dQ_j(x_j) = s_j + t_j \exp(-\tau_j a) \qquad (11.12)$$

and $(A^*\lambda)_j = \lambda_1 + \lambda_2 \xi_j$. Therefore

$$dP = \prod_{j \in I} \frac{\exp(-(\lambda_1 + \lambda_2 \xi_j)a)}{s_j + t_j \exp(-(\lambda_1 + \lambda_2 \xi_j)a)} (s_j \varepsilon_0(dx_j) + t_j \varepsilon_a(dx_j))$$

with the aid of which we compute

$$E_P[X_j] = \frac{t_j a \exp(-(\lambda_1 + \lambda_2 \xi_j)a)}{s_j + t_j \exp(-(\lambda_1 + \lambda_2 \xi_j)a)} = \frac{a}{\frac{s_j}{t_j} \exp((\lambda_1 + \lambda_2 \xi_j)a) + 1}$$

which corresponds to item numbers 2 and 3 in (11.7) after fiddling with a, s_j and t_j, and redefining $\mu_1 = \lambda_1 a$, $\mu_2 = \lambda_2 a$.

Comment: We could have left a variable and take $\Omega = \times_{j \in I}[0, a_j]$ in such a way that $\sum a_j t_j < \infty$. We propose

Exercise 11.1. Repeat the process above and describe possible results.

Exercise 11.2. Consider $(\Omega, \mathcal{F})$ as above and $dQ_j(x_j) = \frac{s_j}{2}\varepsilon_0(dx_j) + \frac{t_j}{2}\varepsilon_a(dx_j) + \frac{s_j}{2}\varepsilon_a(dx_j)$. What conditions must be imposed on s_j, t_j so that $X : \Omega \to C$ have one or two moments? Assume you imposed them and verify that λ_1, λ_2 can be chosen so that (11.8) hold and

$$p_j = \frac{t_j + \left(\frac{s_j}{2}\right)\exp(-(\lambda_1 + \lambda_2\xi_j)\frac{a}{2})}{s_j \cosh(\lambda_1 + \lambda_2\xi_j)\frac{a}{2} + t_j}.$$

11.3 Example 2

Let us now consider $\Omega = [0, \infty)^I$, again with the obvious $\mathcal{F} = \otimes \mathcal{B}([0, \infty))$ and consider the *a priori* $dQ = \otimes dQ_j(x_j)$ with

$$dQ_j(x_j) = l_i \exp(-l_i x_j) dx_j$$

where the l_i are chosen so that $E_Q[\|X\|_1] = \sum E_{Q_j}[X_j] = \sum \frac{1}{l_j} < \infty$. With this the mapping $X : \Omega \to C$ defined as in example 1 is well defined.

The quest for a $P \ll Q$ such that (11.8) holds in the mean, i.e., $E_P[\|X\|_1] = 1$ and $E_P[\sum \xi_i X_i] = m$ leads to finding a member of an exponential family like (11.9), where the right λ^* is to be found minimizing (11.10), but this time

$$Z(\lambda) = \prod_{j \in I} \zeta_j((A^*\lambda)_j) \tag{11.13}$$

with

$$\zeta_j(\tau) = \int_0^\infty \exp(-\tau x_j) l_j \exp(-lx_j) dx_j = \frac{l_j}{l_j + \tau} \tag{11.14}$$

which yields

$$dP = \prod_{j \in I}(l_j + \lambda_1^* + \lambda_2^*\xi_j)\exp(-(l_j + \lambda_1^* + \lambda_2^*\xi_j)x_j)dx_j.$$

A simple integration now leads to

$$P_j = E_P[X_j] = \int_0^\infty x_j(l_j + \lambda_1^* + \lambda_2^*\xi_j)\exp(-(l_j + \lambda_1^* + \lambda_2^*\xi_j)x_j)dx_j$$
$$= \frac{1}{(l_j + \lambda_1^* + \lambda_2^*\xi_j)},$$

which takes care of the first item in the list (11.7).

11.4 Example 3

With $\Omega = [0,\infty)^I$ as above, assume now that each coordinate map $X_j : \Omega \to [0,\infty)$ is to be integer valued (with probability 1). This is achieved by considering *a priori* $dQ = \otimes dQ_j(x_j)$ with $Q_j(x_j)$ charging the non-negative integers only. For example, each Q_j is Poisson with parameter l_j such that $\sum_{j \in I} l_j < \infty$.

We leave for the reader to verify that this time

$$\ln(Z(\lambda)) = -\sum l_j \left(1 - \exp(-(\lambda_1 + \lambda_2 \xi_j))\right)$$

and minimization of $\Sigma(\lambda)$ leads to system

$$\sum l_j \exp(-(\lambda_1^* + \lambda_2^* \xi_j)) = 1$$

$$\sum l_j \xi_j \exp(-(\lambda_1^* + \lambda_2^* \xi_j)) = m$$

from which $\exp(-\lambda_1^*) = \left(\sum l_j \exp(-\lambda_2^* \xi_j)\right)^{-1}$ which we denote by $(z(\lambda_2^*))^{-1}$.

As further exercise we leave for the reader to verify that

$$p_j = l_j \frac{\exp(-\lambda_2^* \xi_j)}{z(\lambda_2^*)}$$

which is the answer provided by the standard maximum entropy method.

11.5 Example 4

A situation in which σ-finite measures cause no problem is provided by taking $\Omega = [0,\infty)^n$, $\mathcal{F} = \mathcal{B}([0,\infty)^n)$ and $dQ(x) = \bigotimes_j \left(\sum_{k \geq 0} \varepsilon_k(dx_j)\right)$, that is, a product of counting measures concentrated on the class of non-negative integers.

Now, an easy computation verifies

$$Z(\lambda) = \prod_{j=1}^{n} \left(\sum_{k \geq 0} \exp(-(\lambda_1 + \lambda_2 \xi_j)n)\right) = \prod_{j=1}^{n} \left(1 - \exp(-(\lambda_1 + \lambda_2 \xi_j))\right)^{-1}.$$

And the corresponding maxentropic distribution would be

$$dP(x) = \sum_{k_1,\ldots,k_n} P^*(k_1,\ldots,k_n) \varepsilon_{k_1}(dx_1)\ldots\varepsilon_{k_n}(dx_n)$$

with

$$P^*(k_1, ..., k_n) = \prod_{j=1}^{n} (1 - \exp(-(\lambda_1 + \lambda_2 \xi_j))) \exp(-(\lambda_1 + \lambda_2 \xi_j) k_j)$$

with respect to which

$$p_j = E_p[X_j] = \sum_{k_1,...,k_n} P^*(k_1, ..., k_n) = \frac{1}{\exp(\lambda_1 + \lambda_2 \xi_j) - 1}$$

which is the fourth item in (11.7).

11.6 Example 5

If we consider $\Omega = \times_{j \in I} [0, a_j]$ with $\sum a_j < \infty$, we impose the uniform a priori $dQ = \bigotimes_{j \in I} \frac{dx_j}{a_j}$ and turn the crank, we obtain

$$Z(\lambda) = \prod_{j \in I} \frac{1 - \exp(-(\lambda_1 + \lambda_2 \xi_j) a_j)}{a_j (\lambda_1 + \lambda_2 \xi_j)}$$

and once the good λ^* has been found minimizing (11.10), we obtain

$$dP = \bigotimes \frac{a_j(\lambda_1^* + \lambda_2^* \xi_j) \exp(-(\lambda_1^* + \lambda_2^* \xi_j) x_j)}{1 - \exp(-(\lambda_1^* + \lambda_2^* \xi_j) a_j)} dx_j$$

as the maxentropic measure on Ω and being a product measure it yields

$$E_P[X_j] = p_j = \frac{1}{\lambda_1^* + \lambda_2^* \xi_j} - \frac{a_j}{\exp(-(\lambda_1^* + \lambda_2^* \xi_j) a_j) - 1}$$

which we leave for the reader to verify. Notice that only when $|I| < \infty$, we are allowed to let $a_j \to a$. This played a role in Chapter 7.

11.7 Example 6

This time, instead of considering measures that change countable sets of Ω, we shall consider countable model for Ω itself. Assume $\Omega = \times_{j \in I} \{0, h_j, ..., N_j h_j\}$ that is, each $x \in \Omega$ is a list in which $x_j = X_j(x) \in \{0, h_j, ..., N_j h_j\}$ where the N_j are positive integers.

On each $\{0, h_j, ..., N_j h_j\}$ we consider binomial distribution as a priori measure, i.e.,

$$Q_j(X_j = k h_j) = \binom{N_j}{k} p_j^k q_j^{N_j - k}$$

if $k \leq N_j$, and equal to zero otherwise.

Again the product structure of the setup makes computing $Z(\lambda)$ easy. We obtain

$$Z(\lambda) = \prod_j (p_j \exp(-(\lambda_1 + \lambda_2 \xi_j)) + q_j)^{N_j}$$

and the corresponding computation, once the good values λ_1^* and λ_2^* that minimize the corresponding version of (11.10) have been found, we can compute p_j to obtain

$$p_j = E_P[X_j] = N_j \frac{p_j}{q_j} \left(\frac{p_j}{q_j} + \exp(\lambda_1 + \lambda_2 \xi_j) \right)^{-1}$$

providing another framework in which items 2 or 3 (11.7) can be obtained.

Reference

[1] Gzyl, H. *"Maxentropic characterization of some probability distributions"*. Studies in Appl. Math. 105 (2000) pp. 235-243.

Chapter 12

Is an image worth a thousand words?

This chapter was contributed by my friend and former classmate Ignacio Ferrín who is a practicing astronomer. In case you feel like commenting or adding to the material, he can be reached at ferrin@ciens.ula.ve

12.1 Problem setup

The title of this chapter follows a known adage which, as far as I know, has not been "proved". Here we shall see that it is nevertheless true.

The image (12.1) that goes with the text, is a photograph of an astronomical observatory and contains an enormous amount of information, which is not apparent to a careless observer.

We may say that in physics there are direct and inverse problems. In a direct problem one uses the laws of physics to predict a phenomenon or study the behavior of system. In an inverse problem one has experimental or observational data, and has to explain it. Preferably assembling a theory that explains the given data. Or at least to use a known theory to explain the data. The data for us will be the photograph and some information provided by the photographer.

The observatory in the photograph is in an international observatory list, and has been assigned a code number. The picture was taken with an Olympus OM-1 camera, with a $f/d = 2.8$ lens. Here f is the focal distance and d is the lens diameter. The film was an ISO 100/ASA21.

With this information, the problem consists of answering the questions posed below. The answers must be quantitative and based on sound reasoning. The information in the picture is complete.

To solve the problem you need a marked ruler, a protractor, a calculator, and you have to measure distances on the picture (so better photocopy it).

Fig. 12.1 The original picture.

You will also need a metric tape to measure some lengths at home: height of chairs, height and width of doors. You will have to access the web and perhaps the references cited at the end.

Assembling the problem took me many years. During this time the list of questions and answers grew, and a few more were added for this chapter. Perhaps you, dear reader, can come up with more. Do not hesitate to contact me for comments.

12.1.1 List of questions for you to answer

1) From which hemisphere was the picture taken?
2) Find the latitude (LAT) of the site.
3) Which is the observatory shown, and what is its code?
4) On which continent is it located?
5) Determine the exposure time T?
6) Compute the focal distance f of the lens.
7) Find the diameter d of the lens.
8) If the camera was triggered at 12:00 P.M. local time what was the date?
9) Find the height E of the building.
10) Estimate the width B of the building.
11) Find the distance D from the building to the camera.
12) Determine the diameter a of the telescope.
13) What angle does it make with respect to the plane of the paper.
14) What is the f/d of the telescope?
15) Is it a refracting or reflecting telescope?
16) Estimate the stellar limiting magnitude SLM
17) What can you say about the cloudiness of that evening?
18) What can you say about the temperature of that evening?
19) Who took the picture?
20) Can one write the answer in less that 1000 words?

12.2 Answers to the questions

12.2.1 Introductory comments

The answers provided are quantitative and scientific. The information provided by the picture is complete: no guess work is necessary. Usually there will be no error estimates for not to complicate the answers needlessly, but in some simple cases, error estimates are provided.

All values computed are either "averages" or "most probable values". You can measure on the picture and obtain your own values, probably a bit different. But most certainly, procedures will have to be similar.

12.2.2 Answers

1) From which hemisphere was the picture taken?

Since there is no polar star in the "center" of the picture, it must have been taken on the southern hemisphere.

2) Compute the latitude (LAT) of the site.

We know that the latitude is negative. Refer to figure 12.1. The latitude equal to the height of the pole P (see figure 12.2) above the horizon.

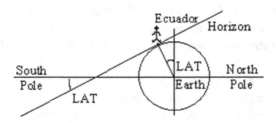

Fig. 12.2   The latitude (LAT) is equal to the height of the pole P above the horizon.

To be able to determine it, we have to identify the stars in the picture, and to know their distance to P. For that I have used the Norton's Star Atlas, [2].

Any other, and there are many, would have been all right. In figure 12.2 I have identified some of the stars that have left and are like trace. If we had a stellar trace tangent to the horizon, the distance of that trace to P would be the latitude of the place absorption, such trace is unavailable.

We have to relativize. The star that will do the job for us is η in the Mensa constellation. The distance to the pole P will be

$$\text{LAT} = \text{distance } (P - \eta\text{Mensa}) \times \frac{\text{linear distance } (P\text{–horizon}) \text{ on the picture}}{\text{linear distance } (P\text{–}\eta\text{Mensa}) \text{ on the picture}}$$

The quotient of the linear distances measures on the picture comes out to be 1.44, and in Hofleit ([1]) "Catalogue of Bright Stars" the declination of that star is $-75°05' = -75.08°$. Thus $90° - 75.08° = 14.92°$ which yields LAT $= -21.51°$.

Since the next question relates to the position of the observatory, we compute LAT with the aid of different stars. We obtain $-21.43°$, $-21.76°$, $-21.55°$ and $-21.22°$. These yield $-21.49° \pm 0.20°$.

3) Which is the observatory on the picture and what is its code?

The statement suggests that observatories have some sort of international code assigned to them. I know two lists of international observatories. One of them can be seen at

$$\text{http}://\text{cfa}-\text{www.harvard.edu/ian/cbat.html}$$

the same list is included in "ceres", an integrated software for the study of asteroids, see [5].

My first reaction was, all right, let us go to the list to identify the observatory by LAT. But, surprise, the observatories are not listed by LAT, but by their longitude (or LON). Dead end? No!

With the aid of a good Geographical Atlas, like the "Great World Atlas" [3], I surveyed the $-21.49°$ LAT. There we can see that Africa lies within the band $(+3°, 35°)$, Madagascar lies with the $(+44°, 48°)$ band, Australia within the $(+117°, 148°)$ and Latin America within the band $(-70°, -41°)$.

Back to the list to find out which observatories lie with these longitude bands have LAT $-21.49°$ approximately.

In Africa we have (with (LON/LAT/CODE)): Johannesburg $(27.88/-26.0/76, 78, 81)$, Bloemfontein $(26.40/-29.1/74)$ and Pretoria$(28.23/-25.7/79)$. In Africa we are not.

In Madagascar we do not have observatories and in Australia we have Siding Spring $(+149.1/-31.2/413)$, Sydney $(+151.2/-34.0/420)$ and Brisbane $(+152.9/-27.5/425)$. Then again, our target is not in Australia.

In Latin America we have some near hits: Rio de Janeiro $(-43.2/-22.6/880)$, Itajubá $(-45.58/-22.5/874)$, Campinas $(-46.83/-22.94/870)$, Tarija $(-64.6/-21.5/820)$ and Cordoba$(-64.55/-31.38/821)$.

So our best bet will be Tarija in Bolivia, Code number 820.

4) On which continent is it located?
   South America.

5) Determine the exposure time T.
   Evidently the shutter stayed open for a long while for the stars have left clearly marked traces. In figure 12.2 I have marked the angle subtended by the trace of α Mensa. Now, the time that a star takes to circle the celestial sphere is not 24 hours. That would be the day it takes the Sun to circle the sphere. A "sidereal day" is 23h56m04s $= 23.9344$ hours. Therefore the exposure time is

$$T = 23.944 \frac{\alpha}{360°}.$$

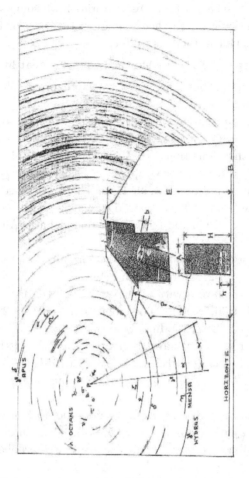

Fig. 12.3 Here we define variables for the analysis. We identified some of the stars in the picture. The South Pole is identified with a letter P. The star pointed with the arrow is the most weak start that the author could identify on the original photo.

With the aid of a protractor I obtained $\alpha = 17.6°$, therefore $T = 1.17$ hours $\approx 1h10m = 70m$.

6) Compute the focal distance f of the lens.

Check with any colleague in optics to understand why the scale (in degrees/cm) factor of a lens of focal distance f (in cm) is given by

$$\text{scale}[\circ/cm] = 57.29/f[cm].$$

For example, a lens with focal distance equal to 1m, would have a scale factor 0.573. In our case we have to find f. For that we have to know scales in the picture. The Olympus OM1 is a 35mm camera producing pictures of size $3.6 \times 2.4 cm$. Therefore the whole picture is an expansion of a frame of that size. If we knew the dimensions (in degrees) of the picture in along its vertical or vertical axes, we would obtain the scale factor. We can make use of the same information we employed for answer No 2, namely

$$\text{scale}[\circ/cm] = \frac{\text{distance}[\circ][P - \eta \text{ Mensa}]}{2.4cm} \times \frac{\text{vertical length of the picture}}{\text{distance}[cm](P - \eta \text{ Mensa})}$$

The two distances entering in the ratio, measured on my photograph, yield the quotient 2.225. The distance in degrees $(P - \eta Mensa)$ is 14.92°. Therefore

$$\text{scale}[\circ/cm] = 13.84[\circ/cm]$$

Therefore $f = 4.14cm$ is the focal distance we want.

7) Find the diameter d of the lens.
Since $f/d = 2.8$, we obtain $d = f/2.8 = 1.48cm$.

8) If the camera was triggered at 12:00 PM local time, what was the date?
In figure 12.3 we show a map of the celestial southern hemisphere copied from Norton's [2]. On the outer border of the map is shown the data at which the sphere has the given appearance at 12:00 PM, local time.
Compare figure 12.3 with figure 12.2, and considering only the beginning of the traces of the stars, we can conclude that if the picture was taken at 12:00 PM, then it must have been taken on December the 2nd. This is due to the fact that the traces of δ and Apus start almost vertically over P. The error is ±2 days.

9) Find the height E of the building.
There are three objects in the picture that allow us to determine a linear scale: (i) the height of the seat of the chair that is visible through the door; (ii) the width of the door and (iii) the height of the door. So I took a tape and measures every chair and door back home. I obtained (a) average height at seats $= h = 0.46m$; (b) average width of doors $= A = 0.76m$ and (c) average height of doors $= H = 1.95m$. Now I measure the ratios E/h, E/A and E/H on the picture and obtain $E_1 = 5.52m$, $E_2 = 4.50m$ and $E_3 = 5.15m$. Since the correct

Fig. 12.4 Southern Hemisphere's Map, used to determine the photo's date. The date is displayed in the external bound (Norton, 1978).

value is known, taking average to obtain $E_a = 5.06m$. The deviation from the extreme values is ±9%, reasonably small given the size and quality of the picture and the measurements made on it. Actual experiment yields errors larger than that.

10) Estimate the width B of the building.

Proceeding as above I obtain the three estimates: $B_1 = 5.67m$, $B_2 = 4.68m$ and $B_3 = 5.29m$. Averaging we obtain $B = 5.20m$.

11) Find the distance D from the building to the camera.

Consider a triangle whose vertical side is the height E of the building, whose base is the unknown distance D to the building. Thus, if β is the angle determined by the base and the line from "my feet" (describe the approximations involved) to the top of the building. That angle β was found in question-answer 5. The angular scale is $13.84°/cm$. Therefore

$$\beta = 13.84°[\circ/cm] \times 2.4cm \times \frac{E[cm]}{\text{vertical length of picture }[cm]}.$$

The rightmost ratio is 0.57 which yields $\beta = 18.80°$. Therefore
$$D = \frac{E[m]}{tg\beta} = 14.86m.$$

12) Determine the diameter a of the telescope.
    Proceeding as in (9) or (10), we obtain three possible values $a_1 = 0.55m$, $a_2 = 0.45m$ and $a_3 = 0.51m$. The average being $0.50m$. Again the maximum deviation from the mean is $\pm 10\%$. Not too bad.

13) What angle does it make with respect to the plane of the paper?
    The projected length of the tube of the telescope can be estimated relative to the chair and door. We obtain $P_1 = 2.33m$, $P_2 = 1.90m$ and $P_3 = 2.17m$, the mean being $P = 2.13m$.
    This is the projected length. It is related to the actual length L by $P = L\cos\gamma$. To obtain $\cos\gamma$ we measure the width b and the length of the upper opening of the tube: $b/a = \cos\gamma$. This yields $b/a = 0.48$ or $\gamma = 61°$.

14) What is f/d for the telescope.
$$f/d = \frac{L}{a} = \frac{P\cos\gamma}{a} = \frac{4.44}{0.50} = 8.88.$$

15) Is it a reflecting or refracting telescope?
    The refracting telescopes (that have a lens) usually have f/d ratios of the order of 15 to correct for chromatic aberration typical of that kind of systems. The reflecting telescopes (that have a mirror and do not show chromatic aberration) can have smaller f/d ratios, usually in the range 6 to 10. A value $f/d = 8.8$ suggests that it is a reflecting telescope.

16) Estimate the stellar limiting magnitude.
    The SLM is the magnitude of the faintest star shown on the picture. The intensity of star is its magnitude. The brightest or faintest star visible to the unaided eye in a clear might have magnitude 6.
    A picture can go deeper than a naked eye. The weakest stars have even larger magnitudes.
    The magnitude measured on a picture depends on the quality of the picture. I do not know what the printers will do to my picture. Anyway I measured on my copy. The stars further away from the pole leave a longer trace, thus the energy is spread over a bigger length than that coming from a similar star nearer to the pole: this increases their SLM.
    Close to the pole, in figure 12.2 I placed an arrow pointing to the faintest star I can see. Hope that you see it. To quantify its SLM

I have used a star atlas that goes deeper than Norton's, it is called Vranometry 2000, [4]. In that atlas, the star marked with the star has an estimated magnitude of 8. Thus near the pole SLM ~ 8, further away it is smaller than 8.

17) What can you say about the cloudiness of the evening?

At least two things: First, no clouds passed by. This is deducted from the uniformity and continuity of the star traces. Had there been cloudiness, during the 70min exposure time, the traces would appear interrupted.

On the other hand, we know the angular spread of the picture

$$\Omega = 13.84°[\circ/cm] \times 2.4[cm] \times 13.84°[\circ/cm] \times 3.6[cm] = 1655[\circ]^2.$$

The celestial sphere subtends an angle of $4\pi[rad]^2$ and $1 rad = 57.3[\circ]$, thus the whole sphere has an angular spread of $41273[\circ]^2$. On any given night, an observer can see half of that much, namely $20.637[\circ]^2$. Therefore our picture only covers 8% of the visible sphere, which stayed cloudless for 1h 10min. That is, if there were clouds and the air was motionless, they would cover only 92% of the globe. Since clouds move about quite a bit, it is reasonable to conclude that it was quite a cloudless evening.

18) What can you stay about the temperature of the night?

Also, at least two things. Since it was December, which in the southern hemisphere means summertime, it means that the temperature was higher than average. Since the temperature drops in the evening and increases in the morning, by midnight it must have gone down somewhat, but it must have been a warm evening.

19) Who took the picture?

The perpetrator of this note, of course.

20) Can you write the answer in less than 1000 words?

I doubt it. It took me over 2500 words to do it. I bet you would not be able to use much less than that. With this I have verified the adage: A picture is worth (more than) a thousand words.

Elementary, dear Watson!

12.3 Bibliographical comments and references

One general book on basic astronomy that the reader may enjoy is [6]. The rest of the references contain material useful for dealing with the matter of this chapter.

References

[1] Hoffleit, D. "Catalogue of Bright Stars". Yale University Observatory, New Haven, Connecticut, 1964.
[2] Norton, A. P. "Norton's Star Atlas". Gall and Inglis Ltd., Edinburgh, Scotland, 1978.
[3] Reader's Digest "Great World Atlas", The Reader's Digest Association, Pleasantville, New York, USA, 1963.
[4] Tirion, W., Rappaport, B., Lovi, G. "Uranometría 2000", Willmann Bell, Inc., Richmond, Virginia, USA, 1988.
[5] Zephyr Services "Ceres: An Integrated Software Package for Minor Planet's Researchers", 1992 Zephyr Services, 1900 Murray Ave., Pittsburgh, PA, 15217, USA. Developed for the Institute for Theoretical Astronomy, St. Petersburg, Russia.
[6] Cooper, W.A. and Walker, E.N. "Getting the measure of the stars". IOP Publishing, Philadelphia, USA, 1989.

Appendix A

Basic topology

If the notions of approximation and probability are fundamental to much of mathematics, topology and measure theory are the framework in which they are realized.

To describe nearness between points in a set S we introduce the notion of topology on S as follows

Definition A.1. A topology $\mathcal{T}$ on a set S is a collection of sets (called open sets) which satisfies

i) Arbitrary unions of sets in $\mathcal{T}$ is in $\mathcal{T}$.
ii) Finite intersections of sets in $\mathcal{T}$ is in $\mathcal{T}$.
iii) $S \in \mathcal{T}, \phi \in \mathcal{T}$.

Definition A.2. The pair $(S, \mathcal{T})$ is called an topological space

Example A.1. If S is a set, and $\mathcal{P}(S)$ denotes the collection of all subsets of S, $\mathcal{P}(S)$ is a topology on S.

Definition A.3. A metric or distance on S is a function $d : S \times S \to [0, \infty)$ satisfying

i) $d(x,y) = 0$ if and only if $x = y$.
ii) $d(x,y) = d(y,x)$ for all $x, y \in S$.
iii) $d(x,y) \leq d(x,z) + d(z,y)$ for all $x, y, z \in S$.

Definition A.4.

a) An open ball $B(x_0, \gamma)$ of radius γ around x_0, is defined by

$$B(x_0, \gamma) = \{x \in S : d(x_0, x) < \gamma\}$$

b) A set O in S is open if and only if (by definition) for any $x \in S$, there is a $B(x, \gamma)$ totally contained in O.

Comment: Part b) of the previous definition provides us with one of the standard ways of defining topologies on sets: just describe what the open sets are. This suggests that to describe a topology we may need a smaller collection of sets.

Definition A.5. A family $\mathcal{B}$ of subsets of S is a basis for a topology $\mathcal{T}$ if

i) $\phi \in \mathcal{B}$.
ii) $S = \bigcup_{B \in \mathcal{B}} B$.
iii) If B_1 and $B_2 \in \mathcal{B}$, then $B_1 \cap B_2 \supset B_0$ for some subset $B_0 \subset \mathcal{B}$.

Once we have the notion of topology we have

Definition A.6. Let $(S_1, \mathcal{T}_1)$ and $(S_2, \mathcal{T}_2)$ be topological spaces and let $f : S_1 \to S_2$. We say that f is continuous whenever $f^{-1}(\mathcal{T}_2) = \{f^{-1}(O) : O \in \mathcal{T}_2\} \subset \mathcal{T}_1$.

Nomenclature: Let $f : (S, \mathcal{T}) \to (\mathbb{R}, \mathcal{T}(\mathbb{R}))$ be continuous, where $\mathcal{T}(\mathbb{R})$ is the topology on $\mathbb{R}$ derived from the distance $d(x-y) = |x-y|$. We usually shorten it by saying $f \in \mathcal{C}(S)$.

Exercise A.1. When the topologies are derived from metrics, translate Definition A.6 into the usual $\varepsilon - \delta$ definition.

Exercise A.2. Consider $(\mathbb{N}, \mathcal{P}(\mathbb{N}))$, i.e. every $\{n\}$ is an open set. Make sense of the statement: the set $\{n, n+1, ...\}$ is a neighborhood of ∞ and define continuity of a real number sequence considered as a continuous map $(\mathbb{N}, \mathcal{P}(\mathbb{N})) \to (\mathbb{R}, \mathcal{T}(\mathbb{R}))$ where $\mathcal{T}(\mathbb{R})$ is the topology induced by the distance $d(x, y) = |x - y|$.

We can transport topologies

Exercise A.3. Let $f : S_1 \to S_2$ be a map and $\mathcal{T}_2$ be a topology on S_2. Verify that $\mathcal{T}(f) = f^{-1}(\mathcal{T}_2)$ is a topology on S_1. It is the smallest topology on S_1 making f continuous.

Exercise A.4. Repeat for a finite family $f_i : S_1 \to S_2, i \in I, |I| < \infty$.

Exercise A.5. (Important for defining weak topologies on function spaces.) Consider an arbitrary family $\mathcal{F}$ of functions $f : S_1 \to S_2$ and $\mathcal{T}_2$ is a topology on S_2. Then

$$\mathcal{B} = \left\{ \bigcap_{f \in \mathcal{I}} f^{-1}(O_i) : \mathcal{I} \text{ finite subfamily of } \mathcal{F}, O_i \text{ open sets in } S_2 \right\}$$

is a basis for a topology on S_1.

Comment: The topology generated by $\mathcal{B}$ is called the weak topology generated by the f's. Weak means that it is contained in any other topology which respect any $f \in \mathcal{F}$ is continuous.

Akin to the notion of continuous functions is

Definition A.7. Let $(S, \mathcal{T})$ be a topological space, consider $\mathbb{R}$ which the usual topology (defined by the absolute value metric) and let $f : S \to \mathbb{R}$ we say that

i) f is *lower semicontinuous* if $f^{-1}(t, \infty) = \{x \in S : f(x) > t\}$ is open,
ii) f is *upper semicontinuous* if $f^{-1}(\infty, t) = \{x \in S : f(x) < t\}$ is open,

for any $t \in \mathbb{R}$ is both cases. If f is both upper and lower semicontinuous, then it is continuous. Verify it!

Definition A.8.

a) A set $F \subset (S, \mathcal{T})$ is *closed* whenever F^c is open.
b) A set $K \subset (S, \mathcal{T})$ is *compact*, if from any covering of K by open sets, we can extract a finite covering. (A covering of K is a collection of sets which contain K.)

Let us now examine a result and some of its consequences important in optimization.

Theorem A.1. *Let $(S_1, \mathcal{T})$ and $(S_2, \mathcal{T})$ be topological spaces. Let $f : S_1 \to S_2$ be continuous and surjective. If S_1 is compact so is S_2.*

Proof. Let U be an open covering of S_2. Then $f^{-1}(U)$ is an open covering of S_1. Since S_1 is compact there exists a finite covering $f^{-1}(U_1), ..., f^{-1}(U_n)$ of S. Since f is onto $U_1, ..., U_n$ is an open covering S_2 (just use $f(f^{-1}(U)) = U$), thus S_2 is compact. □

Corollary A.1. *Let $f : S_1 \to S_2$ be continuous. Let K be compact in S_1, then $f(K)$ is compact in S_2.*

Corollary A.2. *In Corollary A.1 let S_2 be $\mathbb{R}$ (with the usual topology). Let $f : S_1 \to \mathbb{R}$ be continuous and $K \subset S$, be compact, then f assumes its maximum and its minimum in K.*

To prove the facts about extreme of functions it is useful to know.

Theorem A.2. *Let K be compact in $(S, \mathcal{T})$. Then every infinite subset of K has limit point.*

As an application consider the following.

Theorem A.3. *Let $f : S \to \mathbb{R}$ be lower semicontinuous and S be compact. Then S achieves its infimum on S.*

Proof. Consider the covering of S given by the level sets $L_t = \{f > t : -\infty < t < \infty\}$. By compactness finitely many suffice to cover S. Thus there is a t_0 such that $S = \{f > t_0\}$ or f is bounded from below. □

If x_n is a sequence such that $f(x_n) \downarrow \inf\{f(x) : x \in S\}$, there is a point x^* in S and a subsequence $x_{n(k)}$ such that $x_{n(k)} \to x^*$ and $f(x_{n(k)}) \downarrow \inf\{f(x) : x \in S\}$.

Let us close with some standard definitions.

Definition A.9.

a) A topological space $(S, \mathcal{T})$ is said to be *locally compact* whenever every point is contained in an open set having compact closure.
b) $(S, \mathcal{T})$ is *σ-compact* if it can be written as a union of compact sets.

Definition A.10.

a) A subset $D \subset S$ is said to be dense (for $\mathcal{T}$) whenever any open set $O \in \mathcal{T}$ intersects D in a non-empty subset.
b) $(S, \mathcal{T})$ is said to be *separable* if it contains a dense subset.

Definition A.11. A topological space $(S, \mathcal{T})$ is *Hausdorff* whenever for any pair of points x_1, x_2 in S, there exist open sets $O_1, O_2 \in \mathcal{T}$ containing x_1, x_2 respectively, such that $O_1 \cap O_2 = \emptyset$.

Comment: This property ensures uniqueness of limits of sequences. (Verify it). Metric spaces have this property.

Appendix B

Basic measure theory and probability

B.1 Some results from measure theory and integration

Consider the simple function $f(t) = \sum_{i=1}^{n} c_i I_{A_i}(t)$, where the sets A_i are disjoint subsets of $\mathbb{R}$, and the c_i are all different. If we want to give meaning to $\int f(t)dt$ as area under the graph, we would have to set

$$\int f(t)dt = \sum_i c_i m(A_i) = \sum_i c_i m(\{f = c_i\}) \qquad (B.1)$$

where $m(A_i) = m(\{f = c_i\})$ is "the length" of the set A_i. At this point we notice the need to solve two issues

i) To which subsets $A \subseteq \mathbb{R}$ can we assign a length?
   Regarding the right-hand side of (B.1) as a definition of the left-hand side:
ii) Which functions $f : \mathbb{R} \to \mathbb{R}$ can be integrated?

The intuitive road is to notice that if (a, b) is an interval, $m(a, b) = b - a$ has some obvious interesting properties. The same happens to sets obtained from (a, b) by set theoretic operations. This leads to

Definition B.12. A σ-algebra of subsets of a set S is a collection $\mathcal{F} \subset \mathcal{P}(S) :=$ the collection of all subsets of S, such that

i) $\phi \in S$.
ii) Arbitrary countable unions of subsets in $\mathcal{F}$ are in $\mathcal{F}$ (i.e. $A_i \in \mathcal{F}$, $i \geq 1$, then $\cup A_i \in \mathcal{F}$).
iii) $\forall A \in \mathcal{F}$, $A^c \in \mathcal{F}$.

Nomenclature: A pair $(S, \mathcal{F})$ consisting of a set S and a σ-algebra of subsets of S is called a *measurable space*.

Definition B.13. Given a pair of measurable spaces $(S_1, \mathcal{F}_1)$ and $(S_2, \mathcal{F}_2)$, a function $f : S_1 \to S_2$, is said to be measurable (written $f \in \mathcal{F}_1/\mathcal{F}_2$) whenever $f^{-1}(B) \in \mathcal{F}_1$ for any $B \in \mathcal{F}_2$.

We leave it to the reader to verify the following list of facts

1) Given S, $\mathcal{P}(S)$ is a σ-algebra.
2) If S is uncountable, $\mathcal{F} := \{A \subseteq S : \text{either } A \text{ or } A^c \text{ is countable}\}$, then $\mathcal{F}$ is a σ-algebra.
3) Intersection of σ-algebras is a σ-algebra
4) Given a set S, $\mathcal{C}$ = collection of subsets of S.

   i) The family of σ-algebras containing $\mathcal{C}$ is not empty. Denote it by Σ.
   ii) $\sigma(\mathcal{C}) = \bigcap_{\mathcal{F} \in \Sigma} \mathcal{F}$ is the smallest σ-algebra containing $\mathcal{C}$.

5) Given a set Ω, a collection $\{(S_\alpha, \mathcal{F}_\alpha) : \alpha \in I\}$ of measurable spaces and a collection of functions $f_\alpha : E \to S_\alpha$. Notice that $f^{-1}(\mathcal{F}_\alpha)$ is a σ-algebra of subsets of Ω.

   i) $\bigcap_\alpha f^{-1}(\mathcal{F}_\alpha)$ is a σ-algebra on Ω.
   ii) Form the collection $\mathcal{C} = \{f^{-1}(A_i) : A_i \in \mathcal{F}_{\alpha_i}, i \in I, |I| < \infty\}$ $\sigma(\mathcal{C})$ is a σ-algebra on $\mathcal{C}$. Usually denoted $\sigma(f_\alpha : \alpha \in I)$ and called the σ-algebra generated by the family of functions. It is the smallest σ-algebra with respect to which all f_α are measurable.

6) A variation in the previous theme is the following: consider $\Omega = \{\omega : I \to \mathbb{R}, \omega \text{ continuous}\}$ where I is an interval in $\mathbb{R}$. Usually $I = [0, 1]$ or $I = [0, \infty)$. Define, for every $t \in I$ in $X_t : \Omega \to \mathbb{R} : X_t(\omega) = \omega(t)$. The construction described in (5-ii) allows us to give meaning to $\sigma(X_t : t \in I)$ the cylinder σ-algebra on Ω is generated by the coordinate maps X_t.

Nomenclature:

i) Given a topological space $(S, \mathcal{T})$, the **Borel σ-algebra** $\mathcal{B}(S)$ on S is the smallest σ-algebra generated by (the open sets) $\mathcal{T}$.
ii) In the context of Definition B.13, if $f : (S, \mathcal{B}(S)) \to (\mathbb{R}, \mathcal{B}(\mathbb{R}))$ is measurable, one usually writes $f \in b\mathcal{B}(S)$.

Also $f \in \mathcal{B}(S)_+$, $f \in \mathcal{B}(S)$ is used to mean that f is measurable and positive or f is measurable and bounded.

Definition B.14. Let $(S, \mathcal{F})$ be a measurable space. A (positive) measure m on $(S, \mathcal{F})$ is a function $m : \mathcal{F} \to [0, \infty)$, such that if $A_i \in \mathcal{F}$ for $i \geq 1$ is a countable disjoint family, then

$$m(\cup A_i) = \sum m(A_i).$$

Exercise B.6. Prove that $m(\phi) = 0$, and that $A_1 \subset A_2$, $m(A_1) \subseteq m(A_2)$.

The collection $(S, \mathcal{F}, m)$ is called a measure space. When $m(S) < \infty$, the measure is said to be finite.

Definition B.15. When m is a finite measure on $(S, \mathcal{F})$ such that $m(S) = 1$ we say that m is a *probability* (measure) on $(S, \mathcal{F})$.

There are a few extensions of Definition B.13 which appear in a natural way in practice.

1) When $m : \mathcal{F} \to \mathbb{R}$ is such that for any countable disjoint collection $\sum |m(A_i)| < \infty$ and the additivity condition in the definition holds we say that m is a finite signed measure.
2) Similarly, if $m : \mathcal{F} \to V$, where V is a Banach space (see below) and such that for any countable, disjoint collection $\sum \|m(A_i)\| < \infty$, and Definition B.13 is satisfied, then m is said to be a V-valued measure.
3) A measure space is called σ-finite when $S = \cup K_n$, $K_n \in \mathcal{F}$, the family is disjoint and $m(K_n) < \infty$ for each n.

Definition B.16. Consider a locally compact Hausdorff space S, and let $\mathcal{B}(S)$ denote the σ-algebra of Borel sets of S.

A measure μ defined on $(S, \mathcal{B}(S))$ is called a *Borel measure*. It is called a *regular Borel measure* whenever

i) For every $B \in \mathcal{B}(S)$, $\mu(B) = \inf\{\mu(O) : B \subset O, O \text{ open}\}$.
ii) For every $B \in \mathcal{B}(S)$, $\mu(B) = \sup\{\mu(K) : K \subset E, K \text{ compact}\}$.

Definition B.17. We say that the measure space $(S, \mathcal{F}, m)$ is complete whenever $A \subset E$ and $\mu(E) = 0$, the $A \in \mathcal{F}$ (and $m(A) = 0$, of course). We say that a property holds a.e. (almost everywhere)-P whenever the set where it does not hold is a set of P-measure zero.

Constructing Borel measures is a rather convoluted process involving items i) and ii) of Definition B.16. The standard and fundamental case consists of the construction of Lebesgue measure on Euclidean spaces (our usual $\mathbb{R}^n$'s).

A hypercube in $\mathbb{R}^n$ is a set

$$B = \{x \in \mathbb{R}^n : a_i < x_i < b_i, i = 1, ..., n\} = \times(a_i, b_i), i = 1, ..., n$$

since time immemorial, the (Euclidean) volume of such set is taken to be

$$vol(B) = \prod_{i=1}^{n}(b_i - a_i).$$

If we denote by $T_a(A) = \{x + a : x \in A\}$ the translate by a of $A \subset \mathbb{R}^n$, an easy computation shows that $vol(T_a(B)) = vol(B)$.

The construction process leads to

Theorem B.4. *There exist a σ-algebra $\mathcal{F}$ on $\mathbb{R}^n$, containing the Borel sets $B(\mathbb{R}^n)$ and a complete measure m on $(\mathbb{R}^n, \mathcal{F})$ such that*

i) $m(B) = vol(B)$ for any hypercube B.
ii) $\mathcal{F} = \{C \subset S : \exists A \in \mathcal{F}_\sigma \text{ and } B = G_\delta \text{ with } A \subset C \subset B \text{ and } m(B - A) = 0\}$.
iii) m is translation invariant: $m(T_a A) = m(A)$ for all $a \in \mathbb{R}^n$, $A \in \mathcal{F}$.
iv) If μ is any positive, translation invariant measure on $(\mathbb{R}^n, B(\mathbb{R}^n))$ such that $\mu(K) < \infty$ for any compact $K \subset \mathbb{R}^n$, then there exists $c > 0$ such that $\mu(A) = cm(A)$ for any $A \in \mathbb{R}^n$.

Comment: A is $\mathcal{F}_\sigma(G_\delta)$ if it is a countable union (intersection) of closed (open) sets.

Nomenclature: The σ-algebra appearing in Theorem B.4 is called the σ-algebra of Lebesgue measurable sets and the measure m appearing there is called the Lebesgue measure on $\mathbb{R}^n$.

Let us briefly describe the process leading to the definition of the integral $\int f dm$ of a measurable function f with respect of a measure μ on a measurable space $(S, \mathcal{F})$.

Step 1: Let $f(t) = \sum_i c_i I_{A_i}(t)$ a positive simple function: a finite linear combination of functions which take constant value c_i on a set A_i. Recall we use $I_A(t)$ for the function (indicator of A) taking values 1 or 0 depending on whether $t \in A$ or $t \in A^c$. Put

$$\int f(t)m(dt) = \int f(t)dm(t) := \sum c_i m(A_i) \qquad (B.2)$$

which is (B.1) again. Notice that we tolerate $\int f(t)dm(t) = +\infty$.

Step 2: For any positive, measurable f, approximate it by the family $\{f^{(n)}(t)\}$ of simple functions

$$f^{(n)}(t) = \sum_{0 \leq k \leq 2^n - 1} \left(\frac{k}{2^n}\right) I_{\{\frac{k}{2^n} \leq f < \frac{k+1}{2^n}\}}$$

for which

$$\int f^{(n)}(t) dm(t) = \sum_{0 \leq k \leq 2^n - 1} \left(\frac{k}{2^n}\right) m(\{\frac{k}{2^n} \leq f < \frac{k+1}{2^n}\})$$

and denote $f^n(t) \uparrow f(t)$ and

$$\int f(t) m(dt) = \lim_n \int f^{(n)}(t) m(dt).$$

To make things independent of the particular type of simple functions, chosen, one can set

$$\int f(t) m(dt) = \sup\{\int g(t) m(dt) : g(t) \text{ simple function}, 0 \leq g(\cdot) \leq f(\cdot)\}$$
(B.3)

and verify that this defines the left-hand side properly.

Step 3: For any function $f(t)$, decompose it in its positive and negative parts according to $f = \frac{1}{2}(f + |f|) - \frac{1}{2}(|f| - f) = f^+ - f^-$. Both f^+ and f^- are positive. Assume that $\int f^-(t) dm(t) < \infty$ or $\int f^+(t) dm(t) < \infty$

$$\int f(t) dm(t) := \int f^+(t) dm(t) - \int f^-(t) dm(t),$$

i.e., the subtraction is well defined (the result could be $+\infty$ or $-\infty$). Finally, we say

Definition B.18. Let $(\Omega, \mathcal{F}, m)$ be a measure space. We say that f is an *integrable* function whenever

$$\int |f(t)| \, dm(t) = \int f^+(t) m(dt) + \int f^-(t) m(dt) < \infty.$$

Definition B.19. If f is a positive function and $E \in \mathcal{F}$, then

$$\int_E f dm = \sup\{\int I_E g dm : g \text{ simple}, 0 \leq g\}.$$

Exercise B.7. (Properties of the integral)

a) If $0 \leq f \leq g$ then $\int f dm \leq \int g dm$.
b) If $A \subseteq B$ and $f \geq 0$ then $\int_A f dm \leq \int_B f dm$.
c) If $\int f dm$ is defined and c is a real number $\int cf dm = c \int f dm$.

d) If $f(x) \equiv 0$, then $\int f dm = 0$ even if $m(S) = +\infty$.
e) If $m(X) = 0$, then $\int f dm = 0$ even if $f \equiv +\infty$.
f) If f is integrable and $E \in \mathcal{F}$, $\int_E f dm = \int I_E f dm$. (Deduce b) from this and a)).
g) If f and g are integrable, $\int (f+g) dm = \int f dm + \int g dm$.

Three important convergence (approximation) results that are constantly referred to are

Theorem B.5. *(Lebesgue's Monotone Convergence Theorem) Let $\{f_n\}$ be a sequence of measurable functions on $(\Omega, \mathcal{F}, m)$, and assume that*

i) $0 \leq f_1 \leq f_2 \leq ...$
ii) $f_n(x) \to f(x)$ as $n \to \infty$ for every x.

Then f is measurable and

$$\int f_n dm \to \int f dm, \text{ as } n \to \infty.$$

Theorem B.6. *(Fatou's Lemma) Let f_n be positive, measurable functions. Then*

$$\int (\liminf f_n) dm \leq \underline{\lim} \int f_n dm.$$

Comment: Sometimes $\underline{\lim}$ and $\overline{\lim}$ are used for $\liminf$ and $\limsup$ respectively.

Theorem B.7. *(Lebesgue's Dominated Convergence Theorem) Let $f_n(x)$ be a sequence of measurable functions such that $|f_n| < g$ for some positive, integrable g, then*

$$\int \lim_{n \to \infty} f_n d\mu = \lim_{n \to \infty} \int f_n d\mu.$$

These results will explain the continuity of the integration process as a real (or complex) valued linear operator on certain spaces.

An important construction is that of product space and product integral. For that let $(S_1, \mathcal{F}_1, m_1)$ and $(S_2, \mathcal{F}_2, m_2)$ be two measure spaces. We shall denote by $\mathcal{F}_1 \otimes \mathcal{F}_2$ the smallest σ-algebra of subsets of $S_1 \times S_2$ containing the "rectangles" $A_1 \times A_2$, where $A_i \in \mathcal{F}_i$. A measure on $\mathcal{F}$ can be constructed from the obvious $m(A_1 \times A_2) = m_1(A_1) m_2(A_2)$. We had an inkling of how this process goes in $\mathbb{R}^n$. Just assume on faith that it work. An important result at this point, which covers the exchange for order of integration in elementary calculus is.

Theorem B.8. *(Fubini-Tonnelli) Let $(S_1, \mathcal{F}_1, m_1)$ and $(S_2, \mathcal{F}_2, m_2)$ be σ-finite measure spaces and let $(S = S_1 \times S_2, \mathcal{F} = \mathcal{F}_1 \times \mathcal{F}_2, m = m_1 \times m_2)$ be their product space as described above. Let $f : S_1 \times S_2 \to \mathbb{R}$ be an m-integrable function, then*

$$\int f(s_1, s_2) m_2(ds_2) \text{ and } \int f(s_1, s_2) m_1(ds_1)$$

are respectively m_1-integrable and m_2-integrable and

$$\int_S f(x_1, x_2) m(dx_1, dx_2) = \int_{S_1} m_1(dx_1) \int_{S_2} f(x_1, x_2) m_2(dx_2)$$
$$= \int_{S_2} m_2(dx_2) \int_{S_1} f(x_1, x_2) m_1(dx_1).$$

Exercise B.8. Verify that once Lebesgue measure is defined, we can define the integral of $f : (S, \mathcal{F}, m) \to \mathbb{R}_+$ by

$$\int_S f(s) m(ds) = \int_0^\infty |\{s : f(s) \geq x\}| dx.$$

Nice, elegant and useful, but the elementary properties of the integral are a pain in the neck to verify.

Let us close this section with a result that we have quoted many times.

Definition B.20. Let $f : (S, \mathcal{F}) \to (X, \mathcal{G})$ be a measurable mapping between the indicated measurable spaces and let m be a measure on $(S, \mathcal{F})$. The measure n on $(X, \mathcal{G})$ defined by

$$n(A) = m(f^{-1}(A)) = (m \circ f^{-1})(A) = f(m)(A) \qquad (B.4)$$

is called the measure on $(X, \mathcal{G})$ **transported or induced** by f.

Theorem B.9. *(Change of Variables) With the notations of Definition B.20, let $g : (X, \mathcal{G}) \to (\mathbb{R}, \mathcal{B}(\mathbb{R}))$ be either bounded or positive and assume m is a positive measure. Then*

$$\int_X g \, dn = \int_S g(f) \, dm. \qquad (B.5)$$

Proof. Use (B.4) to verify (B.5) for simple functions and carry on from there. □

The other concept we needed repeatedly is that of absolute continuity.

Definition B.21. A measure P on a complete measure space $(S, \mathcal{F}, Q)$ is said to be *absolutely continuous* with respect to Q, written $P \ll Q$, whenever $Q(A) = 0$ implies $P(A) = 0$ for any $A \in \mathcal{F}$.

Theorem B.10. *(Radon-Nikodym)* Let $P \ll Q$ as in Definition B.21. Then there exists a measurable $\rho : S \to [0, \infty)$ such that

$$P(A) = \int_A \rho(x) dQ(x). \tag{B.6}$$

Comment: $\rho(x)$ is called the Radon-Nikodym derivative of P with respect to Q and written $\rho(x) = \frac{dP}{dQ}(x)$.

B.2 Some probabilistic jargon

From the mathematical point of view, probability theory is pretty much like measure theory and integration (real analysis) except when dealing with conditional probabilities and independence, two concepts which did not play a role throughout these notes.

Instead of measurable spaces $(\Omega, \mathcal{F})$ we speak of *sample spaces* and the elements of $\mathcal{F}$ are called *events*. They describe the possible questions we ask about the system under study. The elements in Ω are called *elementary events* or *experiments*.

For reasons of convenient convention, a class of measures is singled out, and called *probabilities*. These are measure P on $\mathcal{F}$ (or $(\Omega, \mathcal{F})$) such that $P(\Omega) = 1$.

Also, measurable real-valued functions on $(\Omega, \mathcal{F})$ are called *random variables*, and if V is vector space on which a σ-algebra is defined, any $X : (\Omega, \mathcal{F}) \to V$ is called a V-valued random variable.

Collections $\{X_\alpha : \alpha \in I\}$ where I is an index set, are called *stochastic process*. When the index set has a geometric interpretation, the name *random field* is usually employed.

Sometimes a class of random variables remains fixed for a given problem. When the class is finite $\{X_1, ..., X_n\}$ one usually forgets about the $(\Omega, \mathcal{F}, P)$ space, and moves onto $(\mathbb{R}^n, \mathcal{B}(\mathbb{R}^n))$ on which the measure $F(dx)$ is defined by

$$F_X(A_1, ..., A_n) = P(X_1 \in A_1, X_2 \in A_2, ..., X_n \in A_n)$$

where $X = (X_1, ..., X_n)$ and in terms of which, the expected value of any function $H(X_1, ..., X_n)$ is given by

$$E[H(X)] = \int_{\mathbb{R}^n} H(x_1, ..., x_n) dF_X(dx). \tag{B.7}$$

We leave it to the reader to verify that this is an application of (B.4) and (B.5).

From (B.7) it is clear that knowing $E[H(X)]$ for all $H \in \mathcal{B}(\mathbb{R}^n)$ would amount to knowing $F_X(dx)$. But sometimes we may reduce the class of functions considerably.

Definition B.22.

a) Let $X : \Omega \to \mathbb{R}^n$ be vector-valued random variable. The *characteristic function* of X is defined by

$$\varphi(k) = E[\exp(i \langle k, X \rangle)] = \int \exp(i \langle k, X \rangle) F_X(dx) \qquad \text{(B.8)}$$

for $k \in \mathbb{R}^n$. It obviously is the Fourier transform of $F_X(dx)$.

b) When $X : \Omega \to \mathbb{R}^n$, we would consider the Laplace transform/exponential generating function of $F_X(dx)$ defined by

$$\zeta(\lambda) = \mathcal{L}(F)(\lambda) = E[\exp(-\langle \lambda, X \rangle)] = \int_{\mathbb{R}^n} \exp(-\langle \lambda, X \rangle) F_X(dx). \qquad \text{(B.9)}$$

Comments: Whereas $\varphi(k)$ is defined on all $\mathbb{R}^n$, $\zeta(\lambda)$ is usually defined on a smaller region of $\mathbb{R}^n$, certainly including $\mathbb{R}^n$. When X ranges over a bounded set or even over all $\mathbb{R}^n$, we may still consider (B.9) but restricted to

$$\mathcal{D}(F) = \{\lambda \in \mathbb{R}^n : \zeta(\lambda) < \infty\}. \qquad \text{(B.10)}$$

These functions characterize not only $\mathbb{R}^n$-valued variables, but more complicated objects that we introduce now.

Consider a separable Banach space V with its collection of Borel sets. (We shall consider two possible natural topologies on Banach space, the strong and the weak topologies, and each will provide us with a collection of Borel sets.)

A random V-valued variable is a measurable mapping $X : \Omega \to V$. When considering the Borel sets specified by the strong topology, it suffices to require that

$$\{\omega \in \Omega : \|X - a\| < r\} = X^{-1}(B(a, r))$$

is in $\mathcal{F}$ for any $a \in V$ and $r > 0$. When dealing with the weak topology, it suffices to require that

$$\{\omega \in \Omega : \ell(X) \in A\} \in \mathcal{F}$$

for any $\ell \in V^*$ and A being an interval in $\mathbb{R}$.

This time the characteristic functions and Laplace transforms of (the distribution of) X are defined as above.

Definition B.23. Let X be a V-valued random variable. The characteristic function of X is defined on V^* by

$$\varphi(\lambda) = E[\exp(i \langle \lambda, X \rangle)]$$

whereas the *Laplace transform* of X is defined by

$$\zeta(\lambda) = E[\exp(-\langle \lambda, X \rangle)]$$

and defined in $\mathcal{D}(X) = \{\lambda \in V^* : \zeta(\lambda) < \infty\}$.

Another important class of random variables, which we have used quite a bit are various random measures. A large class of collections of measures are Banach spaces on their own, thus random measures would be Banach space valued random variables, but it is nicer for applications to think of them just as random measures or random fields.

We have both a probability space $(\Omega, \mathcal{F}, P)$ and a measurable space $(S, \mathcal{B})$.

Definition B.24. A random measure on $(S, \mathcal{B})$ consists of a kernel $X : \Omega \times S \to \mathbb{R}$, such that

i) $\forall \omega \in \Omega$, $X(\omega, \cdot) : S \to \mathbb{R}$ is a measure, i.e. for any disjoint collection $A_i \in \mathcal{B}$, $X(\omega, \cup A_i) = \sum_i X(\omega, A_i)$.

ii) $\forall A \in \mathcal{B}$, $X(\cdot, A) : \Omega \to \mathbb{R}$ is a random variable.

Comment: In practice the identity in i) holds a.e. P and the set of identity depends on the collection A_i. In most cases one can produce a unique set of measure zero of which i) holds for any collection A_i.

If $(S, \mathcal{B})$ is such that any measure m on $\mathcal{B}$ *is* determined by the integrals $\int f(t) m(dt)$ when f ranges over an appropriate large class of functions, say the continuous functions, then knowing the distribution of the random variables $X(f) : \Omega \to \mathbb{R}$ defined by

$$(f, X) \equiv X(f) := \int f(t) X(dt)$$

yields the distribution of X. And since random variables are characterized by either their characteristic functions or Laplace transforms we have

Definition B.25. Let $(\Omega, \mathcal{F}, P)$ be a probability space and $(S, \mathcal{B}, m)$ be a measure space. We say that a random measure (as in Definition B.23) is a *Gaussian random measure* with (co)variance $m(dt)$ whenever

i) For disjoint A_i in $\mathcal{B}$, the $X(A_i)$ are independent.

ii) For any $f : S \to \mathbb{R}$ such that $\int f^2(t) m(dt) < \infty$ and $\alpha \in \mathbb{R}$

$$E[\exp(i\alpha(f, X))] = \exp\left(-\frac{\alpha^2}{2} \int f^2(t) m(dt)\right). \qquad (B.11)$$

Definition B.26. With the same notations as above, assume that m is finite. We say that X is a *Poisson random measure* whenever

i) For disjoint A_i in $\mathcal{B}$, the $X(A_i)$ are independent.
ii) For positive $f \in \mathcal{B}$ and $\alpha > 0$,

$$E[\exp(-\alpha(f, X))] = \exp\left(-\int (1 - e^{-\alpha f(t)}) m(dt)\right). \qquad (B.12)$$

B.3 Brief description of the Kolmogorov extension theorem

Let $\mathcal{I}$ be an infinite index set and let $\Phi = \{J \subset \mathcal{I} : |J| < \infty\}$ be the collection of finite subsets of $\mathcal{I}$. Let us be given a measurable space $(S_\alpha, \mathcal{F}_\alpha)$ for each $\alpha \in \mathcal{I}$. The $(S_\alpha, \mathcal{F}_\alpha)$ may be different or not. Consider $S = \times_{\alpha \in \mathcal{I}} S_\alpha := \{\omega : \mathcal{I} \to \bigcup_\alpha S_\alpha : \omega(\alpha) \in S_\alpha \forall \alpha\}$ and define the obvious coordinate maps

$$X_\alpha : S \to S_\alpha, \quad X_\alpha(\omega) = \omega(\alpha)$$

and for each $J \in \Phi$, $J = (\alpha_1, ..., \alpha_n)$

$$X_J : S \to \times_{\alpha \in J} S_\alpha = S_{\alpha_1} \times ... \times S_{\alpha_n}$$

$$\omega \to (X_{\alpha_1}(\omega), X_{\alpha_2}(\omega), ..., X_{\alpha_n}(\omega)).$$

Similarly, for $J \subset K$, both in Φ

$$X_J^k : \times_{\alpha \in K} S_\alpha \to \times_{\alpha \in J} S_\alpha \equiv \mathcal{F}_J$$

is defined by discarding the necessary coordinates.

Let $\mathcal{F}$ be the smallest σ-algebra on S with respect to which all coordinate maps X_α are measurable. We will write the usual coordinate maps X_α are measurable. We will write the usual

$$\mathcal{F} = \sigma(X_\alpha : \alpha \in \mathcal{I}) = \bigotimes_{\alpha \in \mathcal{I}} \mathcal{F}_\alpha.$$

Note that for $\{\alpha\} \subset J_1 \subset J_2 \subset J_3$ in Φ we have the consistency conditions

$$X_\alpha = X_\alpha \circ X_{J_1}^{J_2} = X_\alpha \circ X_{J_1}$$

$$X_{J_1} = X_{J_1} \circ X_{J_2}^{J_3}$$

$$X_{J_1}^{J_3} = X_{J_1}^{J_2} \circ X_{J_2}^{J_3}.$$

Assume now that for every $J \in \Phi$, we are given a probability measure P_J on $\mathcal{F}_J = \times_{\alpha \in J} \mathcal{F}_\alpha$, and moreover, assume that these P_J satisfy the following consistency requirement:

If $J_1 \subset J_2$, then

$$P_{J_1} = X_{J_1}^{J_2}(P_{J_2}) \tag{B.13}$$

Theorem B.11. *(Kolmogorov Extension Theorem) With the notations introduced above, assume that the consistency condition (B.13) holds, then there exists a unique probability measure on $(S, \mathcal{F})$ such that for any $J \in \Phi$,*

$$P_J = X_J(P). \tag{B.14}$$

Comments: When $J = \{\alpha\}$ we obtain $P_\alpha(A) = P(X_\alpha \in A)$, for $A \in \mathcal{F}_\alpha$, the distribution of X_α, and so on.

Had we started with probability spaces $(S_\alpha, \mathcal{F}_\alpha, Q_\alpha)$, we could form $P_J = \bigotimes_{\alpha \in J} Q_\alpha$. We end up with a setup which we used quite a bit in Chapter 8. In this case all X_α are independent with respect to P, with marginals P_J.

B.4 Basic facts about Gaussian process in Hilbert spaces

We shall mention in Theorem C.12 of Appendix C that when the measure space $(S, \mathcal{B}, m)$ has a countable basis, then $L_2(S, m)$ is a separable Hilbert space. In this case it is easy to put together an $(\Omega, \mathcal{F}, P)$ and define a map

$$\Phi : L_2(S, m) \to L_2(\Omega, P)$$
$$f \to \Phi(f) = \langle f, X \rangle = X(f)$$

in such a way that $\Phi(f)$ is a Gaussian random variable with

$$E[\exp(i\alpha X(f))] = \exp(-\frac{\alpha^2}{2} \|f\|_2^2). \tag{B.15}$$

Consider $\Omega = \mathbb{R}^\mathbb{N}$ with $\mathcal{F} = \mathcal{B}(\Omega)$ and P the product of infinitely many $dP_i(\xi_i) = \frac{\exp(-\frac{\xi_i^2}{2})}{\sqrt{2\pi}} d\xi_i$. This time, as observed at the end of the previous section, the coordinate maps $X_i : \Omega \to \mathbb{R}$ $X_i(\xi) = \xi_i$ if $\xi = (\xi_1, \xi_2, ...)$ are P-independent.

Consider now a complete orthonormal system ψ_i in $L_2(S, m)$. Then $f = \sum f_i \psi_i$ with $\|f\|^2 = \sum_i |f_i|^2$.

Define now
$$\Phi(f) \equiv X(f) := \sum f_i X_i = \langle f, X \rangle \tag{B.16}$$

and notice that the sum converges in $L_2(\Omega, P)$ since $\|f\|^2 < \infty$ and the X_i are orthonormal in $L_2(\Omega, P)$. Also, independence and completing squares leads to (B.15).

If $A \in \mathcal{B}$ is such that $m(A) = \int I_A^2 dm < \infty$, notice that (B.15)-(B.16) asserts that $X(A)$ is a Gaussian random variable. If $f(t) = \sum a_i I_A(t)$ is a simple function in $L_2(S, m)$ $\|f\|^2 = \sum a_i^2 m(A_i)$ and it makes sense to write

$$X(f) = \int f(t) X(dt) = \sum a_i X(A_i) = \langle f, X \rangle$$

and we are being more or less consistent.

We shall close this section by showing how to put a measure P_g on $(\Omega, \mathcal{F})$ $P_g \ll P$ such that for any $h \in L_2(S, m)$

$$E_{P_g}\{\langle h, X \rangle\} = \langle h, g \rangle. \tag{B.17}$$

This result is just a variant of the famous Cameron Martin change of variables formula.

Lemma B.1. *Let $g \in L_2(S, m)$ and let $(\Omega, \mathcal{F}, P)$ and X be as above. Define $\hat{\gamma} \in \Omega$ the sequence appearing in $g = \sum g_i \psi_i$, and define $T_g : \Omega \to \Omega$ by $T_g(\xi) = \xi + \gamma$. Then $P_g = P \circ T_g^{-1} \ll P$ and*

$$\frac{dP_g}{dQ} = \exp\left(\langle g, X \rangle - \frac{1}{2}\|g\|^2\right). \tag{B.18}$$

Proof. Just check what happens for each component measure $dP_i(\xi_i) = \frac{\exp\left(-\frac{\xi_i^2}{2}\right)}{\sqrt{2\pi}} d\xi_i$ on $\mathbb{R}$ when we map $T_{\gamma_i} : \mathbb{R} \to \mathbb{R}$ by $\xi_i \to \xi_i + \gamma_i$. A completion of squares rapidly leads to

$$\frac{dP_\gamma}{dP_i} = \exp\left(\gamma_i \xi_i - \frac{1}{2}\gamma_i^2\right) \quad \text{or} \quad dP_\gamma(\xi_i) = \frac{\exp\left(-\frac{(\xi_i - \gamma_i)^2}{2}\right)}{\sqrt{2\pi}} d\xi_i.$$

Therefore, taking products

$$\frac{dP_g}{dP} = \exp\left(\sum \xi_i \gamma_i - \frac{1}{2}\gamma_i^2\right) = \exp\left(\langle g, X \rangle (\xi) - \frac{1}{2}\|\gamma\|^2\right).$$

□

Corollary B.3. *Identity (B.17) follows from (B.18).*

Proof. Use the fact that $\int_{-\infty}^{\infty} \xi_i \frac{\exp(-\frac{(\xi_i - \gamma_i)^2}{2})}{\sqrt{2\pi}} d\xi = \gamma_i.$ □

Appendix C

Banach spaces

C.1 Basic stuff

Definition C.27. A *norm* on a vector space V is a positive valued function $\|\cdot\| : V \to \mathbb{R}_+$ satisfying

i) $\|x\| = 0$ if and only if $x = 0$.
ii) $\|x + y\| \leq \|x\| + \|y\|$ for all $x, y \in V$ (Triangle inequality).
iii) $\|\alpha x\| = \alpha \|x\|$ for all $x \in V$, $\alpha \in \mathbb{R}$ (or $\mathbb{C}$ if V is a vector space over the complex numbers).

Comment: When condition i) fails, $\|\cdot\|$ is called a *seminorm*.

Example C.2. Let $V = \{x : \mathbb{N} \to \mathbb{R}\}$ be the space of infinite sequences of real numbers. Subspaces of V acquire different names according to the following norms

$$\ell_\infty = \{x \in V : \|x\|_\infty := \sup_n |x_n| < \infty\}$$

(see the comment below) and

$$\ell_p = \{x \in V : \|x\|_p := \left(\sum |x_n|^p\right)^{\frac{1}{p}} < \infty\} \; for \; p \geq 1.$$

Comment: The triangle inequality in the ℓ_p case has a name. It is called Minkowski inequality. It asserts that

$$\left(\sum |x_i + y_i|^p\right)^{\frac{1}{p}} \leq \left(\sum |x_i|^p\right)^{\frac{1}{p}} + \left(\sum |y_i|^p\right)^{\frac{1}{p}}. \tag{C.1}$$

Consider V to be the class of measurable functions on a measure space $(S, \mathcal{B}, m)$. This time set

$$L_\infty(S, m) = \{f \in V : \|f\|_\infty := \sup_n |f(x)| < \infty\}$$

$$L_p(S, m) = \{f \in V : \|f\|_p := \left(\int |f(t)|^p \, dm(t)\right)^{\frac{1}{p}} < \infty\} \text{ for } p \geq 1.$$

Comment: For not to be bothered with inessential values of f, $\|\cdot\|_\infty$ is usually defined by

$$\|f\|_\infty := \operatorname*{ess\,sup}_n f = \inf\{n : m\{|f| > n\} = 0\}$$

which allows f to go wild on sets of measure zero.

Comment: When considering the L_p spaces, we also want $\|f\|_p = 0$ to imply $f = 0$ in order for $\|\cdot\|_p$ to be a norm. this is achieved by identifying two functions f_1 and f_2 such that $\|f_1 - f_2\| = 0$.

Definition C.28. The set $B(x,r) = \{y \in V : \|x - y\| < r\}$ is called an open ball (in V) around x of radius r. The smaller topology on V generated by the open balls is called the *strong* topology on V. Here too we say that V is separable whenever it contains a countable dense subset.

Definition C.29. A measure μ on $(S, \mathcal{B})$ has a countable basis if and only if, for each $\varepsilon > 0$, any collection $\mathcal{C} \subset \mathcal{B}$ of subsets of finite measure such that

$$A, B \in \mathcal{C}, \quad A \neq B \Rightarrow m((A - B) \cup (B - A)) \geq \varepsilon$$

is countable.

The next result relates the separability of $L_p(S, m)$ to the existence of countable bases for $(S, \mathcal{B}, m)$.

Theorem C.12. *Let $1 \leq p < \infty$. The space $L_p(S, \mathcal{B}, m)$ is separable if and only if $(S, \mathcal{B}, m)$ has countable basis.*

Exercise C.9. Look up a proof of, or even better prove, that Lebesgue measure on $(\mathbb{R}^n, \mathcal{B}(\mathbb{R}^n))$ has countable basis.

Comment: In $L_2(S, m)$ separability amounts to the existence of a basis, i.e., of a countable collection whose finite linear combinations are dense in $L_2(S, m)$.

The verification of the triangle inequality hinges on the full version of Minkowski's inequality.

Theorem C.13. *For $f, g \in L_p(S, m)$*

$$\|f + g\|_p \leq \|f\|_p + \|g\|_p .$$

Two important inequalities, also cropping up everywhere are contained in

Theorem C.14. *(Hölder Inequality)* Let $1 \leq p, q < \infty$ be related by $\frac{1}{p} + \frac{1}{q} = 1$. If $f \in L_p(S,m)$ and $g \in L_q(S,m)$, then $f, g \in L_1(S,m)$ and

$$\int |f(t)g(t)|\, m(dt) \leq \left(\int |f(t)|^p\, dm(t)\right)^{\frac{1}{p}} \left(\int |g(t)|^q\, dm(t)\right)^{\frac{1}{q}}$$

which for $p = q = 2$ is called the *Cauchy-Schwartz-Bunyakovsky inequality*.

To go from normed spaces to Banach spaces, we need to be sure that limit points are present.

Definition C.30.

a) A sequence $\{x_n : n \geq 1\}$ in a normed space is called a *Cauchy sequence* $\lim_{n \to \infty} \|x_n - x_{n+k}\| \to 0$ for any $k \geq 1$.
b) A normed vector V space is called a *Banach* space whenever any Cauchy sequence in V has a limit in V.

Theorem C.15. *(Riesz-Fischer)* The normed spaces $L_p(S, \mathcal{B}, m)$ with $1 \leq p < \infty$ are complete, i.e., they are Banach spaces.

C.2 Continuous linear operator on Banach spaces

Throughout V and W will denote Banach spaces with norms $\|\cdot\|_V$ and $\|\cdot\|_W$. When there is no confusion, we shall drop the subscripts.

Definition C.31. A linear mapping $A : V \to W$ is said to be *bounded* if there exists a positive K such that $\|Ax\| \leq K \|x\|$.

Exercise C.10. Prove that $A : V \to W$ is continuous (in the strong topologies) on V and W if and only if it is bounded.

Exercise C.11. Define $\|A\| = \sup_{x \neq 0} \frac{\|Ax\|}{\|x\|}$ and verify that the collection $\mathcal{L}(V,W)$ of all continuous, bounded linear operators between V and W is a Banach space with norm $\|A\|$.

Two standard notions of convergence of operators are contained in

Definition C.32. Let $\{A_n : n \geq 1$ and A be in $\mathcal{L}(V,W)$. We say that

a) $A_n \to A$ *uniformly* whenever $\|A_n - A\| \to 0$ as $n \to \infty$.
b) $A_n \to A$ *strongly* whenever $\|A_n x - Ax\|_W \to 0$ as $n \to \infty$ for all $x \in V$.

Two central results about linear operators are

Theorem C.16. *(Uniform boundedness theorem/Banach-Steinhaus theorem)* Let $\{A_n : n \geq 1\}$ be a family of operators in $\mathcal{L}(V, W)$. If $\sup_n\{\|A_n x\|_W : n \geq 1\} \leq M(x) < \infty$, then there is $M < \infty$ such that $\sup_n\{\|A_n\|_W : n \geq 1\} \leq M < \infty$.

Comment: In this statement W only needs to be a normed space.

Theorem C.17. *(Open mapping the theorem)* Let $A \in \mathcal{L}(V, W)$ and assume that A is onto, then A maps every open set in V onto an open set of W.

Operators (and functions in general) need not always be defined on the whole V. One needs the standard

$$\mathcal{D}(A) = \{x \in V : Ax \in W\} := \text{domain of } A$$
$$\mathcal{R}(A) = \{y \in W : y = Ax, x \in \mathcal{D}(A)\} := \text{range of } A$$
$$\ker(A) = \{x \in V : Ax = 0\} := \text{kernel of } A \equiv \text{null space of } A$$
$$G(A) = \{(x, Ax) : x \in \mathcal{D}(A)\} := \text{graph of } A.$$

Definition C.33. A linear operator $A : \mathcal{D}(A) \to W$ is said to be *closed* whenever $G(A)$ is a *closed linear subspace* of $V \times W$.

Exercise C.12. A is closed whenever for $x_n \to x$ in V, $x_n \in \mathcal{D}(A)$ and $Ax_n \to y \in W$ it follows that $x \in \mathcal{D}(A)$ and $y = Ax$.

Theorem C.18. Let $A : V \to W$ be linear and closed ($\mathcal{D}(A) = V$ for emphasis), then A is bounded.

Comment: This implies that when $\mathcal{D}(A) \subset V$ and A is closed but unbounded, $\mathcal{D}(A)$ cannot be a closed linear subspace with respect to $\|\cdot\|_V$.

Corollary C.4. Let $A : V \to W$ be closed and invertible. Then A^{-1} is closed.

Definition C.34. An operator A in $\mathcal{L}(V, W)$ is said to be compact if it maps bounded sets in V onto compact sets in W.

Equivalently, A is compact if for any bounded sequence x_n in V, Tx_n has a convergent subsequence in W. Also, A is compact if it maps *weakly convergent sequences* in V (see below) onto strongly convergent sequences in W.

Example C.3. Suppose $\mathcal{R}(A)$ is finite dimensional. Then A is compact. This example covers many of the applications we have dealt with above. Consider $V = L_p(S,m)$ and $k_i(\cdot) = k(i,\cdot) \in L_q(S,m)$ for $i = 1,...,n$ with $p^{-1} + q^{-1} = 1$ and define $A: V \to \mathbb{R}^n$ by

$$(Ax)_i = \int k(i,t)x(t)dm(t).$$

This A is compact according to what we said above.

Theorem C.19.

a) Let $A_n \in \mathcal{L}(V,W)$ for $n \geq 1$. If $A_n \to A$ in $\mathcal{L}(V,W)$ and the A_n are compact, so is T.
b) Let $A \in \mathcal{L}(V,W)$ and $B \in \mathcal{L}(W,U)$. If either of B, A is compact, so is BA.
c) If $A \in \mathcal{L}(V,W)$ is compact, so is $A^* \in \mathcal{L}(W^*,V^*)$ (see next section).

C.3 Duality in Banach spaces

When W is equal to $\mathbb{R}$ (or $\mathbb{C}$ when V is a vector space over the complex field) any $T \in \mathcal{L}(V,\mathbb{R})$ is called a linear functional, and $\mathcal{L}(V,\mathbb{R})$ (or $\mathcal{L}(V,P)$) is denoted by V^*.

The canonical example of linear, unbounded functional is: Take $V = \ell_\infty$. For $x = (x_1,...,x_n,...)$ set $T(x) = \sum x_i$. Note that for $u_n = (1,1,...,1,0,...)$ with all components equal to zero after the n-th, we have $Tu_n = n$, while $\|u_n\|_\infty = 1$, that is T is unbounded and defined in $\mathcal{D}(T) = \ell_1$. If instead of ℓ_∞ we consider ℓ_1, then $|T(x)| \leq \sum |x_i| \leq \|x\|_1$.

Defining $\|T\|$ as above, V^* becomes a Banach space of its own, which is called the dual of V.

Example C.4. Use Minkowski inequality. Denote by ℓ_p^n, for $1 \leq p \leq \infty$, the Banach space obtained by imposing the p-norm $\|x\|_p = \left(\sum_1^n |x_i|^p\right)^{\frac{1}{p}}$. Let $q^{-1} + p^{-1} = 1$. Then $\ell_q^n = (\ell_p^n)^*$ for any finite n.

Verify as well that $\ell_1^* = \ell_\infty$ (i.e., any linear functional on ℓ_1 can be identified with a bounded sequence) but ℓ_∞^* is not ℓ_1. Actually $c_0^* \sim \ell_1$ where $c_0 \subset \ell_\infty$ is the space of bounded sequences $\{x_n : n \geq 1\}$ such that $x(n) \to 0$ as $n \to \infty$.

Theorem C.20. *(Riesz-Representation theorem).* Let $(S, \mathcal{B}, m)$ be a σ-finite measure space and let $V = L_p(S, m)$ for $1 \leq p < \infty$. Let $q^{-1} + p^{-1} = 1$. Then

i) for each $\lambda \in L_q(S, m)$ $T_\lambda(x) = \int \lambda(t) x(t) m(dt)$ defines an element of L_p^*.

ii) given a $T \in L_p^*$, there exists $\lambda \in L_q$ such that $T(x) = \int \lambda(t) x(t) m(dt)$ with $\|T\|$ either $\|\lambda\|_q$ if $p < \infty$ or $\|\lambda\|_\infty$ ($= \text{ess sup }\lambda$) if $p = \infty$.

Exercise C.13. Note that $(\mathbb{N}, \mathcal{P}(\mathbb{N}), \mu)$ where $\mu(\{n\}) = 1$ is the counting measure, is a σ-finite measure space. Make sure you deduce the results of the previous example from Theorem C.20.

Example C.5. Let S be a compact subset of some $\mathbb{R}^n$, and $V = \mathcal{C}(S)$ be the class of real valued continuous functions on S with the $\|x\|_\infty$ norm. V is a Banach space. $V^* = \mathcal{C}(S)^*$ is the class of (signed) regular Borel measures on S, that is, each $T \in V^*$ can be described by

$$T(x) = \int x(t) \mu(dt) := \langle \mu, x \rangle$$

where μ has finite total variation described by

$$\|\mu\| = \sup_{\|x\|_\infty \leq 1} \left| \int x(t) \mu(dt) \right|.$$

Comment: A similar result holds for complex valued functions and the class of complex valued regular Borel measures.

This example is a particular case of the following situation. Let $(S, \mathcal{T})$ be a locally compact Hausdorff space, and denote by $\mathcal{C}_0(S)$ the class and functions vanishing at infinity (i.e. $\forall \varepsilon > 0$, there exists compact $K \subset S$ such that $|f(x)| < \varepsilon$ off K). Denote by $\mathcal{C}$ the σ-algebra of subsets of S which makes all $\mathcal{C}_0(X)$ measurable. Measures m defined on $\mathcal{C}$ are called *Baire* measures. Note as well that $\mathcal{C}_0(S)$ is a Banach space under the norm $\|x\|_\infty = \sup_{t \in S} |x(t)|$. We have,

Theorem C.21. *(Riesz)* With the notations introduced above, let $T \in \mathcal{C}_0(S)^*$ be a positive functional ($T(x) \geq 0$ whenever $x \geq 0$), then there exists a Baire measure μ on $(S, \mathcal{C})$ such that

$$T(x) := \langle \mu, x \rangle = \int x(t) \mu(dt).$$

Corollary C.5. *Now let S be compact and the rest as above. Let $T \in C(X)^*$. Then there exists a finite, signed Baire measure μ such that*

$$T(x) := \langle \mu, x \rangle = \int x(t)\mu(dt)$$

and $\|T\| = |\mu|(X)$ the total variation of μ on X.

Definition C.35.

a) Let V be a topological space, and V^* its dual. The weak topology on V is the smallest topology making all elements of V^* continuous

b) x_n converges weakly, written $x_n \left(\overset{w}{\to}\right) x$, to x in V if and only if $\lambda(x_n) = \langle \lambda, x_n \rangle \to \langle \lambda, x \rangle = \lambda(x)$ for any $\lambda \in V^*$.

Exercise C.14. Verify that in the finite dimensional examples given above, the weak and strong topologies coincide.

Example C.6. Let $V = \ell_p$ consider the sequence $x^{(n)} = (x_1^{(n)}, ..., x_j^{(n)}, ...)$ in which $x_n^{(n)} = 1$ and $x_j^{(n)} = 0$ for all other j. Then $x^{(n)} \in \ell_p$. Note that for any $\lambda \in V^* = \ell_q$, $\|\lambda\|_q^q = \sum |\lambda_j|^q < \infty$ implies that $\lambda_j \to 0$ as $j \to \infty$. Thus $\lambda(x^{(n)}) = \lambda_n \to 0$ as $n \to \infty$, or $x_n \left(\overset{w}{\to}\right) 0$. But $\|x^{(n)} - 0\| = 1$, that is $x^{(n)}$ does no converge strongly to zero.

Comment: Nevertheless, $x^{(n)} \left(\overset{w}{\to}\right) x$ in ℓ_p if and only if $\sup_n \|x^{(n)}\| \leq M < \infty$ and $x_k^{(n)} \to x$ (components converge pointwise) as $k \to \infty$.

In many variational problems in Banach spaces, by appropriate dualization one passes to a variational problem in the dual of the given space. In this case it is important to know when some bounded sets are compact (in some appropriate topology). To take care of this note that given an $x \in V$, it induces a linear functional Tx on V in the obvious way

$$T_x(\lambda) = \langle \lambda, x \rangle = \lambda(x).$$

This provides way of defining a weaker topology on V^*, naturally called the weak-*-topology. Again this is the smallest topology on V^* with respect to which all T_x defined above are continuous. We have

Exercise C.15. A sequence $\{\lambda_n : n \geq 1\}$ in V^* converges in the weak-*-topology to $\lambda \in V^*$ whenever for any $\lambda \in V$, $\langle \lambda_n, x \rangle \to \langle \lambda, x \rangle$.

Exercise C.16. Verify that $\mathcal{I} : V \to V^{**}$ is an isometric embedding of V into V^{**}. (Note that $\|T_x\| = \sup_{\|\lambda\| \leq 1} |\langle \lambda, x \rangle| = \|x\|$.)

We come to a (weak) version of

Theorem C.22. *(Banach-Alaoglu) Let $B(r) = \{x : \|x\| < r\}$ be an open ball around 0 in V. Then*

$$K = \{\lambda \in V^* : |\langle \lambda, x \rangle| \leq 1 \text{ for every } x \in B(r)\}$$

i.e., the closed unit ball in V^*, then K is weak--compact.*

Definition C.36. We say that a space is *reflexive* when the embedding in the previous exercise is an isomorphism.

Exercise C.17. Assume V is reflexive. Is the unit ball in the weak topology weakly compact?

We finally come to the applications of duality theory to the characterization of ranges and kernels of bounded operators. The following results have as much geometric flavor as you can get.

Consider a Banach space V and its dual V^* and subspaces V_1 and V_1^* of each, not necessary closed.

Definition C.37. The annihilators of V_1 and V_1^* are defined by

$$V_1^\perp = \{\lambda \in V^* : \langle \lambda, x \rangle = 0, \forall x \in V_1\}$$
$$(V_1^*)^\perp = \{\lambda \in V : \langle \lambda, x \rangle = 0, \forall x \in V_1^*\}.$$

The following exercise is a simple theorem

Exercise C.18.

a) $(V_1^\perp)^\perp = $ is the norm closure of V_1 in V.
b) $((V_1^*)^\perp)^\perp = $ is the norm closure of V_1^* in V^*.

Theorem C.23. *Let $A \in \mathcal{L}(V, W)$. Then there is a unique $A^* \in \mathcal{L}(W^*, V^*)$ called the adjoint of A satisfying*

$$\langle \lambda, Ax \rangle = \langle A^*\lambda, x \rangle$$

for all $\lambda \in V^*$, $x \in V$. Moreover $\|A\|$ (in $\mathcal{L}(X, Y)$) $= \|A^*\|$ (in $\mathcal{L}(W^*, V^*)$).

The following theorem is so simple, that we spell out the proof

Theorem C.24. *Let $A \in \mathcal{L}(V, W)$ and $A^* \in \mathcal{L}(W^*, V^*)$ its adjoint. Then $\ker(A^*) = \mathcal{R}(A)^\perp$ and $\ker(A) = \mathcal{R}(A^*)^\perp$.*

Proof. In each of the following columns, each statement is equivalent to the preceding one

$$\lambda \in Ker(A^*) \qquad x \in Ker(A)$$
$$A^*\lambda = 0 \qquad Ax = 0$$
$$\langle A^*\lambda, x \rangle = 0, \forall x \in V \qquad \langle \lambda, Ax \rangle = 0, \forall \lambda \in V^*$$
$$\langle \lambda, Ax \rangle = 0, \forall x \in V \qquad \langle A^*\lambda, x \rangle = 0, \forall \lambda \in V^*$$
$$\lambda \in \mathcal{R}(A)^\perp \qquad x \in \mathcal{R}(A^*)^\perp$$

We leave for the reader to work out the details. □

Corollary C.6.

a) $\ker(A^*)$ *is weak-*-closed in* V^*.
b) $\mathcal{R}(A)$ *is dense in* W *if and only if* A^* *is one-to-one.*
c) A *is one-to-one if and only if* $\mathcal{R}(A^*)$ *is weak-*-dense in* V^*.

We are almost through building up notation to examine the solvability of

$$(T - \lambda I)x = y$$

where $T \in \mathcal{L}(V) := \mathcal{L}(V, V)$ is a compact operator. For applications to inverse problems T will be either A^*A or AA^* acting on Hilbert spaces.

Definition C.38. The spectrum $\sigma(T)$ of $T \in \mathcal{L}(V)$ is by definition

$$\sigma(T) = \{\lambda \in \mathbb{C} : T - \lambda I \text{ is \underline{not} invertible}\}$$

Comment: $\lambda \in \sigma(T)$ whenever one of the following holds

a) $\mathcal{R}(T - \lambda I)$ is not all of X.
b) $T - \lambda I$ is not one-to-one.

Further properties of compact operators are contained in

Theorem C.25.

a) $A \in \mathcal{L}(V, W)$ *is compact and* $\mathcal{R}(A)$ *is closed in* W, *then* $\dim(\mathcal{R}(A)) < \infty$.
b) *If* $T \in \mathcal{L}(V)$ *and* T *is compact and* $\lambda \neq 0$, *then* $\dim(\ker(T - \lambda I)) < \infty$.
c) *If* $\dim(V) = \infty$, $T \in \mathcal{L}(V)$ *is compact, then* $0 \in \sigma(T)$.
d) *If* $T \in \mathcal{L}(V)$ *and* T *is compact and* $\lambda \neq 0$, *then* $\mathcal{R}(T - \lambda I)$ *is closed in* T.

Definition C.39. The scalars $\lambda \in \sigma(T)$ are called *eigenvalues* of T.

Theorem C.26. *Let $T \in \mathcal{L}(V)$ be compact. Let $E(r) = \{\lambda \in \sigma(T) : |\lambda| > r\}$. Then:*

a) *$E(r)$ is finite and*
b) *If $\lambda \in E(r)$, $\mathcal{R}(T - \lambda I) \neq V$.*

And to finish

Theorem C.27. *Let $T \in \mathcal{L}(V)$ be compact. If $\lambda \neq 0$ the following four numbers are equal and finite*

1) $\alpha = \dim(\ker(T - \lambda I))$.
2) $\beta = \dim(X/\mathcal{R}(T - \lambda I))$.
3) $\alpha^* = \dim(\ker(T^* - \lambda I))$.
4) $\beta^* = \dim(X^*/\mathcal{R}(T^* - \lambda I))$.

Also

a) *If $\lambda \neq 0$ and $\lambda \in \sigma(T)$, then λ is an eigenvalue of T and T^*.*
b) *(Riesz-Schauder Theorem) The spectrum $\sigma(T)$ is at most countable and has at most one limit point which is 0.*

Let us close this subsection with some basics about projections. Throughout V will be a Banach space

Definition C.40. *A linear mapping $P : V \to V$ is a projection whenever $P^2 = P$.*

Lemma C.2. *Let $P : V \to V$ be a projection. Then*

a) $\mathcal{R}(P) = \ker(I - P) = \{x : Px = x\}$.
b) $\ker(P) = \mathcal{R}(I - P)$.
c) $\mathcal{R}(P) \cap \ker(P) = 0$ *and* $V = \mathcal{R}(P) \oplus \ker(P)$.
d) *If V_1 and V_2 are subspaces such that $V_1 \cap V_2 = 0$ and $V_1 + V_2 = V$, there exists a projection P on V such that $V_1 = \mathcal{R}(P)$ and $V_2 = \ker(P)$.*

And a bit more elaborate is

Theorem C.28. *Let P be a continuous projection on V, then $V = \mathcal{R}(P) \oplus \ker(P)$ (This is (c) in Lemma C.2 plus closedness of the subspaces.) And conversely, given $V = V_1 \oplus V_2$ (again like in (d) above plus closedness of the subspaces) then the projection with range V_1 and kernel V_2 is continuous.*

C.4 Operators on Hilbert spaces. Singular values decompositions

Hilbert spaces are the infinite dimensional analogues of Euclidean spaces. This time we have

Definition C.41. A scalar product on a vector space V is a mapping $\langle \cdot, \cdot \rangle : V \times V \to \mathbb{R}$ (or $\mathbb{C}$) satisfying

a) $\langle x, y \rangle = \langle y, x \rangle$ (or $\overline{\langle y, x \rangle}$ when the scalars are complex).
b) $\langle x, ay_1 + by_2 \rangle = a \langle x, y_1 \rangle + b \langle x, y_2 \rangle$.
c) $\langle x, x \rangle \geq 0$ and $\langle x, x \rangle = 0$ iff $x = 0$.

Furthermore, the norm in V is obtained from the scalar product by $\|x\|^2 = \langle x, x \rangle$, and V is assumed to be closed in this norm.

This time the notions of perpendicularity introduced in Definition C.37 coincide with their geometrical counterparts.

Theorem C.29. *(Hilbert-Schmidt) Let A be a self adjoint compact operator on a Hilbert space V. Then there is a complete orthonormal basis $\{\phi_n\}$ for V so that $A\phi_n = \lambda_n \phi_n$ and $\lambda_n \to 0$ as $n \to \infty$.*

Proof. For each eigenvalue of A, chose a set of orthonormal eigenvectors corresponding to that eigenvalue. This is a finite set! The collection $\{\phi_n\}$ of all these eigenvectors is orthonormal since eigenvectors corresponding to different eigenvalues are orthonormal. Let V_1 be the closure of the span of $\{\phi_n\}$. Note that self adjointness of A implies $A : V_1 \to V_1$ and $A : V_1^\perp \to V_1^\perp$. Let $\hat{A}$ be the (self-adjoint and compact) restriction of A to $V_1^\perp$. If any $\lambda \neq 0$ is in $\sigma(\hat{A})$, then it is an eigenvalue of A, thus $\hat{A}$ must be the zero operator on $V_1^\perp$ (this is because $\hat{A}$ is selfadjoint and $\|\hat{A}\| = \sup\{|\lambda| : \lambda \in \sigma(\hat{A})\}$). Therefore $V_1^\perp = \{0\}$ for if $\varphi \neq 0$ is in $V_1^\perp$, $\hat{A}\varphi = A\varphi = 0$ implies that $\varphi \in V_1$. Thus $V_1 = V$.

The last part of the theorem is just the Riesz-Schauder Theorem. □

As an application of the Hilbert-Schmidt Theorem we have the result that played a central role in Chapter 4.

Theorem C.30. *(Singular values decomposition) Let A be a compact operator on the Hilbert space V. Then there exists a (not necessarily complete) orthonormal set $\{\psi_n : n \geq 1\}$ and $\{\phi_n : n \geq 1\}$ and positive real numbers*

$\{\lambda_n : n \geq 1\}$ such that
$$A(\cdot) = \sum_{n\geq 1} \lambda_n(\psi_n, \cdot)\varphi_n$$
converges in norm. The numbers λ_n are called the singular values of A.

Proof. Note that A (and A^*) are compact, then so is A^*A which is self adjoint. By the previous theorem, there exists an orthonormal set $\{\psi_n : n \geq 1\}$ and strictly positive μ_n (because A^*A is positive, i.e., $(x, A^*Ax) \geq 0$) so that $A^*A\psi_n = \mu_n \psi_n$.

Let λ_n be the positive square root of μ_n and set $\phi_n = \frac{A\psi_n}{\lambda_n}$. Verify that ϕ_n are orthonormal and that
$$Ax = \sum \lambda_n(\psi_n, x)\phi_n$$
for any $x \in V$. □

C.5 Some convexity theory

Just in case, a nonempty subset $\mathcal{C}$ of a vector space is said to be convex if for any two x, y in $\mathcal{C}$, $\alpha x + (1-\alpha)y$ is in $\mathcal{C}$ for $0 < \alpha < 1$.

Also, $f : \mathcal{C} \to (-\infty, +\infty)$ is said to be convex whenever $\mathcal{C}$ is convex and for any $x, y \in \mathcal{C}$, $0 < \alpha < 1$, the following holds
$$f(\alpha x + (1-\alpha)y) \leq \alpha f(x) + (1-\alpha)f(y).$$

Theorem C.31. *(Hahn-Banach) Let V_1 be a subspace of a Banach space V, and let $\lambda : V_1 \to \mathbb{R}$ be a linear functional such that*
$$|\lambda(x)| \leq p(x), \quad x \in V_1$$
where $p : V \to [0, \infty)$ is a given convex function. Then λ can be extended to a functional $\lambda : V \to \mathbb{R}$ such that
$$|\lambda(x)| \leq p(x), \quad x \in V.$$

Exercise C.19. (Application)

a) Let V, V^* be dual vector spaces and let $x_1, ..., x_n \in V$, $\lambda \in V^*$. Suppose you know the numbers
$$\lambda(x_i) = y_i, \quad i = 1, ..., n.$$
Prove that given the vector $y \in \mathbb{R}^n$ with components y_i, there exists a $\lambda \in V^*$ such that
$$\lambda(x_i) = y_i.$$

b) Let $V = \mathcal{C}([0,1])$ and let $x_i(t) = t^i$ for $i = 1, ..., n$. Prove that there exists a signed Borel measure $m(dt)$ on $([0,1], \mathcal{B}([0,1]))$ such that

$$\int_0^1 t^i m(dt) = m_i, \quad i = 1, ..., n$$

where m_i are given numbers. Question: what condition must be imposed on the m_i if you insist on the $m(dt)$ to be a positive Borel measure?

Comment: (Actually only the fact that V is a vector space is needed for the theorem to hold.) Many results about minimization of convex functions on convex sets depend on the following separation results.

Theorem C.32. *Let A and B be disjoint, nonempty, convex sets in a Banach space V.*

a) *If A is open, there exists $\lambda \in V^*$ and γ in $\mathbb{R}$ such that $\langle \lambda, x \rangle < \gamma \leq \langle \lambda, y \rangle$, for every $x \in A$ and every $y \in B$.*
b) *If A is compact and B is closed, then there exists $\lambda \in X^*, \gamma_1, \gamma_2 \in \mathbb{R}$ such that $\langle \lambda, x \rangle < \gamma_1 < \gamma_2 < \langle \lambda, y \rangle$ for $x \in A$ and $y \in B$.*

Corollary C.7.

a) *If V is a Banach space, then V^* separates points of V.*
b) *If V_1 is a subspace of V and $x_0 \in V$ is not in the closure $\overline{V}_1$ of V, there exists $\lambda \in V^*$ such that $\langle \lambda, x_0 \rangle = 1$ and $\langle \lambda, x \rangle = 0$ on V_1.*

An application of these results is the Krein-Milman theorem.

Definition C.42.

a) Let K be a subset of a vector space V, a nonempty subset E of K is called an *extreme set* for K whenever for $x, y \in K$ and $0 < t < 1$, $tx + (1-t)y \in E$, then $x, y \in E$. Sometimes E is called a face of K.
b) An *extreme point* for K is an extreme set consisting of one point only.
c) Given a subset E of V, the *convex hull* of E is the *smallest convex* set containing E. The closed hull of E is the closure of its convex hull.

Theorem C.33. *(Krein-Milman) Let K be a Banach space. If K is a compact, convex subset of V, then K is the closed convex hull of the set of its extreme points.*

Let us now examine the issue of integration of Banach space valued functions.

Definition C.43. Let V be a Banach space and V^* its dual. Let $(S, \mathcal{F}, m)$ be a measure space and $X : S \to V$ be a measurable. Assume that for any $\lambda \in V^*$, the real valued function $\langle \lambda, X(s) \rangle$ is integrable. If there exists a $y \in V^*$ such that

$$\langle \lambda, y \rangle = \int \langle \lambda, X(s) \rangle \, m(ds)$$

for every $\lambda \in V^*$, we say that X is weakly-m-integrable and write

$$y = \int X(s) m(ds).$$

Comment: If $\int \|X(s)\| \, m(ds) < \infty$, then X is weakly-m-integrable. If we had the analogue of approximation by simple functions, we could define stronger integrability.

Exercise C.20. Verify that when $\|X(\cdot)\|$ is m-integrable, then

$$\|y\| = \left\| \int X(s) m(ds) \right\| \leq \int \|X(s)\| \, m(ds).$$

<u>Hint</u>: Consider the 1-dimensional subspace generated by y, define $\langle \lambda, y \rangle = \|y\|$ and invoke Hahn-Banach appropriately.

Appendix D

Further properties of entropy functionals

D.1 Properties of entropy functionals

Given a (complete) probability space $(\Omega, \mathcal{F}, Q)$ (but it may well be a σ-finite measure space) we can consider the class $\mathcal{A}(Q) = \{m \in \mathcal{M} : m \ll Q\}$, where $\mathcal{M}$ is the class of all (signed) finite measures on $(\Omega, \mathcal{F})$.

If need be, check with Rudin's [5-2] or with Simmon's [1] to verify that $\mathcal{M}$ can be made a Banach space by defining the norm

$$\|m\| = m^+(\Omega) + m^-(\Omega)$$

where m^+, m^- are the upper and lower variations of m given by the Hahn-Jordan decomposition.

Exercise D.21. How does $\mathcal{A}(Q)$ sit in $\mathcal{M}$?

Define the isometric isomorphism

$$\begin{aligned} \mathcal{I} : \mathcal{A}(Q) &\to L_1(Q) \\ m &\to \frac{dm}{dQ} \end{aligned}$$

where (obviously) $\frac{dm}{dQ}$ is defined up to null Q-sets (or as an equivalence class to be really proper). That's why we need Q null sets to be in $\mathcal{F}$. Note that

$$\left\|\frac{dm}{dQ}\right\|_1 = \left\|\frac{dm^+}{dQ}\right\|_1 + \left\|\frac{dm^-}{dQ}\right\|_1 = m^+(\Omega) + m^-(\Omega) = \|m\|.$$

Also, for $\rho \in L_1(Q)$, $dm = \rho dQ$ defines $\mathcal{I}^{-1}(\rho)$. This should convince us that the map $\mathcal{I}$ defined above is really an isometric isomorphism.

Our entropy functionals were defined in subsets of

$$\mathcal{P}(Q) = \left\{P \ll Q : \int_\Omega dP = 1\right\}$$

the class of all probability measures on $(\Omega, \mathcal{F})$ which are absolutely continuous with respect to Q. From what we said above it is clear that we can think of $\mathcal{P}(Q)$ as a closed, convex subset of $L_1(Q)$.

We can now define $S_Q : \mathcal{P}(Q) \to [-\infty, \infty)$ by

$$S_Q(P) = -\int_\Omega \frac{dP}{dQ} \ln\left(\frac{dP}{dQ}\right) dQ$$

or $-\infty$ when $E_P\left[\left|\ln\left(\frac{dP}{dQ}\right)\right|\right] = +\infty$. $S_Q(P)$ is called the entropy of P (with respect to Q).

It is easy to verify that $\varphi(t) = t \ln t$ is continuous on $(0, \infty)$, extendible continuously to $[0, \infty)$ and can be set $\varphi(t) = -\infty$ for $t < 0$, it is strictly convex on $(0, \infty)$ and has a minimum at $\frac{1}{e}$. Thus $S_Q(P)$ is a concave functional.

Proposition D.1. $S_Q(P)$ *is upper semicontinuous on* $\mathcal{P}(Q)$.

Proof. Actually, the integral defining $S_Q(P)$ is defined on the closed subset of $L_1(Q)_+$ obtained as $\mathcal{I}(S_Q(P))$. We have to verify that for every real number r, $\{P \in \mathcal{P}(Q) : S_Q(P) < r\}$ is open. Or equivalently, that $K = \{P \in \mathcal{P}(Q) : -S_Q(P) \leq r\}$ is closed in $L_1(Q)$ for any r.

Let $\rho_n \in K$ tend to $\rho \in L_1(Q)$. There exists at least one subsequence $\rho_{n(k)}$ of ρ_n tending to ρ a.s. $-Q$. Thus $\rho \geq 0$ and $\int_\Omega \rho dQ = 1$. From the continuity of φ on $[0, \infty)$ we have $\varphi(\rho_n) \to \varphi(\rho)$ a.s. Q and therefore, by Fatou's lemma

$$\lim_{k \to \infty} \int_\Omega \varphi(\rho_n) \geq \int_\Omega \varphi(\rho) dQ$$

which implies that $dP = \rho dQ$ is in K. □

Exercise D.22. Verify that $S_Q(P)$ is strictly convex.

Proposition D.2. *Let* $\mathcal{P}_1$ *be a nonempty convex subset of* $\mathcal{P}(Q)$. *If* $\mathcal{I}(\mathcal{P}_1)$ *is compact in* $L_1(Q)$, *there exists a* $P^* \in \mathcal{P}_1$ *which maximizes* $S_Q(P)$ *on* $\mathcal{P}_1$. *Since* $S_Q(P)$ *is strictly convex,* P^* *is unique.*

Proof. Exercise for the reader. □

The next result allows us to do away with the compactness assumption of $\mathcal{P}_1 \subset \mathcal{P}(Q)$.

Theorem D.34. *Assume that* $Q(\Omega) < \infty$. *Let* $\mathcal{P}_1 \subset \mathcal{P}(Q)$ *be a nonempty, convex subset. If* $\mathcal{I}(\mathcal{P}_1)$ *is closed in* $L_1(Q)$, *there exists* $P^* \in \mathcal{P}_1$ *maximizing* $S_Q(P)$ *on* $\mathcal{P}_1$.

Proof. Set $l = \inf\{\frac{Q(\Omega)}{e} - S_Q(P) : P \in \mathcal{P}_1\}$. Since $Q(\Omega) < \infty$ and $t \ln t \geq -\frac{1}{e}$, we have $l \in [0, \infty)$. If $l = \infty$ we got nothing to prove. If $l < \infty$, we shall prove that there is $P^* \in \mathcal{P}_1$ such that $S_Q(P^*) = Q(\Omega) - l$. For that consider

$$\mathcal{P}_1^* = \{P \in \mathcal{P}_1 : \frac{Q(\Omega)}{e} - S_Q(P) \leq 2l\}.$$

Since $\mathcal{P}_1$ is convex and closed, it follows from the lower semicontinuity of $\frac{Q(\Omega)}{e} - S_Q(P)$ that

$$l = \inf\{\frac{Q(\Omega)}{e} - S_Q(P) : P \in \mathcal{P}_1\}.$$

In Rudin's [5-2] you can find the proof that a convex set in a Banach space is closed if and only if it is weakly closed. Recall as well that the weak topology on $L_1(Q)$ is the setwise convergence of probability measures. Therefore $\mathcal{P}_1^*$ is weakly closed and the map $P \to \frac{Q(\Omega)}{e} - S_Q(P)$ is weakly lower semicontinuous.

If we manage to prove that $\mathcal{P}_1^*$ is weakly sequentially compact, then it will be weakly compact and we shall obtain the existence of P^* from Theorem A.2 in Appendix A.

To prove that $\mathcal{P}_1^*$ is weakly sequentially compact it suffices to prove that for $P \in \mathcal{P}_1^*$ and $\rho = \frac{dP}{dQ}$, for every sequence $\{E_n : n \geq 1\}$ of sets in $\mathcal{F}$ having empty intersection

$$\lim_{n \to \infty} \int_{E_n} \rho dQ = \lim_{n \to \infty} P(E_n) = 0$$

uniformly in $P \in \mathcal{P}_1^*$.

Note that if $Q(E_n) = 0$ for some n, we would have nothing to prove. Thus we can assume that $Q(E_n) > 0$ for all n and put

$$c_n := \mu(E_n)^{-\frac{1}{2}}$$

and note that $c_n \uparrow \infty$ as $n \to \infty$. Also, since $\ln t \to \infty$ as $t \to \infty$ (or $-\frac{\varphi(t)}{t} \to \infty$ as $t \to \infty$) we conclude that there exists n_0 such that for $n > n_0$ the following hold: $c_n > y$ and $\varphi(t) = t \ln t > 0$ for $c_n < t < \infty$ for any given $y > 0$. For such y and n,

$$\int_{E_n} \rho dQ \leq \int_{E_n \cap \{\rho \leq c_n\}} \rho dQ + \int_{E_n \cap \{\rho > c_n\}} (\rho - y) dQ + \int_{E_n} y dQ$$

$$\leq c_n Q(E_n) + y Q(E_n) + \int_{E_n \cap \{\rho > c_n\}} \frac{(\rho - y) \varphi(\rho)}{\varphi(\rho)} dQ$$

$$\leq \frac{1}{c_n} + \frac{y}{c_n^2} + \frac{c_n - y}{\varphi(c_n)} \int_{\{\rho > c_n\}} \varphi(\rho) dQ$$

$$\leq \frac{1}{c_n} + \frac{y}{c_n^2} + \frac{c_n - y}{\varphi(c_n)} \left(2l - \frac{Q(\Omega)}{e}\right)$$

since $\frac{\varphi(t)}{t} \to \infty$ as $t \to \infty$ and $c_n \to \infty$ we are done. □

Exercise D.23. To verify the inequality at the third step, verify that for a convex function defined on $(0, \infty)$ and for $x_1 < y_1 \leq x_2$, $x_1 < y_2 \leq x_2$

$$\frac{f(y_1) - f(x_1)}{y_1 - x_1} \leq \frac{f(x_2) - f(y_2)}{x_2 - y_2}.$$

To end the appendix, let us verify a property of the functional

$$K(P_1, P_2) := \int_\Omega \frac{dP_1}{dQ} \ln \left(\frac{\frac{dP_1}{dQ}}{\frac{dP_2}{dQ}}\right) dQ$$

defined on $\mathcal{P}(Q)$. Notice that even though we assume Q to be a σ-finite measure, both $\frac{dP_1}{dQ}$ and $\frac{dP_2}{dQ}$ are probability densities. The number $K(P_1, P_2)$ is usually called: relative entropy, Kullback-Leibler information number, information distance or information for discrimination.

The following was essential for us

Proposition D.3. *With the notations introduced above, $K(P_1, P_2) \geq 0$ and $K(P_1, P_2) = 0$ if and only if $P_1 = P_2$.*

Proof. Let $\Omega_1 \subset \Omega$ be the support of P_1 (or of $\frac{dP_1}{dQ}$). By Jensen's inequality

$$-K(P_1, P_2) = \int_{\Omega_1} \frac{dP_1}{dQ} \ln \left(\frac{\frac{dP_2}{dQ}}{\frac{dP_1}{dQ}}\right)$$

$$\leq \ln \left(\int_\Omega \frac{dP_1}{dQ} \left(\frac{\frac{dP_2}{dQ}}{\frac{dP_1}{dQ}}\right) dQ\right)$$

$$= \ln 1 = 0$$

since $\ln x$ is strictly concave function, we have equality if and only if $\frac{\frac{dP_2}{dQ}}{\frac{dP_1}{dQ}} = 1$
a.e.$-Q$ or $P_1 = P_2$. □

D.2 A probabilistic connection

This section could have been placed in Chapter 5, advising the reader to skip it in case of need for it contains further mathematical connections between probability (large deviations theory) and entropy functionals. The basic identity in Chapter 5, that is,

$$\sup\{S(P,Q) : P \ll Q, E_P[AX] = y\} = \inf\{S(\lambda) : \lambda \in W^*\}$$

offers an unexpected "connection" between inverse problems and the theory of large deviations.

In Chapter 5 we proved the existence of a measure $P(\lambda^*)$ which maximized $S(P,Q)$ and such that $E_{P(\lambda^*)}[AX] = y$.

In this section we assume that a $\lambda^* \in W^*$ exists such that

$$AE_Q\left[\frac{Xe^{-\langle\lambda^*, AX\rangle}}{Z(\lambda^*)}\right] = y \tag{D.1}$$

and describe a context in which both $S(\lambda)$ and $S(P,Q)$ appear in a natural way and in which identity (5.13) holds.

Our presentation is rather sketchy. Full detail is spread out in the first three chapters of Deuschel and Stroock's book [10]. We shall introduce obvious modifications.

Let us denote by P^* the measure appearing in (5.16), i.e., $\frac{dP^*}{dQ} = \frac{\exp(-\langle\lambda^*, AX\rangle)}{Z(\lambda^*)}$ and consider the product space $\hat{C} = C^{N_+}$, $\hat{\mathcal{F}} = \otimes_1^\infty \mathcal{F}$, $\hat{P} = \otimes_1^\infty P^*$, where $\mathcal{F}$ denotes the Borel σ-algebra on C and $N_+ = \{1, 2, ...\}$ stands for the positive integers.

In this set up the coordinate maps $X_n : \hat{C} \to C$ are independent, identically distributed random variables, having mean

$$\hat{E}_{\hat{P}}[X_i] = E_P[X_1] = x^*$$

and $Ax^* = y$.

The measures $P^n = \otimes_1^n P^*$ describe the joint distributions of $(X_1, ..., X_n)$ and we denote by P_n the distribution of

$$\Sigma_n = \frac{1}{n}\sum_{i=1}^n X_i.$$

The Banach space version of the strong law of large numbers goes as follows

Theorem D.35. *If $\int \|\xi\| P^*(d\xi) < \infty$, then there is an $m(P^*)$ in C such that $\Sigma_n \to m(P^*)$ a.s $-\hat{P}$. Moreover $m(P^*)$ denotes the unique $x^* \in B$ such that for any $\mu \in V^*$*

$$\langle \mu, x^* \rangle = \int \langle \mu, \xi \rangle P^*(d\xi).$$

Proof. See Chapter 3 of [10]. □

Actually, for every $\lambda \in W^*$ we have

$$\langle \lambda, Ax^* \rangle = \langle \lambda, \lim A\Sigma_n \rangle \qquad (D.2)$$

$a.e - \hat{P}$. Also $\hat{P}\{\|A\Sigma_n - y\| \leq \varepsilon\} \to 1$ for any $\varepsilon > 0$, but the interesting question is: How does $\hat{P}\{\|A\Sigma_n - y\| > \varepsilon\}$ decay as n becomes large?

The distribution of $A\Sigma_n$ is the image of P_n under $A : V \to W$ and can be recovered from its Laplace transform

$$\hat{E}_{\hat{P}}[e^{-\langle \eta, A\Sigma_n \rangle}] = E_{P^*}[e^{-\langle \eta, AX_1 \rangle/n}]^n = Z_*\left(\frac{\eta}{n}\right)^n$$

since the X_i are all independent relative to $\hat{P}$. Here $\eta \in W^*$ and

$$Z_*(\eta) = \frac{Z(\eta + \lambda^*)}{Z(\lambda^*)}. \qquad (D.3)$$

From $A\Sigma_n \to Ax^* = y$ a.s $- \hat{P}$ we have, for any continuous bounded $f : W \to \mathbb{R}$, $\hat{E}_{\hat{P}}[f(A\Sigma_n)] \to f(y) = \int f(z)\varepsilon_y(dz)$, i.e.

$$P_n \to \varepsilon_{x^*} \text{ and } A(P_n) \to \varepsilon_{Ax^*} = \varepsilon_y$$

weakly. The relevant definitions at this point are

Definition D.44. A function $I : W \to [0, \infty)$ is said to be a rate function if it is lower semi-continuous. We say that the $A(P_n)$ satisfy the full large deviation principle with rate function I if for any Borel set Γ in W

$$-\inf_{int(\Gamma)} I(x) \leq \liminf_{n \to \infty} \tfrac{1}{n} \ln(A(P_n)(\Gamma))$$
$$\leq \limsup \tfrac{1}{n} \ln(A(P_n)(\Gamma)) \leq -\inf_{\bar{\Gamma}} I(x). \qquad (D.4)$$

Definition D.45. I is said to be a good rate function if $\{z \in W : I(z) \leq L\}$ is compact in W for all $L \geq 0$.

If we define for $z \in W$
$$S^*(z) = \inf\{\langle \lambda, z\rangle + \ln(Z_*(\lambda)) : \lambda \in W^*\}$$
it happens that $-S^*(z)$ is a good rate function for $A(P_n)$ and (D.4) becomes
$$\sup\{S^*(z) : z \in int(\Gamma)\} \leq \liminf_{n\to\infty} \tfrac{1}{n}\ln(A(P_n)(\Gamma))$$
$$\leq \limsup_{n\to\infty} \tfrac{1}{n}\ln(A(P_n)(\Gamma)) \qquad (D.5)$$
$$\leq \sup\{S^*(z) : z \in \bar{\Gamma}\}.$$

A few interesting properties of $S^*(z)$ are the following: Notice that $S_{P^*}(\eta, z) = \langle \eta, z\rangle + \ln(Z_*(\eta))$ satisfies $S_{P^*}(0, z) = 0$, therefore $S^*(z) \leq 0$.

Since $\langle \eta, y\rangle = \int \langle \eta, A\xi\rangle P^*(d\xi)$ for any λ, from Jensen's inequality we have $S_{P^*}(\eta, y) \geq 0$. Therefore $S^*(y) = 0$. (Recall that $E_{P^*}[AX] = y$.)

And since $-S^*(z) = \sup\{\langle \eta, z\rangle - \ln(Z_*(-\eta)) : \eta \in W^*\}$ we obtain that $S^*(z)$ is concave and upper semicontinuous. Summing up, we have established:

Lemma D.3. *The function $S^*(z)$ defined on W^* satisfies*

i) $S^*(z)$ *is upper semicontinuous and concave.*
ii) $S^*(z) \leq 0$.
iii) If $y = E_{P^*}[AX]$ then $S^*(y) = 0$.
iv) The large deviations of P_n are controlled by (D.5).

The apparition of $S(P, P^*)$ comes at the end of a more convoluted road. Instead of considering the large deviations of the laws P_n (or equivalently of $A(\Sigma_n)$), we should now consider the large deviations of the empirical measures
$$A(L_n) = \frac{1}{n}\Sigma \varepsilon_{AX_i}$$
as elements of $\mathcal{P}(W) \subset \mathcal{M}(W)$. The class $\mathcal{M}(W)$ of all finite measures on W can be made a Banach space when the total variation norm is put on it, and $\mathcal{P}(W)$ is a compact (in the weak topology on $\mathcal{M}(W)$) convex subset of $\mathcal{M}(W)$, which can be appropriately metrized. Again, for details the reader should refer to [10].

The distributions of ε_{AX} and $A(L_n)$ are elements of $\mathcal{P}(\mathcal{P}(W))$ which can be characterized by their Laplace transform. To define the Laplace transform we need know that

$\mathcal{M}(W)^* = C_b(W, \mathbb{R}) :=$ the class of continuous bounded functions on W.

The Laplace transform of the law $\tilde{P}^*$ of ε_{AX} as element of $\mathcal{P}(\mathcal{P}(W))$ is

$$Z_{\tilde{P}^*}(\varphi) = E_{\tilde{P}^*}[e^{-\langle\varphi,v\rangle}] = \int e^{-\langle\varphi,v\rangle}d\tilde{P}^*(v) = \int e^{-\varphi(AX)}dP^*$$

where the pairing $\langle\varphi, v\rangle$ is just $\int \varphi(\eta)v(d\eta)$ for $v \in \mathcal{P}(W)$. Notice that $\tilde{P}^*$ is the image of P^* under the map $V \to \mathcal{P}(W)$ given by $\xi \to \varepsilon_{A(\xi)}$.

To describe the large deviations of $\tilde{P}_n$ we must set for $P_W \in \mathcal{P}(W)$

$$\Sigma_P(\varphi, P_W) = \langle\varphi, P\rangle + \ln(Z_{\tilde{P}^*}(\varphi)) \tag{D.6}$$

and then

$$\Sigma^*(P_W) = \inf\{S(\varphi, P_W) + \ln(Z_{\tilde{P}^*}(\varphi)) : \varphi \in M(W)^*\}. \tag{D.7}$$

The following comments, leading to the identification of $\Sigma^*(P)$ are essentially the steps in the proof of Lemma 3.2.13 of [8]. We saw above that $\Sigma^*(P_W) \leq 0$ and for P_a defined by

$$P_a(\Gamma) = \int_{\mathcal{P}(W)} v(\Gamma)d\tilde{P}^*(v)$$

or equivalently, for any $\varphi \in C_b(W, \mathbb{R}) = M(W)^*$,

$$\langle\varphi, P_a\rangle = \int_{\mathcal{P}(W)} \langle\varphi, v\rangle\, d\tilde{P}^*(v) = \int \varphi(A\xi)dP^*(\xi)$$

we have $\Sigma^*(P_a) = 0$.

We want to verify that for $P \in \mathcal{P}(C)$ and $P_W = A(P)$, then

$$\Sigma^*(P_W) = S(P, P^*) = -\int \ln(\frac{dP}{dP_*})dP.$$

Assume for the time being that $\rho = \frac{dP}{dP^*} \geq \alpha > 0$.

Then

$$\exp\{\int -\varphi dv + S(P, P^*)\} = \exp\left(\int(-\varphi(A) - \ln(\rho))dP\right)$$

$$\leq \int \exp(-\varphi(A) - \ln(\rho))dP$$

$$= \int e^{-\varphi(A)}dP = Z_{P^*}(\varphi)$$

or equivalently

$$S(P_W, P^*) \leq \langle\varphi, v\rangle + \ln(Z_{P^*}(\varphi)) \quad \forall \varphi$$

which amounts to saying that

$$S(P_W, P^*) \leq \Sigma^*(P_W).$$

Assume now that $\Sigma^*(P_W) > -\infty$. This will imply that there is $P \in \mathcal{P}(C)$ such that $P_W = A(P) \ll A(P^*)$.

Note that

$$-\infty < \Sigma^*(P_W) \leq \int \varphi dP_W + \ln(Z_{\tilde{P}^*}(\varphi))$$

which implies that $\Sigma^*(P_W) \leq \int \varphi dP_W$ for $\varphi \in C_b(W, \mathbb{R})$.

Since indicators of Borel sets can be approximated by continuous functions, we can replace φ in the last inequality by $-rI_\Gamma$ for $\Gamma \in$ Borel in W. If we choose Γ such that $\int I_\Gamma(A)dP^* = 0$, we obtain that

$$\infty > -\Sigma^*(P_W) \geq r \int I_\Gamma(\eta)dP_W \text{ for all } r > 0.$$

This yields $P_W(\Gamma) = 0$.

Clearly, $P_W \ll A(P)$. If we set $\rho_0 = \dfrac{dP_W}{dA(P^*)}$ and then put $dP = \rho_0(A)dP^*$, clearly $P_W = A(P)$.

Again, assuming that $\rho_0 \geq \alpha > 0$ and uniformly bounded, and setting $\varphi = -\ln(\rho_0)$ we obtain once more

$$-\infty < \Sigma^*(P) \leq -\int \ln(\rho_0)dA(P) = -\int \rho_0 \ln(\rho_0)dA(P^*) = S(P, P^*).$$

We leave up to the reader to browse in [8] to see how to remove the assumptions of strict positivity plus uniform boundedness on ρ_0.

Before ending this section we mention that P^* was brought in only because of (D.1). All assertions hold true for any $Q \in \mathcal{P}(C)$. To obtain something like (5.13) take a look at exercise 3.2.12 in [10].

And, a presentation of the famous Entropy Concentration Theorem appears in [11]. This is also related to the large deviations theory. It asserts that the "most probable empirical distributions are maxentropic".

D.3 Extended definition of entropy

Here we present a definition of entropy that cover the case in which the measure $Q << P$ is not a probability measure. Since the variational characterization presented below is more natural with the statisticians sign convention that with the physicists convention, we consider

Definition D.46. Let $\mathcal{M} = \{u : \mathbb{R}^n \to [0, \infty), \int u dP < \infty\}$, this is the class of densities of finite measures on $\mathbb{R}^n$ absolutely continuous with respect to P. Define

$$\hat{S}_P[u] = \int u\ln(u)dP - \int udP \ln\left(\int udP\right)$$

whenever $|u\ln(u)|$ is P-integrable, and $-\infty$ otherwise.

Clearly this reduces to $-S(Q,P)$ whenever $Q << P$ is a probability measure. The following characterization is interesting.

Lemma D.4. *Let $V = \{v : \int e^v dP \leq 1\}$, then*

$$\hat{S}_P[u] = \sup\{\int uvdP : v \in V\}.$$

Proof. Consider the Lagrangian function defined on $V \times \mathbb{R}$ for a fixed $u \in \mathcal{M}$,

$$L(u,v,\lambda) = \int uvdP - \lambda\left(\int e^v dP - 1\right).$$

Notice that

$$\sup\{\int uvdP : v \in V\} = \sup_V \inf_{\lambda \geq 0} L(u,v,\lambda).$$

Since $L(u,v,\lambda)$ is concave in v and linear in λ, a saddle point solution exists and (abbreviating) sup inf = inf sup. To compute the right end, consider $v + sh$, for arbitrary h and s small enough. Compute the derivative of $L(u, v+sh), \lambda)$ at $s = 0$ to obtain $\int (u - e^v \lambda)hdP = 0$. That is $v^* = \ln(u/\lambda)$. Substitute in $L(u,v,\lambda)$, equate the derivative with respect to λ to zero to obtain $\lambda^* = \int udP$. Substitute to obtain

$$L(u, v^*, \lambda^*) = \hat{S}_P[u]$$

as desired. □

D.4 Exponetial families and geometry in the space of probabilities

We already know of Jaynes' proposal to solve a problem that appears in a variety of problems in different fields, consisting of finding a probability p defined on some measurable space $(\Omega, \mathcal{B})$ when only the expected values

$$E_p[X_i] = y_i; \quad i = 1, ..., M \tag{D.8}$$

are known. Think of it as a generalized moment problem. To solve this problem by means of the maximum entropy method, one realizes that the class of probability measures

$$\mathcal{C} = \{p \,|\, (1) \text{ holds}\}$$

is a convex set, and the systematic way to pick up points from a convex set consists of maximizing a concave function: Find the $p \in \mathbb{P}$ that maximizes a given concave function defined over $\mathcal{C}$.

To be specific, and this represents no loss of generality, let us consider a finite dimensional problem, that is $\Omega = \{1, 2, ..., N\}$ and $\mathcal{B}$ is the collection of all subsets of Ω. The class of all probability distributions on Ω is the following simplex in $\mathbb{R}^N$

$$\mathbb{P} = \{p = (p_1, ..., p_n) \,|\, \sum p_i = 1, \, p_i \geq 0 \,\forall i\}.$$

The fact that many people find curious about the solution to (D.8) is the apparition of exponential families of the type

$$p_i = \frac{e^{-<\lambda, X>(i)}}{Z(\lambda)} \qquad (D.9)$$

where $<\lambda, X> \equiv \sum_{k=1}^{M} \lambda_k X_k$, and where $Z(\lambda) = \sum_{i=1}^{N} e^{-<\lambda, X>(i)}$ is an obvious normalization factor. How this comes about, and how it is related to the entropy function is rather well known.

Here we present an entirely geometrical way of looking at the problem according to which the exponential parametrization of the points of $\mathbb{P}$ is natural. An exponential parametrization is a mapping $\mathbb{R}^N \to \mathbb{P}$, sending $\xi \in \mathbb{R}^N$ onto $p(\xi)$ given by

$$p_i(\xi) = \frac{e^{-\xi_i}}{\sum_{k=1}^{N} e^{-\xi_i}} \qquad (D.10)$$

then (D.9) represents (D.10) restricted to the manifold satisfying the constraints (D.8), i.e., $\sum p_i X_k(i) = y_k$.

For this a special geometry is introduced on the set of positive vectors thought of as a special subset of $\mathbb{C}^n$. The fundamental item being the definition of a connection (or its associated distribution of horizontal planes) and the notion of parallel transport and geodesics that the connection determines. Our presentation is a simplified version of the material appearing in references [13]-[15], where references to the original literature are given.

D.4.1 *The geometry on the set of positive vectors*

In this section we after briefly recalling some the basic facts about $\mathbb{C}^n$ thought of as a C^*-algebra, we define a special connection and describe its geodesics and parallel transport along them.

D.4.1.1 *The basic set up*

For reasons that shall become clear below, we want to understand in which sense the curve

$$\gamma(t) = a(0)^{1-t} a(1)^t \tag{D.11}$$

where $a(0)$, $a(1)$ are positive vectors, is a geodesic. To begin with, we shall think of $\mathbb{A} \equiv \mathbb{C}^n$ as the set of complex valued functions $x : [1, n] \to \mathbb{C}$ with the natural algebraic structure imposed on them in which $xy(k) \equiv x(k)y(k)$. Actually this structure turns $\mathbb{A}$ into a commutative C^*-algebra. In this algebra we consider

$$G^* = \{x \in \mathbb{A} \mid x \neq 0\}$$

which is a (commutative) group (with respect to the ordinary product of functions), and the class $G^+ \subset G$ which denotes the class of positive elements, or vectors in $\mathbb{C}^n$ with strictly positive components. Note to begin with that G^+ is a homogeneous space for the group action defined by

$$L_g : G^+ \to G^+ \quad L_g(a) = (g^*)^{-1} a g^{-1}, \quad \forall\, a \in G^+$$

for any $g \in G$. Here x^* denote the complex conjugate of x. Since the product is commutative, $L_g(a) = |g|^{-2} a$. If we identify G^+ with the diagonal matrices with strictly positive elements, every $a \in G^+$ defines a scalar product on $\mathbb{C}^n$ by $<x, y>_a = \sum a(i) x^*(i) y(i)$. Now we may interpret the group action as an isometry $\mathbb{C}_a^n \to \mathbb{C}_{L_g(a)}^n$. Here $\mathbb{C}_a^n$ is $\mathbb{C}^n$ with the scalar product $<x, y>_a$. Interpretations aside, for any fixed $a_0 \in G^+$ we can define the projection operator

$$\pi_{a_0} : G \to G^+$$

by means of

$$\pi_{a_0}(g) = L_g(a_0)$$

and notice right away that the fiber (isotropy group) over a_0 defined by

$$I_{a_0} = \{g \in G \mid \pi_{a_0}(g) = a_0\} = \{g \in G \mid |g| = 1\} = \mathbb{T}^n$$

the n-dimensional torus. An interesting issue is whether this consequence of commutativity can be related to the de-coherence inherent to classical mechanics. (Think of the a's as density matrices.)

Since G is clearly an open subset of $\mathbb{C}^n$, its tangent space at any point is $\mathbb{C}^n$, i.e.

$$(TG)_1 = \mathbb{C}^n = \mathbb{A},$$

and it is easy to see that

$$(TI_{a_0})_1 = V_{a_0} = \{0\} \oplus i\mathbb{R}^n$$

which in the non-commutative case corresponds to the anti-hermitian elements in $\mathbb{A}$.

The derivative $(D\pi_{a_0})_1(x)$ of π_{a_0} at 1 in the direction of $x \in \mathbb{A}$ is easy to compute, and it is given by

$$(D\pi_{a_0})_1(x) = -a_0(x + x^*).$$

Clearly

$$(D\pi_{a_0}) : \mathbb{A} \to (TG^+)_{a_0} \equiv \mathbb{R}^n \oplus \{0\}.$$

Note as well that the horizontal space at a_0, defined by

$$H_{a_0} \equiv \{x \in \mathbb{A} \,|\, (a_0)^{-1}x^* a_0 = x\} = \{x \in \mathbb{A} \,|\, x^* = x\} = \mathbb{R}^n \oplus \{0\}.$$

Therefore, we have the obvious splitting

$$\mathbb{A} = H_{a_0} \oplus V_{a_0}.$$

Not only that, the map $(D\pi_{a_0})_1$ is invertible from the left. That is, there exists a mapping $\kappa_{a_0} : (TG^+)_{a_0} \to (TG)_1$, given by

$$\kappa_{a_0}(z) \equiv -\frac{a_0^{-1}}{2} z$$

such that $(TG^+)_{a_0} \stackrel{\kappa}{\to} (TG)_1 \stackrel{(D\pi_{a_0})}{\to} (TG^+)_{a_0}$ is the identity mapping.

The mapping κ is called the **structure 1-form** of the homogeneous space G^+, and it is an $\mathbb{A}$-valued linear mapping defined on $(TG^+)_{a_0}$. All the geometry on G^+ comes from κ.

D.4.2 Lifting curves from G^+ to G and parallel transport

Let us begin with a basic lemma. Here the reason of being of the connection κ will become apparent.

Lemma D.5. *Let $a(t) : [0,1] \to G^+$ be a continuous curve in G^+. There exists a curve $g(t) : [0,1] \to G$, (called **the lifting of** $a(t)$ **to** G) such that*

$$L_{g(t)}(a_0) = \pi_{a_0}(g(t)) = a(t) \tag{D.12}$$

where (from now on) the notation $a(0) = a_0$ will be used now and then.

Proof. Let us verify that the solution to the (transport) equation

$$\dot{g}(t) = \kappa_{a(t)}(\dot{a}(t))g(t) \tag{D.13}$$

satisfies (D.12). Here the commutativity makes things really simple. Equation (D.13) explicitly spelt out is

$$\dot{g}(t) = -\frac{\dot{a}(t)}{2a(t)}g(t); \quad g(0) = 1$$

which can easily be solved to yield

$$g(t) = \left(\frac{a(0)}{a(t)}\right)^{1/2}. \tag{D.14}$$

Note that $\pi_{a_0}(g(t)) = \left(\frac{a(t)}{a(0)}\right)^{1/2} a(0) \left(\frac{a(t)}{a(0)}\right)^{1/2} = a(t)$. $\square$

The **parallel transport** along $a(t)$ (from $a(0)$ to $a(1)$) is defined in

Definition D.47. *Let $a(t)$ be a curve in G^+ and let $g(t)$ be its lifting to G. The parallel transport along $a(.)$ is the mapping $\tau(a(.)) : (TG^+)_{a(0)} \to (TG^+)_{a(1)}$ defined by*

$$\tau(a(.))(X) = L_{g(1)}(X). \tag{D.15}$$

We may now say that $a(t)$ is a **geodesic** if $\dot{a}(0)$ is transported onto $\dot{a}(t)$ by means of the (time) rescaled curve $b(s) := a(st)$, $s \in [0,1]$. From (D.14)-(D.15) it is clear that this amounts to

$$\dot{a}(t) = \frac{a(t)}{a(0)}\dot{a}(0) \Leftrightarrow \frac{\dot{a}(t)}{a(t)} = \frac{\dot{a}(0)}{a(0)} \equiv X.$$

Or equivalently,

Lemma D.6. *The curve $a(t)$ is a geodesic if and only if there exists a (real) vector X such that*

$$a(t) = a(0)e^{tX}.$$

Comment: This means that the lifted geodesic is given by
$$g(t) = a(0)^{1/2} e^{-tX/2}.$$

To tie up with the construction in the previous and following sections, observe that if we specify the initial and final points of the geodesic, the vector X is automatically determined:

$$a(1) = a(0)e^X \Rightarrow X = \ln\left(\frac{a(1)}{a(0)}\right),$$

and the equation of the geodesic can be rewritten as

$$a(t) = a(0)^{1-t} a(1)^t \qquad (D.16)$$

which is nothing but (D.11).

D.4.3 From geodesics to Kullback's divergence

We just saw how geodesics are associated with the parallel transport determined by the connection. Now we introduce a pseudo-norm in each tangent space and examine the distance it defines along geodesics. For any $Z \in \mathbb{A}$ define the seminorm (i.e., $\|Z\| = 0 \not\Rightarrow Z = 0$)

$$\|Z\| = |<1, Z>| \equiv |\sum Z(i)|.$$

Even though this is a rather peculiar choice we shall see that it leads to Kullback's divergence.

To define a pseudo-metric on TG^+, we begin by defining it at $(TG^+)_1$ by $\|X\|_1 \equiv \|X\|$, and transporting it to any other $(TG^+)_{a_0}$ by means of the group action: that is, we set

$$\|X\|_{a_0} = |<a_0^{-1}, X>|$$

and it is easy to verify that this is a consistent definition. Note now that if $a(t)$ is a geodesic joining $a(0)$ to $a(1)$ in G^+, then the "length" of the velocity vector along the geodesic is constant, for

$$\|\dot a(t)\|_{a(t)} = |<a(t)^{-1}, a(t)X>| = \|X\|_1 = \|\ln(a(1)/a(0))\|$$

and therefore, the geodesic distance from $a(0)$ to $a(1)$ is given by

$$d(a(0), a(1)) = \int_0^1 \|\dot a(t)\|_{a(t)} dt = \|X\| = \|\ln(a(1)/a(0))\|.$$

Comment: Note that the choice of 1 for the definition of the norm is not mandatory. For example, for a fixed $q \in G^+$ we may have defined

$$\|X\|_{(q),a(0)} := |<qa(0)^{-1}, X>|$$

and we would have ended up with

$$\|\dot{a}(t)\|_{(q)a(t)} = |<qa(t)^{-1}, a(t)X>| = \|X\|_{(q)} = \|q\ln(a(1)/a(0))\|.$$

which is still symmetric in $a(0), a(1)$. If we chose $q = a(1)$, we would end up with

$$\|X\|_{a(1)} = |\sum a(i)\ln(a_1(i)/a_0(i))|$$

which is not symmetric anymore (the norm depends on the final point of the trajectory) and it corresponds to the **Kullback divergence** between $a(0)$ and $a(1)$. Notice that when restricted to probabilities, $K(a_1, a_0) = \sum a(i)\ln(a_1(i)/a_0(i)) \geq 0$ and $K(a_1, a_0) = 0 \Leftrightarrow a_1 = a_0$. But the last two properties are not necessarily true for arbitray positive vectors, Moreover, K does not satisfy the triangular inequality, that is it is not a true distance.

D.4.4 Coordinates on $\mathbb{P}$

The set of probabilities on a finite sample space $\Omega = \{1, ..., n\}$ can be described as the closure of the manifold $\mathbb{P} = \{\mathbf{p} = (p_1, ..., p_n) \in \mathbb{R}^n | p_i > 0; \Sigma_{i=1}^n p_i = 1\}$ in $\mathbb{R}^n$. Observe now that we can identify rays in G^+ with points in $\mathbb{P}$ via an equivalence relation.

Definition D.48. We say that $a(1), a(2) \in G^+$ are equivalent if and only if $\frac{a(1)}{<1, a(1)>} = \frac{a(2)}{<1, a(2)>}$.

Comment: That is we identify lines in $\mathbb{R}^n_+$ with the point they intersect at $\mathbb{P}$, and $\mathbb{P}$ can be thought of as the projective space in G^+, in other words as the quotient space $G^+/\sim$, where for the time being, $\sim$ is the equivalence relation defined in (D.48).

We saw above that for any two points a_0 and a_1, there is vector field $X = \ln(\frac{a_1}{a_0})$ such that $\gamma(t) = a_0 e^{tX}$ is the geodesic joining a_0 to a_1 in G^+ (and G).

Note that the trace on $\mathbb{P}$ of a geodesic given by (D.16), or if you prefer, the equivalence class of each point of the geodesic, is given by

$$\gamma(t) = \frac{p(t)}{<1, p(t)>} = \frac{p(0)^{1-t}p(1)^t}{\sum (p_0(i))^{1-t}(p_1(i))^t}. \quad (D.17)$$

Thus (D.17) is the projection of a geodesic in $\mathbb{A}$ onto the projective space $\mathbb{P}$ which is class of probabilities on the discrete set $\{1, 2, ..., n\}$.

D.4.4.1 Exponential Families

Let us now examine a bit further in what sense (D.10) is natural in our setup. Set $a(0) = 1$ and let $a(1)$ be any other point in G^+. We now know that there exists a real vector X, actually given by $X = \ln a(1)$, such that $a(t) = e^{tX}$ joins 1 geodesically to $a(1)$, and the trace on $\mathbb{P}$ of this geodesic is $p(t) = a(t)/<1, a(t)>$ given by

$$\frac{e^X}{\sum e^{X(i)}}$$

that is we have a correspondence between vectors in $\mathbb{R}^n$ thought of as tangent vectors to TG^+ and probabilities in $\mathbb{P}$, which we shall now explore further.

We shall consider the mapping

$$\Phi : (TG^+)_1 \simeq \mathbb{R}^n \to \mathbb{P} \simeq G^+/\sim; \quad X \to \Phi(X) = \frac{e^X}{E[e^X]} \equiv \frac{e^X}{\zeta(X)}. \quad (D.18)$$

Let us examine some basic properties of Φ. To begin with to observe that $\Phi(X) = \Phi(X + \alpha 1)$. Thus Φ as defined cannot be a bijective map.

Comment: Note that we chose $1 \in G^+$ to define the norm $\|Z\|$ and to define Φ. This is just for technical convenience. Notice as well that collinear vectors differ in $\|.\|$-norm by $e^{\alpha 1}$ for appropriate α.

To understand the interplay between algebra and probability further, we note that $\mathbb{R}^n \simeq \mathcal{A}^s \simeq (TG^+)_1 = B \oplus H^s$, where $H = KerE[.]$ and H^s is the class of real, centered random variables, and for this we want to think of the expected value as of a linear mapping from $\mathcal{A}^s$ onto a commutative algebra B (which in this case coincides with $\mathbb{R}$). This additive decomposition at the Lie algebras induces a multiplicative decomposition at the group level. That is we can write any positive element in $g \in G^+$ as $g = e^X = e^{<1,X>}e^{X-<1,X>}$. This establishes a mapping from $B \times C$ where $C = \{e^Y \mid E[Y] = 0\}$ onto G^+.

Note now that the projection $g = e^{<1,X>}e^{X-<1,X>} \to \frac{e^{X-<1,X>}}{E[e^{X-<1,X>}]}$ is independent of $e^{<1,X>}$. This motivates the following: To make the map Φ a bijection, we have to restrict its domain. Basically $\mathbb{R}^n$ is an n-dimensional manifold whereas $\mathbb{P}$ is only $(n-1)$-dimensional. Thus if we define

$$\Phi : H^s \to \mathbb{P}; \quad Y \to \Phi(Y) = \frac{e^Y}{E[e^Y]} \quad (D.19)$$

we have a bijection, the inverse mapping being given by

$$q(i) \to Y(i) = \ln q(i) - \sum_i \ln q(i).$$

Note that $E[Y] = 0$, and the last identity may be the basis for a geometric interpretation of the surprisal function $\ln q(i)$ associated with a probability distribution $q(i)$ on $[1,n]$.

To conclude we note that the special role played by the vector 1 can be done away with: We could define expected values with respect to any given (and fixed) $p \in G^+$ by the standard $E_p[X] = \sum p(i)X(i) = <p, X>$. We note again that $e^X = e^{<p,X>}e^{X-<p,X>}$ and if we put $M_p \equiv (KerE_p[.])^s$, or more explicitly:

$$M_p = \{X \in \mathbb{R}^n | \Sigma_{i=1}^n p_i X_i = 0\} = \{X : [1,n] \to \mathbb{R} | E_p[X] = 0\}$$

which is a linear subspace of $\mathbb{R}^n$, on which the following maps are defined

$$\Psi_p(Y) : M_P \to \mathbb{R}; \quad \Psi_p(Y) = \ln E_p[e^Y],$$

$$\Phi_p(Y) : M_p \to \mathbb{P}; \quad Y \to e^{Y-\Psi(Y)}p = \frac{e^Y}{E_p[e^Y]}p.$$

Now it is not hard to see that the collection $\{M_p, \Phi_p\}$ can be used to define an atlas and provide $\mathbb{P}$ with a differentiable manifold structure, but we shall not carry out this point here. And this is a good point to stop for the time being.

D.5 Bibliographical comments and references

Two classics for introductory topology are [1] and [2], and in [3], [4] and [5] you will find all the measure theories that are needed to follow these notes (and more!).

In [5] you will find all basic concepts of probability explained with all necessary proofs included. In [5]-[7] there are variations on the Kolmogorov Extension Theorem, and in [6] quite a bit on measures on Banach space is explained.

The contents of Appendix D are taken almost verbatim from [9].

References

[1] Simmons, G. F. *"Introduction to Topology and Analysis"*. McGraw-Hill, New York, 1963.

[2] Diedonné, J. *"Foundations of Modern Analysis. Vol. I"*. Acad. Press, New York-San Diego, 1960.

[3] Rudin, W. *"Real and Complex Analysis"*. McGraw-Hill, New York, 1966.
[4] Royden, H. *"Real Analysis"*. The Macmillan Comp., New York, 1968.
[5] Bauer, H. *"Probability Theory and Elements of Measure Theory"*. Holt, Runehart and Winston, Inc., New York, 1972.
[6] Gihman, I. and Skorohod, A. *"The Theory of Stochastic Processes"*. Springer-Verlag, Berlin, 1974.
[7] Rao, M. M. *"Stochastic Processes: General Theory"*. Kluwer Acad. Pubs., Dordrecht, 1995.
[8] Kallenberg, O. *"Random Measures"*. Akademic Verlag, Berlin, (& Acad. Press., New York), 1975.
[9] Vajda, I. *"Theory of Statistical Inference and Information"*, Kluwer Acad. Pubs., Dordrecht, 1989.
[10] Deuschel, J. D. and Strook, D. *"Large Deviations"*. Academic Press, Boston, 1989.
[11] Gamboa, F. and Gassiat, E. *"New Bayesian methods for ill posed problems"*. Prep. Orsay. Sept., 1995.
[12] Robert, C. *"An entropy concentration theory: applications in artificial intelligence and descriptive statistics"*. J. Apply. Prob. Vol. 27 (1990), pp. 303-313.
[13] Gzyl, H. and Recht, L *"Geometry on the space of probabilities I: the finite dimensional case"*. Revista Matemtica Iberoamericana. 22, (2006) pp. 545-558.
[14] Gzyl, H. and Recht, L. *"Geometry on the space of probabilities II: projective spaces and exponential families"*. Revista Matemtica Iberoamericana, Vol 3 (2006), pp. 833-849.
[15] Gzyl, H. and Recht, L. *"Intrinsic geometry on the class of probability densities and exponential families"*. Publications Matematiques, Vol. 51 (2007) pp. 309-332.

Appendix E

Software user guide

The software provided with this volume is designed to be used in the MatLab® environment, so you need to have it running in the background.

E.1 Installation procedure

Run the application *lip.exe*, this action causes the execution of the program WinZip®, which is assumed to be installed on the machine. WinZip will decompress the files and the first window will appear.

By default, the files will be decompressed into `c:\Matlab\LIP`. If the user so prefers, she/he can redirect the path by typing in a new one or do it using the "Browse" button to choose one.

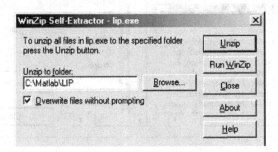

Once you have made up your mind as to where you want your files, press the unzip button.

Once the execution of "Unzip" is complete, a notice announcing that 167 files have unzipped successfully appears. If you want to go on, press OK.

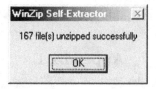

Now you have to begin a Matlab session, that is the Matlab command window must be open. You must now include in the Matlab path the folder that you selected to keep in the files lip.exe. To do that, in the Matlab task bar activate the "File" menu and choose the "Set Path" option. You should now see the window displayed here.

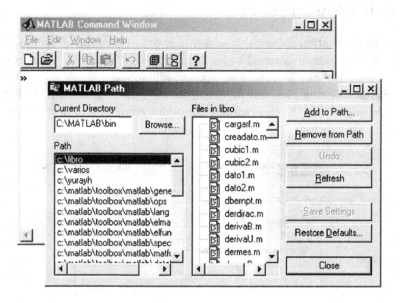

Once the Path Browser window is open, push the button "Add to Path..." on the right. A new window in which you have to make sure that the path to the folder in which you placed lip.exe is displayed. If you proceed by default, you should see the windows as shown. Press the OK button.

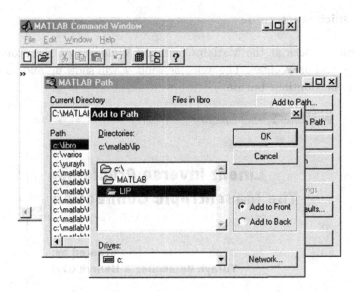

If you do not want to go through the drag of repeating these steps each time you use this software, save the actual configuration by pressing the "Save Settings" button. Now press the "Close" button and you are all set to go.

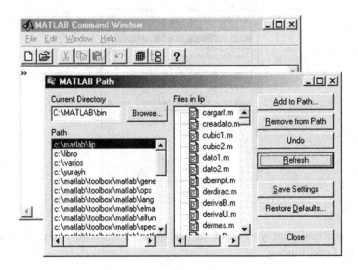

E.2 Quick start guide

Now you are back at the Matlab Command Window. At the prompt >> type vlip and press enter. The welcome window will show up and you have to press the Continue button.

Then the Menu of Basics Selections appear, here you can select the type of problem that you want to solve

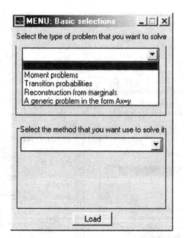

and the method that you want to use to solve it. And then you have to press the Load button.

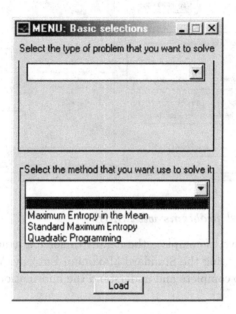

According to your selection you will see different windows appear. If you complete the information required in it you can observe and/or save the results, by pressing the OK button, clear them if you press the Cancel button or you may choose to go back to the Menu and change your selection of problem. Now you have different windows according to your selection in the Menu.

E.2.1 *Moment problems with MEM*

The user will have to complete the details to solve a generalized moment problem with the Maximum Entropy Method in the Mean. We refer the user to Section 7.2, to complete and understand the information required in this window.

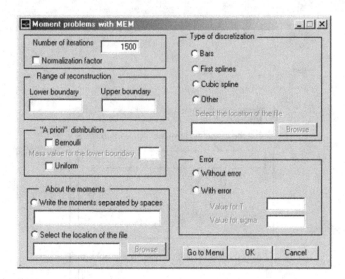

E.2.2 Moment problems with SME

The user will have to complete the details to solve a generalized moment problem in [0,1], using the Standard Maximum Entropy. We refer the user to Section 7.3, to complete and understand the information required in this window.

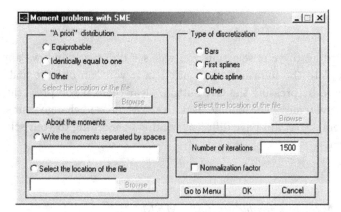

E.2.3 Moment problems with Quadratic Programming

This is an additional tool that can be used to solve the generalized moment problem in [0,1] with a quadratic programming method, with or without projection, according to the material explained in Section 7.1.2, using the different discretization matrices as explained in Chapter 7.

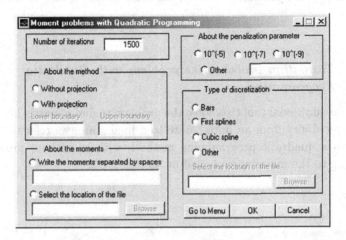

E.2.4 Transition probabilities problem with MEM

The user will have to complete the details to find a matrix that solve the problem of transition probability from an initial state to a final state, with the Maximum Entropy Method in the Mean. We refer the user to Section 7.6, to complete and understand the information required in this window.

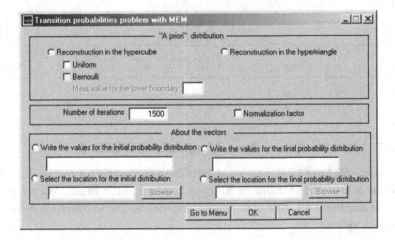

E.2.5 Transition probabilities problem with SME

Sorry!!, but this software does not cover this case. This implies that if you select this option you can see a warning message and return to the Menu.

E.2.6 Transition probabilities problem with Quadratic Programming

This is an additional tool that solve the problem to find a matrix for a transition probability from an initial state to a final state, we refer to Section 7.6, with a quadratic programming method, with or without projection, according to the material explained in Section 7.1.2.

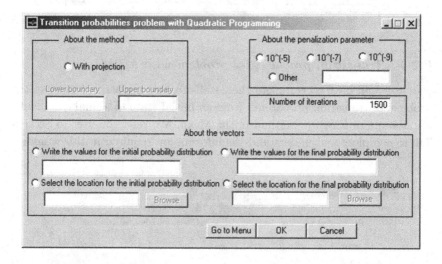

E.2.7 Reconstruction from Marginals with MEM

The user will have to complete the details to reconstruct a matrix from its marginals using the Maximum Entropy Method in the Mean. We refer the user to Section 9.9.1, to complete and understand the information required in this window.

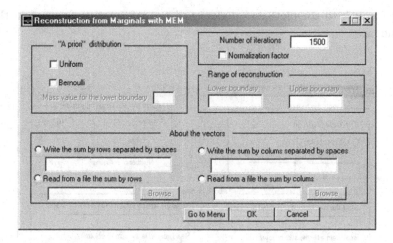

E.2.8 Reconstruction from Marginals with SME

The user will have to complete the details to reconstruct a matrix from its marginals using the Standard Maximum Entropy Method. We refer the user to Section 9.9.2, to complete and understand the information required in this window.

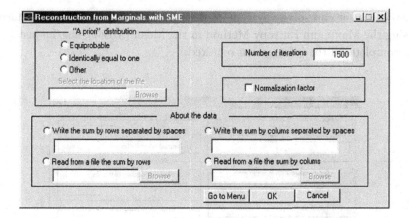

E.2.9 Reconstruction from Marginals with Quadratic Programming

The user will have to complete the details to reconstruct a matrix from its marginals using a Quadratic Programming Method. We refer the user to

Section 9.7, to complete and understand the information required in this window.

[Screenshot: Reconstruction from Marginals with Quadratic Programming dialog with About the method (Without projection / With projection, Lower boundary, Upper boundary), About the penalization parameter (10^(-5), 10^(-7), 10^(-9), Other), Number of iterations 1500, About the data (Write the sum by rows / columns separated by spaces, Read from a file the sum by rows / columns, Browse), Go to Menu, OK, Cancel.]

E.2.10 A generic problem in the form $Ax = y$, with MEM

Here the user can solve any type of discrete problem in the $Ax = y$ form using the Maximum Entropy Method in the Mean (MEM), she/he just has to complete all the information required.

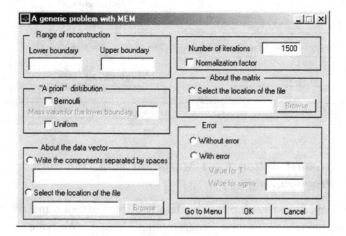

E.2.11 A generic problem in the form $Ax = y$, with SME

Here the user can solve any type of discrete problem in the $Ax = y$ form using the Standard Maximum Entropy Method (SME), just complete all the information required.

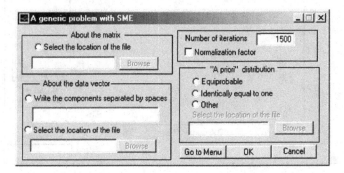

E.2.12 A generic problem in the form $Ax = y$, with Quadratic Programming

This is an extra tool for this version that permit one to solve any type of discrete problem in the $Ax = y$ form, using Quadratic Programming, just complete all the information required.

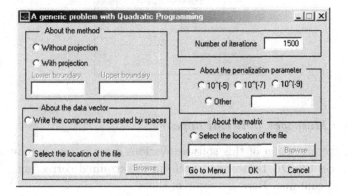

E.2.13 The results windows

The results can be displayed in two different windows according to the type of problem selected. In one of them you can visualize the result in a plot area, and you have the numerical values of the result on the right.

324 Linear Inverse Problems: The Maxentropic Connection

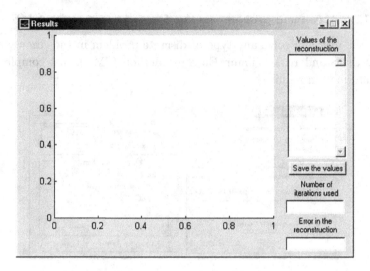

The other is a small window, and the results are displayed in the Command window of Matlab.

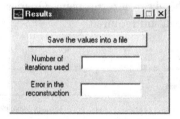

E.2.14 Messages that will appear

During the execution of this software various error messages may appear, if you press the OK button and perform the required correction, you will be able to continue. Here are some of them:

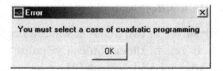

Software user guide

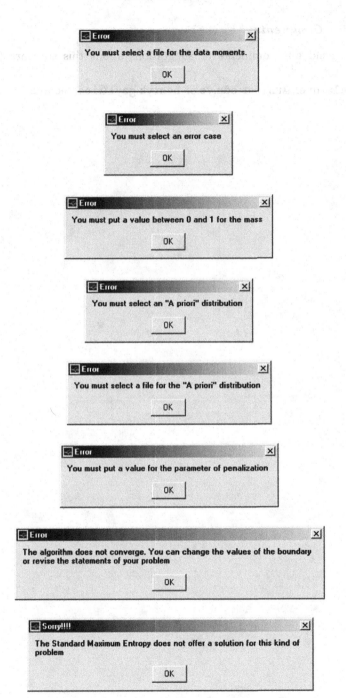

E.2.15 *Comments*

Please, send any comments or suggestions about this software (or this book), to:

yvelasquez@unimet.edu.ve or henryk.gzyl@iesa.edu.ve

Series on Advances in Mathematics for Applied Sciences

Editorial Board

N. Bellomo
Editor-in-Charge
Department of Mathematics
Politecnico di Torino
Corso Duca degli Abruzzi 24
10129 Torino
Italy
E-mail: nicola.bellomo@polito.it

F. Brezzi
Editor-in-Charge
IMATI - CNR
Via Ferrata 5
27100 Pavia
Italy
E-mail: brezzi@imati.cnr.it

M. A. J. Chaplain
Department of Mathematics
University of Dundee
Dundee DD1 4HN
Scotland

C. M. Dafermos
Lefschetz Center for Dynamical Systems
Brown University
Providence, RI 02912
USA

J. Felcman
Department of Numerical Mathematics
Faculty of Mathematics and Physics
Charles University in Prague
Sokolovska 83
18675 Praha 8
The Czech Republic

M. A. Herrero
Departamento de Matematica Aplicada
Facultad de Matemáticas
Universidad Complutense
Ciudad Universitaria s/n
28040 Madrid
Spain

S. Kawashima
Department of Applied Sciences
Engineering Faculty
Kyushu University 36
Fukuoka 812
Japan

M. Lachowicz
Department of Mathematics
University of Warsaw
Ul. Banacha 2
PL-02097 Warsaw
Poland

S. Lenhart
Mathematics Department
University of Tennessee
Knoxville, TN 37996–1300
USA

P. L. Lions
University Paris XI-Dauphine
Place du Marechal de Lattre de Tassigny
Paris Cedex 16
France

B. Perthame
Laboratoire J.-L. Lions
Université P. et M. Curie (Paris 6)
BC 187
4, Place Jussieu
F-75252 Paris cedex 05, France

K. R. Rajagopal
Department of Mechanical Engrg.
Texas A&M University
College Station, TX 77843-3123
USA

R. Russo
Dipartimento di Matematica
II University Napoli
Via Vivaldi 43
81100 Caserta
Italy

Series on Advances in Mathematics for Applied Sciences

Aims and Scope

This Series reports on new developments in mathematical research relating to methods, qualitative and numerical analysis, mathematical modeling in the applied and the technological sciences. Contributions related to constitutive theories, fluid dynamics, kinetic and transport theories, solid mechanics, system theory and mathematical methods for the applications are welcomed.

This Series includes books, lecture notes, proceedings, collections of research papers. Monograph collections on specialized topics of current interest are particularly encouraged. Both the proceedings and monograph collections will generally be edited by a Guest editor.

High quality, novelty of the content and potential for the applications to modern problems in applied science will be the guidelines for the selection of the content of this series.

Instructions for Authors

Submission of proposals should be addressed to the editors-in-charge or to any member of the editorial board. In the latter, the authors should also notify the proposal to one of the editors-in-charge. Acceptance of books and lecture notes will generally be based on the description of the general content and scope of the book or lecture notes as well as on sample of the parts judged to be more significantly by the authors.

Acceptance of proceedings will be based on relevance of the topics and of the lecturers contributing to the volume.

Acceptance of monograph collections will be based on relevance of the subject and of the authors contributing to the volume.

Authors are urged, in order to avoid re-typing, not to begin the final preparation of the text until they received the publisher's guidelines. They will receive from World Scientific the instructions for preparing camera-ready manuscript.

SERIES ON ADVANCES IN MATHEMATICS FOR APPLIED SCIENCES

Published*:

Vol. 69   Applied and Industrial Mathematics in Italy
         eds. M. Primicerio, R. Spigler and V. Valente

Vol. 70   Multigroup Equations for the Description of the Particle Transport
         in Semiconductors
         by M. Galler

Vol. 71   Dissipative Phase Transitions
         eds. P. Colli, N. Kenmochi and J. Sprekels

Vol. 72   Advanced Mathematical and Computational Tools in Metrology VII
         eds. P. Ciarlini et al.

Vol. 73   Introduction to Computational Neurobiology and Clustering
         by B. Tirozzi, D. Bianchi and E. Ferraro

Vol. 74   Wavelet and Wave Analysis as Applied to Materials with Micro or
         Nanostructure
         by C. Cattani and J. Rushchitsky

Vol. 75   Applied and Industrial Mathematics in Italy II
         eds. V. Cutello et al.

Vol. 76   Geometric Control and Nonsmooth Analysis
         eds. F. Ancona et al.

Vol. 77   Continuum Thermodynamics
         by K. Wilmanski

Vol. 78   Advanced Mathematical and Computational Tools in Metrology
         and Testing
         eds. F. Pavese et al.

Vol. 79   From Genetics to Mathematics
         eds. M. Lachowicz and J. Miękisz

Vol. 80   Inelasticity of Materials: An Engineering Approach and a Practical Guide
         by A. R. Srinivasa and S. M. Srinivasan

Vol. 81   Stability Criteria for Fluid Flows
         by A. Georgescu and L. Palese

Vol. 82   Applied and Industrial Mathematics in Italy III
         eds. E. De Bernardis, R. Spigler and V. Valente

Vol. 83   Linear Inverse Problems: The Maximum Entropy Connection
         by H. Gzyl and Y. Velásquez

*To view the complete list of the published volumes in the series, please visit:
http://www.worldscibooks.com/series/samas_series.shtml